METHODS IN MOLECULAR BIOLOGY

For further volumes:
http://www.springer.com/series/7651

For over 35 years, biological scientists have come to rely on the research protocols and methodologies in the critically acclaimed *Methods in Molecular Biology* series. The series was the first to introduce the step-by-step protocols approach that has become the standard in all biomedical protocol publishing. Each protocol is provided in readily-reproducible step-by step fashion, opening with an introductory overview, a list of the materials and reagents needed to complete the experiment, and followed by a detailed procedure that is supported with a helpful notes section offering tips and tricks of the trade as well as troubleshooting advice. These hallmark features were introduced by series editor Dr. John Walker and constitute the key ingredient in each and every volume of the *Methods in Molecular Biology* series. Tested and trusted, comprehensive and reliable, all protocols from the series are indexed in PubMed.

Thermomorphogenesis

Methods and Protocols

Edited by

Meng Chen

Department of Botany and Plant Sciences, Institute for Integrative Genome Biology, University of California, Riverside, CA, USA

Humana Press

Editor
Meng Chen
Department of Botany and Plant
Sciences, Institute for Integrative
Genome Biology
University of California
Riverside, CA, USA

ISSN 1064-3745 ISSN 1940-6029 (electronic)
Methods in Molecular Biology
ISBN 978-1-0716-3813-2 ISBN 978-1-0716-3814-9 (eBook)
https://doi.org/10.1007/978-1-0716-3814-9

Cover Illustration Caption: Image provided by Dr. De Fan.

This Humana imprint is published by the registered company Springer Science+Business Media, LLC, part of Springer Nature.
The registered company address is: 1 New York Plaza, New York, NY 10004, U.S.A.

Paper in this product is recyclable.

Preface

Welcome to this volume of Methods in Molecular Biology dedicated to the captivating field of thermomorphogenesis, a discipline that unravels the intricacies of warm temperature responses in plants. The knowledge of how plants perceive, interpret, and respond to elevated temperatures has advanced dramatically in the last few years, with far-reaching implications for agriculture, ecology, and our fundamental understanding of plant development. The principles of thermomorphogenesis also provide the basis for addressing the challenges of creating resilient plants in a changing climate. The publication of this methods book is a testament that the study of thermomorphogenesis has become a major discipline within the field of plant molecular biology.

This methods book brings together a compendium of methodologies for investigating thermomorphogenesis, addressing three major sections on the following topics: thermomorphogenesis responses, temperature sensing in thermomorphogenesis, and temperature signaling in thermomorphogenesis. Although the methods focus on the model plant species *Arabidopsis thaliana*, most of the methodologies should be readily applicable to other plant species. The book is directed primarily at the active research scientist, student, or technician. In addition to being an aid for the research laboratory, many of the 23 chapters will be valuable in graduate and undergraduate laboratory courses in plant molecular biology.

Riverside, CA, USA *Meng Chen*

Contents

Preface .... *v*
Contributors .... *ix*

Part I Thermomorphogenesis Responses

1 A Guide to Quantify Arabidopsis Seedling Thermomorphogenesis at Single Timepoints and by Interval Monitoring 3
Philipp Janitza, Zihao Zhu, Muhammad Usman Anwer, Martijn van Zanten, and Carolin Delker

2 Dissection of Daytime and Nighttime Thermoresponsive Hypocotyl Elongation in *Arabidopsis* 17
De Fan and Meng Chen

3 Light-Dependent High Ambient Temperature-Induced Senescence Assay Using Whole Seedlings 25
Chanhee Kim and Giltsu Choi

4 Assessing Temperature Responses in Roots 37
Sanghwa Lee and Wolfgang Busch

5 A Luciferase Imaging-Based Assay for Studying Temperature Compensation of the Circadian Clock 43
Hongtao Zhang and Stacey L. Harmer

6 Investigating the Effects of Temperature on Pathogen Propagation in *Arabidopsis* 55
Zhan Li, Yang Liu, and Jian Hua

7 Assessing the High Temperature Effects on Stomatal Production 65
Rini Rahiman and On Sun Lau

8 Characterizing the Synergistic Response Induced by Warm Temperatures and Shade 75
Yogev Burko

Part II Temperature Sensing in Thermomorphogenesis

9 Measurement of Phytochrome B Thermal Reversion Rates In Vivo 85
Cornelia Klose and Andreas Hiltbrunner

10 Characterization of Thermoresponsive Photobody Dynamics 95
Juan Du and Meng Chen

11 Phase Separation of Phytochrome B in HEK293T Cells 105
Di Chen and Shangwei Zhong

12 Isolation of Phytochrome B Photobodies 113
Yongmin Kwon, Chanhee Kim, and Giltsu Choi

13 Analysis of Phase Separation of EARLY FLOWERING 3 123
Maolin Peng, Stephanie Hutin, Aleksandra Mironova, Chloe Zubieta, and Philip A. Wigge

14 In Vitro Determination of Temperature-Dependent DNA Binding of the Evening Complex Using Electrophoretic Mobility Shift Assays 135
Stephanie Hutin, Philip A. Wigge, and Chloe Zubieta

15 Monitoring Real-Time Temperature Dynamics of a Short RNA Hairpin Using Förster Resonance Energy Transfer and Circular Dichroism 149
Martin Balcerowicz, Philip A. Wigge, Marco Di Antonio, and Betty Chung

PART III TEMPERATURE SIGNALING IN THERMOMORPHOGENESIS

16 Characterization of PIF4 Phosphorylation by SPA1 161
Sanghwa Lee and Enamul Huq

17 Chromatin Immunoprecipitation to Investigate H2A.Z Dynamics in Response to Environmental Changes 169
J. Paola Saldierna Guzmán, Mark Zander, and Björn C. Willige

18 Assessing the Function of CBF1 in Modulating the Interaction Between Phytochrome B and PIF4 183
Jing Peng, Xiaojing Dong, Shuhua Yang, and Jigang Li

19 GST Pull-Down Assay to Study PIF4 Binding In Vitro 195
Abhishesh Bajracharya, Berry Dickey, and Yongjian Qiu

20 Investigating Circadian Gating of Temperature Responsive Genes 213
Rachel I. Strout, Calum A. Graham, Antony N. Dodd, and Dawn H. Nagel

21 RuBisCO Depletion and Subcellular Fractionation for Enhanced Florigen Detection in *Arabidopsis* 227
Katarzyna Gawarecka and Ji Hoon Ahn

22 Identifying Causal Genes for Thermo-Responsive Rosette Growth in *Arabidopsis* by Genome-Wide Association Study 239
Zhixue Wang and Jian Hua

23 Mathematical Modeling of Photo- and Thermomorphogenesis in Plants 247
Gabriel Rodriguez-Maroto, Pablo Catalán, Cristina Nieto, Salomé Prat, and Saúl Ares

Correction to: Chromatin Immunoprecipitation to Investigate H2A.Z Dynamics in Response to Environmental Changes............................. C1
J. Paola Saldierna Guzmán, Mark Zander, and Björn C. Willige

Index *263*

Contributors

JI HOON AHN • *Department of Molecular Life Sciences, Korea University, Seoul, South Korea*

MUHAMMAD USMAN ANWER • *Institute of Agricultural and Nutritional Sciences, Martin Luther University Halle-Wittenberg, Halle (Saale), Germany; Molecular Genetics and Physiology of Plants, Ruhr University Bochum, Bochum, Germany*

SAÚL ARES • *Grupo Interdisciplinar de Sistemas Complejos (GISC), Madrid, Spain; Centro Nacional de Biotecnologia (CNB), CSIC, Madrid, Spain*

ABHISHESH BAJRACHARYA • *Department of Biology, University of Mississippi, Oxford, MS, USA*

MARTIN BALCEROWICZ • *Division of Plant Sciences, University of Dundee at the James Hutton Institute, Dundee, UK*

YOGEV BURKO • *Department of Vegetables and Field Crops, Agricultural Research Organization, Volcani Institute, Bet Dagan, Israel*

WOLFGANG BUSCH • *Plant Molecular and Cellular Biology Laboratory, Salk Institute for Biological Studies, La Jolla, CA, USA*

PABLO CATALÁN • *Grupo Interdisciplinar de Sistemas Complejos (GISC), Madrid, Spain; Department of Mathematics, Universidad Carlos III de Madrid, Madrid, Spain*

DI CHEN • *Peking University Institute of Advanced Agricultural Sciences, Shandong Laboratory of Advanced Agriculture Sciences in Weifang, Weifang, China; State Key Laboratory of Protein and Plant Gene Research, School of Life Sciences, Peking University, Beijing, China*

MENG CHEN • *Department of Botany and Plant Sciences, Institute for Integrative Genome Biology, University of California, Riverside, CA, USA*

GILTSU CHOI • *Department of Biological Sciences, Korea Advanced Institute of Science and Technology, Daejeon, South Korea; Department of Biological Sciences, KAIST, Daejeon, South Korea*

BETTY CHUNG • *Department of Pathology, University of Cambridge, Cambridge, UK*

CAROLIN DELKER • *Institute of Agricultural and Nutritional Sciences, Martin Luther University Halle-Wittenberg, Halle (Saale), Germany*

MARCO DI ANTONIO • *Imperial College London, Chemistry Department, Molecular Science Research Hub, London, UK; The Institute of Chemical Biology (ICB), Molecular Science Research Hub, London, UK; The Francis Crick Institute, London, UK*

BERRY DICKEY • *Department of Biology, University of Mississippi, Oxford, MS, USA*

ANTONY N. DODD • *Department of Cell and Developmental Biology, John Innes Centre, Norwich Research Park, Norwich, UK*

XIAOJING DONG • *State Key Laboratory of Plant Environmental Resilience, College of Biological Sciences, China Agricultural University, Beijing, China*

JUAN DU • *Department of Botany and Plant Sciences, Institute for Integrative Genome Biology, University of California, Riverside, CA, USA*

DE FAN • *Department of Botany and Plant Sciences, Institute for Integrative Genome Biology, University of California, Riverside, CA, USA*

KATARZYNA GAWARECKA • *Department of Molecular Life Sciences, Korea University, Seoul, South Korea*

CALUM A. GRAHAM • *Department of Cell and Developmental Biology, John Innes Centre, Norwich Research Park, Norwich, UK*
STACEY L. HARMER • *Department of Plant Biology, College of Biological Sciences, University of California, Davis, CA, USA*
ANDREAS HILTBRUNNER • *Institute of Biology II, Faculty of Biology, University of Freiburg, Freiburg, Germany; Signaling Research Centers BIOSS and CIBSS, University of Freiburg, Freiburg, Germany*
JIAN HUA • *School of Integrative Plant Science, Plant Biology Section, Cornell University, Ithaca, NY, USA; Plant Biology Section, School of Integrative Plant Science, Cornell University, Ithaca, NY, USA*
ENAMUL HUQ • *Department of Molecular Biosciences and The Institute for Cellular and Molecular Biology, The University of Texas at Austin, Austin, TX, USA*
STEPHANIE HUTIN • *Laboratoire de Physiologie Cellulaire and Végétale, Univ. Grenoble Alpes/CNRS/CEA/INRA/IRIG, Grenoble, France; Laboratoire Physiologie Cellulaire et Végétale, Univ. Grenoble Alpes, CNRS, CEA, INRAE, IRIG-DBSCI, Grenoble, France*
PHILIPP JANITZA • *Institute of Agricultural and Nutritional Sciences, Martin Luther University Halle-Wittenberg, Halle (Saale), Germany*
CHANHEE KIM • *Department of Biological Sciences, Korea Advanced Institute of Science and Technology, Daejeon, South Korea; Department of Biological Sciences, KAIST, Daejeon, South Korea*
CORNELIA KLOSE • *Institute of Biology II, Faculty of Biology, University of Freiburg, Freiburg, Germany*
YONGMIN KWON • *Department of Biological Sciences, KAIST, Daejeon, South Korea*
ON SUN LAU • *Department of Biological Sciences, National University of Singapore, Singapore, Singapore*
SANGHWA LEE • *Plant Molecular and Cellular Biology Laboratory, Salk Institute for Biological Studies, La Jolla, CA, USA; Department of Molecular Biosciences and The Institute for Cellular and Molecular Biology, The University of Texas at Austin, Austin, TX, USA; Plant Molecular and Cellular Biology Laboratory, Salk Institute for Biological Studies, La Jolla, CA, USA*
JIGANG LI • *State Key Laboratory of Plant Environmental Resilience, College of Biological Sciences, China Agricultural University, Beijing, China*
ZHAN LI • *Guangdong Provincial Key Laboratory of Plant Molecular Breeding, State Key Laboratory for Conservation and Utilization of Subtropical Agro-Bioresources, South China Agricultural University, Guangzhou, China; School of Integrative Plant Science, Plant Biology Section, Cornell University, Ithaca, NY, USA*
YANG LIU • *School of Integrative Plant Science, Plant Biology Section, Cornell University, Ithaca, NY, USA*
ALEKSANDRA MIRONOVA • *Laboratoire de Physiologie Cellulaire and Végétale, Univ. Grenoble Alpes/CNRS/CEA/INRA/IRIG, Grenoble, France*
DAWN H. NAGEL • *Department of Botany and Plant Sciences, University of California, Riverside, CA, USA*
CRISTINA NIETO • *Centro Nacional de Biotecnologia (CNB), CSIC, Madrid, Spain; Instituto Nacional de Investigacion y Tecnologia Agraria y Alimentaria (INIA), CSIC, Madrid, Spain*
JING PENG • *State Key Laboratory of Plant Environmental Resilience, College of Biological Sciences, China Agricultural University, Beijing, China*

MAOLIN PENG • *Leibniz-Institut für Gemüse-und Zierpflanzenbau, Theodor-Echtermeyer-Weg 1, Großbeeren, Germany*

SALOMÉ PRAT • *Centro Nacional de Biotecnologia (CNB), CSIC, Madrid, Spain; Centro de Investigación en Agrigenomica (CRAG), CSIC-IRTA-UAB-UB, Barcelona, Spain*

YONGJIAN QIU • *Department of Biology, University of Mississippi, Oxford, MS, USA*

RINI RAHIMAN • *Department of Biological Sciences, National University of Singapore, Singapore, Singapore*

GABRIEL RODRIGUEZ-MAROTO • *Grupo Interdisciplinar de Sistemas Complejos (GISC), Madrid, Spain; Department of Mathematics, Universidad Carlos III de Madrid, Madrid, Spain*

J. PAOLA SALDIERNA GUZMÁN • *Department of Soil and Crop Sciences, College of Agricultural Sciences, Colorado State University, Fort Collins, CO, USA*

RACHEL I. STROUT • *Department of Botany and Plant Sciences, University of California, Riverside, CA, USA*

MARTIJN VAN ZANTEN • *Plant Stress Resilience, Institute of Environmental Biology, Utrecht University, Utrecht, The Netherlands*

ZHIXUE WANG • *State Key Laboratory of Crop Genetics & Germplasm Enhancement and Utilization, College of Agriculture, Nanjing Agricultural University, Nanjing, China; Plant Biology Section, School of Integrative Plant Science, Cornell University, Ithaca, NY, USA*

PHILIP A. WIGGE • *Leibniz-Institut für Gemüse-und Zierpflanzenbau, Theodor-Echtermeyer-Weg 1, Großbeeren, Germany; Leibniz-Institut für Gemüse- und Zierpflanzenbau, Großbeeren, Germany; Institute of Biochemistry and Biology, University of Potsdam, Potsdam, Germany; Leibniz Institute for Vegetable and Ornamental Crops (IGZ), Grossbeeren, Germany*

BJÖRN C. WILLIGE • *Department of Soil and Crop Sciences, College of Agricultural Sciences, Colorado State University, Fort Collins, CO, USA*

SHUHUA YANG • *State Key Laboratory of Plant Environmental Resilience, College of Biological Sciences, China Agricultural University, Beijing, China*

MARK ZANDER • *Waksman Institute of Microbiology, Department of Plant Biology, School of Environmental and Biological Sciences, Rutgers, The State University of New Jersey, Piscataway, NJ, USA*

HONGTAO ZHANG • *Department of Plant Biology, College of Biological Sciences, University of California, Davis, CA, USA*

SHANGWEI ZHONG • *Peking University Institute of Advanced Agricultural Sciences, Shandong Laboratory of Advanced Agriculture Sciences in Weifang, Weifang, China; State Key Laboratory of Protein and Plant Gene Research, School of Life Sciences, Peking University, Beijing, China*

ZIHAO ZHU • *Institute of Agricultural and Nutritional Sciences, Martin Luther University Halle-Wittenberg, Halle (Saale), Germany; Leibniz Institute of Plant Genetics and Crop Plant Research (IPK) Gatersleben, Seeland, Germany*

CHLOE ZUBIETA • *Laboratoire de Physiologie Cellulaire and Végétale, Univ. Grenoble Alpes/CNRS/CEA/INRA/IRIG, Grenoble, France; Laboratoire Physiologie Cellulaire et Végé tale, Univ. Grenoble Alpes, CNRS, CEA, INRAE, IRIG-DBSCI, Grenoble, France*

Part I

Thermomorphogenesis Responses

Chapter 1

A Guide to Quantify Arabidopsis Seedling Thermomorphogenesis at Single Timepoints and by Interval Monitoring

Philipp Janitza, Zihao Zhu, Muhammad Usman Anwer, Martijn van Zanten, and Carolin Delker

Abstract

Temperature-induced elongation of hypocotyls, petioles, and roots, together with hyponastic leaf responses, constitute key model phenotypes that can be used to assess a plant's capacity for thermomorphogenesis. Phenotypic responses are often quantified at a single time point during seedling development at different temperatures. However, to capture growth dynamics, several time points need to be assessed, and ideally continuous measurements are taken. Here we describe a general experimental setup and technical solutions for recording and measuring seedling phenotypes at single and multiple time points. Furthermore, we present an R-package called "*rootdetectR*," which allows easy processing of hypocotyl, root or petiole length, and growth rate data and provides different options of data presentation.

Key words Thermomorphogenesis, Phenotyping, Growth rate analyses, RootDetection, *RootdetectR*

1 Introduction

Increases in ambient temperature, below the heat stress level, affect virtually all aspects of plant growth and development. In seedlings, warm temperatures cause elongation of hypocotyls, petioles, and roots while causing a reduction in leaf area (reviewed in [1, 2]). In combination with hyponasty, this suite of phenotypes is part of the morphological adaptations to warm ambient temperatures which are collectively termed thermomorphogenesis [3, 4]. Elongation and hyponasty of stems and petioles are considered to aid cooling of the leaf blades by facilitating efficient transpiration, presumably to maintain high photosynthesis efficiency [5, 6]. Commonly, thermomorphogenesis phenotypes of *Arabidopsis* seedlings are quantified based on photographs taken from seedlings grown on sterile agar medium on vertically orientated plates. These types of single

Meng Chen (ed.), *Thermomorphogenesis: Methods and Protocols*, Methods in Molecular Biology, vol. 2795, https://doi.org/10.1007/978-1-0716-3814-9_1,

time point or "end point" analyses have been—and remain to be—an essential tool in the identification of the regulatory components orchestrating thermomorphogenesis, especially because seedling response capacity is a good predictor for temperature responses later in the plants' life cycle, including being predictive for flowering time [7]. Furthermore, tedious monitoring of growth rates in dense time series over several days, including dark periods, has revealed differences in the regulation of temperature-induced growth (rates) during development and in different photoperiods [8–12]. While single timepoint analyses for a few genotypes are easily conducted and can be manually analyzed using, for instance, the freeware package ImageJ, large-scale screenings of, e.g., mapping populations or generation of time-series for multiple genotypes are still challenging in terms of data acquisition, processing, and analysis. Furthermore, analysis of growth rates requires imaging of growth at multiple time points during the day and ideally also during the night period. Illumination of plants by infrared (IR) light emitting diodes (LEDs) and use of cameras that are modified to capture IR images can provide an easy solution to circumvent perturbing the given light/dark rhythm [13, 14].

Here, we describe the basic setup and equipment needed for single and multiple timepoint analyses of hypocotyl, root, or petiole lengths. We also present a simple method for data acquisition and subsequent data analyses with a custom-made R package called "*rootdetectR.*"

2 Materials

Any standard protocol for seed sterilization and media preparation for *Arabidopsis* culture can be used for growth assays as long as the medium is transparent. Here, we will give an example using *Arabidopsis thaliana salts* (ATS) medium [15] with liquid surface sterilization of seeds, but commercially available ½ strength Murashige Skoog (MS) medium and alternative seed sterilization methods (e.g. chlorine gas) can similarly be used.

2.1 Arabidopsis *Sterile Culture*

1. 1 M KNO_3: Dissolve 101 g in 1 L deionized water and autoclave.
2. 1 M $MgSO_4$: Dissolve 246.5 g $MgSO_4x7H_2O$ in 1 L deionized water and autoclave.
3. 1 M $Ca(NO_3)_2$: Dissolve 236 g 1 L deionized water and autoclave.
4. 20 mM Fe-Na-EDTA: Dissolve 8.42 g Fe-EDTAx3H_2O in 1 L deionized water and autoclave.

5. 1 M KPO pH 5.5: Dissolve 15.79 g K_2HPO_4 and 123.8 g 1 M KH_2PO_4 in 1 L deionized water and autoclave.
6. Micronutrient stock solution: Dissolve 70 mM H_3BO_3 (4.32 g/L), 14 mM $MnCl_2x4H_2O$ (2.77 g/L), 0.5 mM $CuSO_4$ (80 mg/L), 1 mM $ZnSO_4x7H_2O$ (288 mg/L), 0.2 mM $Na_2MoO_4x2H_2O$ (48 mg/L), 10 mM NaCl (580 mg/L), and 0.01 mM $CoCl_2x6H_2O$ (2.38 mg/L) in deionized water and autoclave.
7. Solid ATS medium according to [15]: Mix 5 mM KNO_3 (5 mL 1 M stock/L), 2.5 mM KPO (2.5 mL 1 M stock/L), 2 mM $MgSO_4$ (2 mL 1 M stock/L), 2 mM $Ca(NO_3)_2$ (2 mL 1 M stock/L), 0.05 mM Fe-Na-EDTA (2.5 mL 20 mM stock/L), and micronutrients: 70 μM H_3BO_3, 14 μM $MnCl_2$, 0.5 μM CuSO4, 1 μM $ZnSO_4$, 0.2 μM Na_2MoO_4, 10 μM NaCl, and 0.01 μM $CoCl_2$ (1 mL micronutrient stock solution/L). Sucrose may be added to the medium (0.5–1% w/v) depending on the specific experimental setup (*see* **Note 1**). Add deionized water to a total volume of 1 L. Stir thoroughly and transfer to autoclave-safe flasks containing plant-media gelling agent (e.g., 0.5% (w/v) Agargel™, Sigma Aldrich). Autoclave medium according to the manufacturer instruction of your autoclave. After autoclaving, be very cautious to avoid boiling retardation! Once the medium has cooled down to ~80 °C, close lids tightly and invert 3–4 times to evenly distribute the gelling agent. Let medium cool down further to 50–60 °C and pour it into clear plastic petri dishes under a sterile hood. For square plates use 50–75 mL of medium, for small round plates approximately 25–40 mL of medium (depending on the exact size of the plates). Let the medium in plates cool down until fully solidified, ideally with lids open under a sterile hood to avoid condensation. Plates can be stored at 4 °C for several weeks.
8. Petri dishes for cultivation: Use either square plates with vents (e.g., 120 × 120 × 17 mm, Greiner Bio-one) or smaller round plates, depending on the phenotype to be assayed. Square plates are commonly used for root growth assays as they are easy to handle and provide space for root growth analysis over several days. However, phenotyping of hypocotyl length can also be performed using, e.g., smaller round plates (e.g., 94 × 16 mm) if appropriate stands are available to avoid moving of plates during cultivation.
9. Plaster tape (e.g., Leucopore, *see* **Note 2**).
10. 70% ethanol.
11. Sodium hypochlorite solution (also known as "bleach"): mix 7 volume units of 12% sodium hypochlorite with 3 volume units of deionized water and add 0.01% (v/v) of Triton-X. *The ready-to-use* bleach solution can be stored at 4 °C for several weeks.

12. Sterile deionized water to wash and stratify seeds.
13. 1.5 mL reaction tubes, 1 mL micropipette with corresponding sterile 1 mL pipette tips (alternatively sterilized glass Pasteur pipettes with narrow tips can be used).
14. Black permanent marker.
15. Optional: print-out labels containing QR codes (*see* **Note 3).**
16. Optional: shaker for 1.5 mL reaction tubes.
17. Sterile hood for plate preparation and controlled environment/growth cabinet/phytotron for plant cultivation, with option to position plates near-vertically, at an angle of ideally ~70 (up to 90) degrees.
18. Light meter and/or data logger to measure and/or record light and temperature regime.

2.2 Single Time Point Analysis

1. Digital camera with fixture to maintain equal distance/focus for all pictures in one set, or alternatively a standard flat-bed scanner connected to a PC.
2. Black background (cardboard, fine-threaded cloth, etc.).
3. Ruler or similar object with millimeter (mm) scale.
4. Optional: computer with camera operating software.

2.3 Infrared Time-Lapse Imaging of Multiple Time Points (See Note 4)

1. Digital SLR camera with modified 830 IR long pass filter [13] (*see* **Note 5**), including a SD memory card with sufficient capacity, or a connection to a laptop for storage of images. Optional: External power supply for camera (useful for long-term use of camera).
2. 2 IR-LEDs (840 nm) with power supply (*see* **Note 6**).
3. Black background (cardboard or cloth).
4. Adjustable plate holder (acrylic) or similar fixture to keep petri dishes in fixed place (Fig. 1).
5. SLR camera remote control with timer, or a laptop and software that can control the camera.
6. Tape or cable binders to fix power supplies in place.

2.4 Data Acquisition and Analysis

While numerous software options to capture, analyze, and present data are available, we present an option using software that is presently available, free of charge, and thus accessible to everyone.

1. Java-enabled computer or digital device able to process large image files with the following software installed: RootDetection (www.labutils.de) or equivalent measurement software such as ImageJ (https://imagej.net/ij/index.html).
2. Table calculation software (e.g., LibreOfficeCalc).
3. R, Rstudio, and the Rpackage "rootdetectR" (*see* **Note 7**).

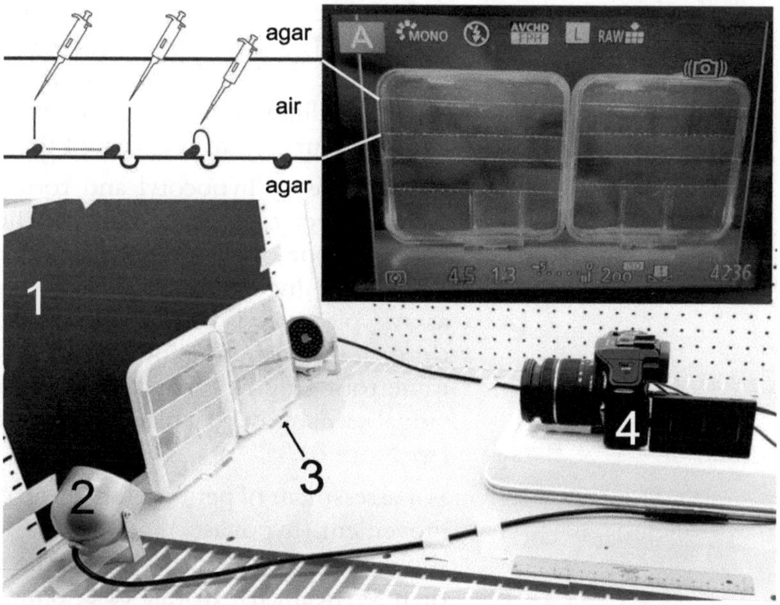

Fig. 1 Components and setup for multiple timepoint analyses of thermomorphogenesis by IR photography. To capture images at multiple time points (e.g., hourly), plates are positioned close to vertically in front of a black background (1). In addition to the general illumination within the controlled environment, IR lamps (2) are positioned to illuminate plates and allow image capture during the day and dark period. Plates are secured in place (e.g., in acrylic adjustable plate holders) (3). Pictures are captured by a digital SLR camera with modified light filters (4). For analysis of hypocotyl or root thermomorphogenesis, seeds can be placed on the surface of the growth medium and seedlings grow along a vertical axis while in contact with the medium. When analyzing petiole hyponastic rhythms, seedling shoots should be devoid of contact to the medium. Here, strips of growth medium are removed and seeds are placed in small indentations in the agar surface (see picture inset and cartoon)

3 Methods

3.1 Preparation of Arabidopsis *Sterile Culture*

1. Prepare one plate of solid ATS medium for each genotype/temperature combination to be analyzed.
2. For seed sterilization, transfer *Arabidopsis* seeds to 1.5 mL reaction tubes and treat with 70% ethanol for 3–5 min while gently shaking. After ethanol removal, incubate seeds in sodium-hypochlorite solution for 7–10 min followed by 3–5 wash steps with sterile deionized water.
3. Stratify seeds at 4 °C in sterile deionized water for 3–4 days prior to plating to facilitate uniform germination. Stratification after plating (sowing) is also possible, but is discouraged in case of multiple time point analyses due to excessive condensation water buildup.

4. Label plates either with custom-generated QR code stickers (*see* **Note 3**) or by using a black permanent marker on the back of the plate. Make sure that the label will be out of range of the actual seedlings after cultivation, but it should be visible when taking pictures from the back of the plate.
5. For analysis of hypocotyl and root elongation, seeds can be plated directly on the surface of the medium. Seeds can be placed in one or multiple lines or even be randomly dispersed when only hypocotyls are to be analyzed. For root growth assays, place seeds on a single horizontal line within the top third of the plate to allow sufficient space to accurately determine root lengths. To facilitate accurate measurements, individual seeds should be placed with sufficient space to each other (*see* **Note 8**).
6. For assessment of petiole elongation or if rhythmicity of petiole movement (hyponasty) is to be analyzed [11], seedling shoots should be devoid of contact with the media surface throughout their vertical axis. In this case, remove one or two strip(s) of 2–3 cm of cultivation medium under a sterile hood prior to plating. Place seeds in small indentations on the surface of the cut to prevent excessive movement of seedlings (Fig. 1, *see* **Note 9**).
7. Cultivate plates in a close-to-vertical orientation (we recommend an angle of ~70–90 degrees) in a controlled environment setting (phytotron, growth cabinet, etc.). Always check light and temperature conditions prior to—and ideally during—the experiment to ensure the designated settings are met (e.g., similar light intensities of control and warm temperature conditions). Most commonly, thermomorphogenesis phenotypes are determined either after seedlings have been grown constantly for 5–8 days in control (e.g., 20–21 °C) or warm (e.g., 27–29 °C) temperature or seedlings are first grown for 1–4 days in control temperature conditions after which the warm temperature samples are shifted to higher temperatures for another 4–8 days while control seedlings remain at the colder temperature. In our experience, temperature effects on hypocotyl length can be captured already a few hours after germination and within 24 h after shifting the seedlings to the warm temperature [9, 10], while hypocotyl elongation seizes once the first true leaves emerge. To avoid introducing artificial variation into the data, mark ungerminated seeds or late germinated seedlings after 2–4 days and exclude these from later analyses.

3.2 Single Time Point ("End Point") Analyses

1. Set up the photography unit. Place a digital camera in a fixture, so that the area of a dummy plate is fully covered. Place black cardboard or fine-threaded black cloth as background and fix it with tape to aid contrast. Mark the position of the plate (or use a mold, *see* **Note 10**) to ensure a fixed position of all plates to be processed and place a ruler or similar size standard with mm markings next to the plate.
2. Take one picture of a dummy plate including the ruler with mm markings as template for the size standard.
3. Take pictures of all individual plates by placing them face down (i.e., pictures of seedlings are taken through the back of the plate). That way, residual condensation water does not run down the seedlings, which could cause reflections or distortions on the photos. Make sure that the camera settings are optimized for a strong contrast between the dark background and light roots and/or shoot organs (*see* **Note 11**, Fig. 2). If you make use of the QR code labeling, make sure that the code is in focus and clearly visible on the back of the plate. Do not change the camera position during the picture capture of plates from one experiment to ensure that the size standard of the dummy plate refers accurately to all plates of one set.

   An alternative to a photography setup is the use of a standard flat-bed scanner to image the plates (use at least 600 dpi), if a single time point is assayed. Discard the plates afterward in this case.
4. Measure your trait(s) of interest using RootDetection, ImageJ, or equivalent software (*see* **Note 12**). When using RootDetection, photos need to be as ".jpeg" files (Fig. 2). To measure root, hypocotyl, or petiole length, sample names ("Label") can be extracted and assigned to each measurement directly (*see* **Note 13**). RootDetection also generates a copy of each picture, which includes all measurements highlighted for documentation. Make sure to generate the size standard by measuring 3–5 times 10 mm on the ruler of the picture of the dummy plate and label these measurements as "10 mm."

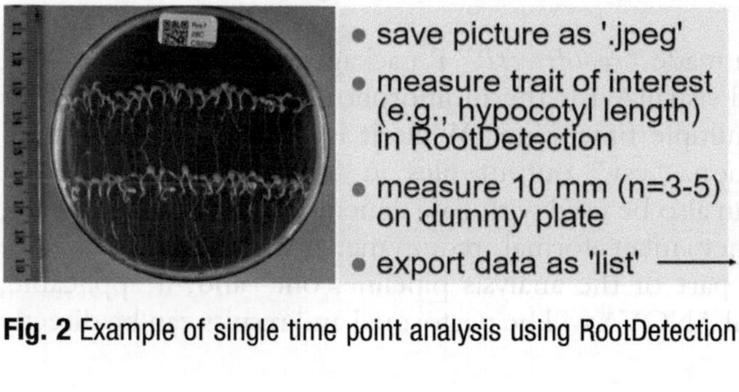

Nr	Filename	RootNr	Label	Desc	LengthPx	LengthMM
1	IMG_1707.JPG	1	Col;20C		80.709	80.709
2	IMG_1707.JPG	2	Col;20C		76.026	76.026
3	IMG_1707.JPG	3	Col;20C		42.45	42.45
4	IMG_1707.JPG	4	Col;20C		72.45	72.45
5	IMG_1707.JPG	5	Col;20C		53.712	53.712
6	IMG_1707.JPG	6	Col;20C		47.854	47.854
7	IMG_1707.JPG	7	Col;20C		56.921	56.921
8	IMG_1707.JPG	8	Col;20C		91.198	91.198
9	IMG_1707.JPG	9	Col;20C		54.231	54.231
10	IMG_1707.JPG	10	Col;20C		58.669	58.669
11	IMG_1707.JPG	11	Col;20C		66.261	66.261
12	IMG_1707.JPG	12	Col;20C		48	48
13	IMG_1707.JPG	13	Col;20C		49.244	49.244
14	IMG_1707.JPG	14	Col;20C		62.169	62.169
15	IMG_1707.JPG	15	Col;20C		67.417	67.417

Fig. 2 Example of single time point analysis using RootDetection

3.3 Analysis of Multiple Time Points

1. Position plates in plate holders and set up IR lights and the SLR camera (including power supplies and remote control if physically wired to camera). If the camera has a rotatable display, position the display to easily view captured photos without having to move the camera (Fig. 1).
2. Check whether battery of camera and remote control (if applicable) is sufficient for the entire time of the designated experiment (usually 3–7 days)!
3. Take test pictures to ensure correct focus of the camera, and correct lighting by the IR lights (avoid shadows as much as possible) and adjust camera settings if necessary (example settings for a Panasonic G5 camera modified with a IR long pass filter (830 nm): lens H-PS14042, aperture f4.5, shutter speed 1.3″, and ISO 200). Important: make sure to test camera settings under day and night conditions! While the 830 nm IR-filter modification blocks out most of the visible light, there may still be some minor differences in the pictures obtained during day and night periods. Select settings that allow good visualization of seedlings in day and night conditions. If using different IR filters, make sure the wavelength threshold still blocks out most of the light used at daytime in your controlled environment.
4. Take picture with size standard (ruler) containing mm markings containing a dummy plate or only a ruler.
5. Set up remote control of the camera with required intervals (e.g., hourly or 2-hourly).
6. Run experiment for the designated time period, make sure to record the zeitgeber time (ZT) of the start of the experiment in your experimental documentation.
7. Transfer picture files to computer and measure trait of interest. If growth rates are to be determined, make sure that data for individual roots at different timepoints can be associated (*see* **Note 14**). When using RootDetection, export measurements as "list" for further processing using "*rootdetectR*." If you used an alternate program but want to make use of the data processing options in "*rootdetectR*," make sure to convert your table to the required format (*see* Subheading 3.4).

3.4 Data Processing, Analysis, and Presentation

The custom made "*rootdetectR*" R package is designed to process, analyze, and visualize the thermomorphogenesis data obtained for single or multiple timepoints. While it is designed to work with RootDetection ".csv" output files in "list" configuration (see Fig. 2), it can also be used with data generated with other software, provided they meet formal requirements outlined below (see Fig. 3). As part of the analysis pipeline, one- and, if applicable, two-factorial ANOVA will be performed and results can be directly

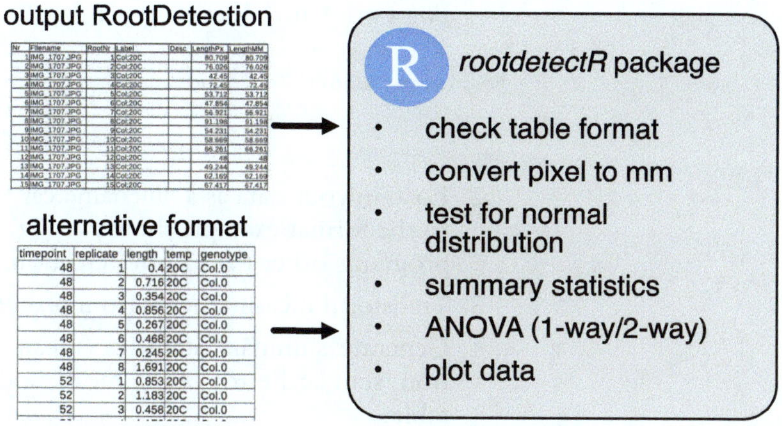

Fig. 3 Overview of the data processing pipeline of "*rootdetectR*"

plotted into graphs, even if the data does not necessarily meet the requirements for an ANOVA. Thus, it is necessary to critically inspect the generated file regarding the test for normality, carefully inspect boxplots for distribution, and/or test for homogeneity of variances, before placing any emphasis on the statistical results!

The steps below highlight the general approach followed in the data processing, analysis, and visualization. For individual commands and the full range of options provided by the package, please refer to the "*rootdetectR*" manuals (i.e., "vignettes") which are provided in two variants: a short introduction containing the most usual and important commands and an advanced manual with a comprehensive overview of possible options.

1. To install the "*rootdetectR*" package and get the vignette (i.e., detailed user manual of the package), type the following commands into the GUI of R or RStudio:

```
install.packages(c("knitr", "rmarkdown", "dev tools"))

library(devtools)

install_github("PhilippJanitza/rootdetectR", build_vignettes
= TRUE)

library(rootdetectR)
```

To view the detailed manual of the package and the available commands, use the following commands:

```
vignette("Advanced_Introduction", package = "rootdetectR")
```

For a brief introduction type:

```
vignette("Short_Introduction", package = "rootdetectR")
```

2. Load in your data as a "filename.csv" file. Make sure to adhere to the format example shown in Fig. 3 if data was obtained by programs other than RootDetection.
3. Transform measurements from pixel to mm (*see* **Note 15**).
4. Generate summary statistics (mean, median, standard deviation, standard error of the mean, and sample size) and save as table.
5. Test for normal distribution, perform ANOVA(s).
6. Plot data and export graphs. There are numerous options available for data visualization. Some examples of "*rootdetectR*" outputs are shown in Fig. 4.

4 Notes

1. Keep in mind that sucrose can affect the regulation of growth responses via the PIF4/TOR/brassinosteroid pathway [16–18].
2. Parafilm (or similar) can be used to seal plates but can result in the buildup of ample condensation water. We recommend using plaster tape (e.g., Leucopore) that can circumvent water buildup, but is more prone to drying-out of the medium when the controlled environment used has low humidity.
3. You can generate custom QR-code labels that can be printed on commercially available stickers (e.g., AVERY). This function is implemented in the RootDetection measurement software but there are also other tools available. These can be read by RootDetection during data acquisition and facilitate easy labeling of data. However, it is also possible to enter Label information manually (see RootDetection manual on www.labuitls.de).
4. Equipment described will allow the capture of 2–3 plates (depending on the space available in the controlled environment) at one ambient temperature setting. Monitoring of growth at different ambient temperatures in parallel will require at least two sets of equipment (i.e., one set per temperature).
5. While there are IR-capable cameras available (e.g., security cameras), they usually have a low resolution and are not well suited to allow reliable measurements on *Arabidopsis* seedling organs. Custom modification of standard digital cameras for IR

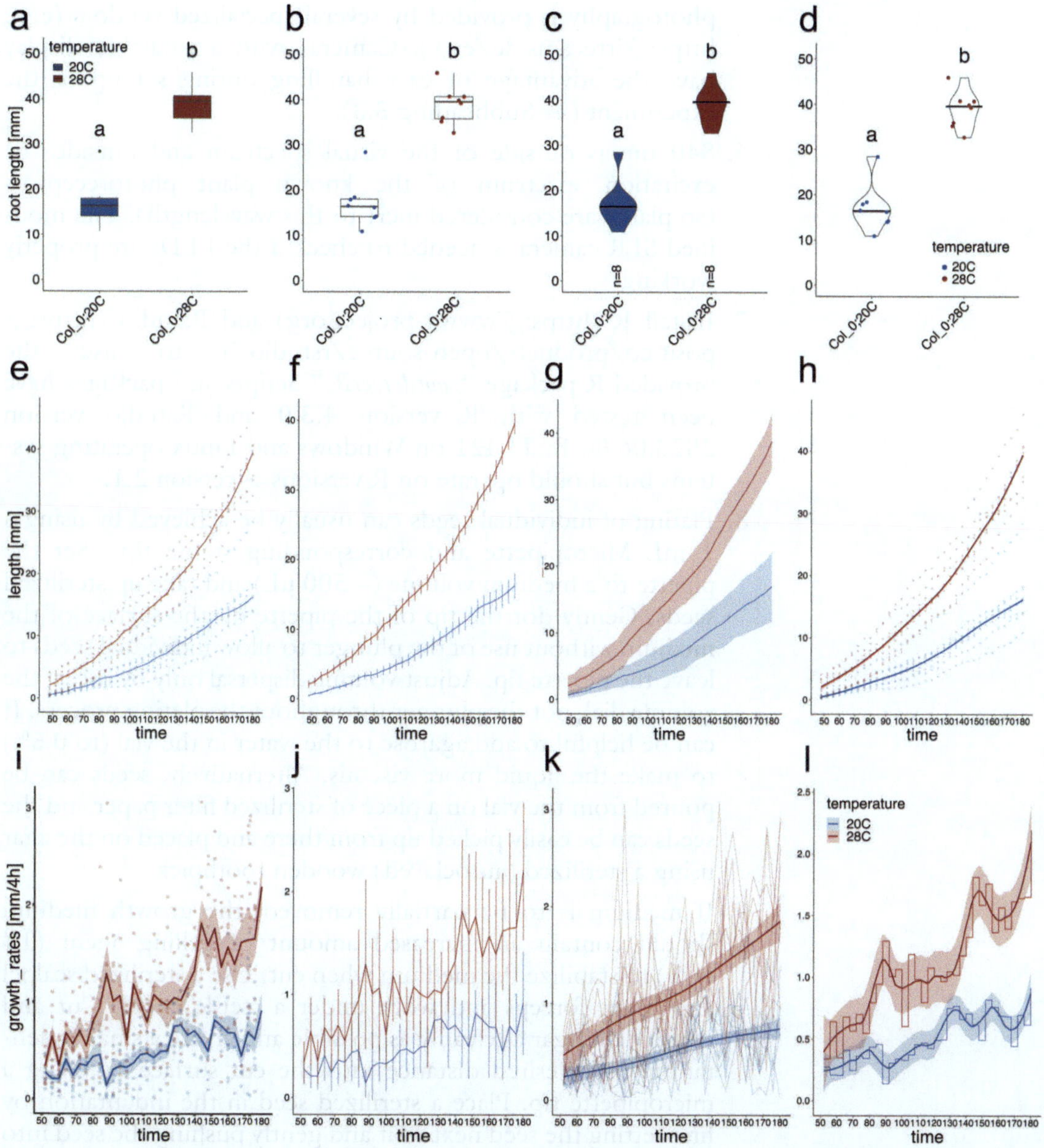

Fig. 4 Example output options generated by "*rootdetectR*" There are numerous output options for single (**a–d**) and multiple (**e–l**) timepoint analyses of which only some typical examples are shown here. Data of a single time point can be plotted as (**a**) a simple boxplot with medians, interquartile ranges, and outliers, (**b**) as boxplots with individual data points superimposed, (**c**) as violin plots with medians, or (**d**) as violin plots with individual data points. For multiple time points, absolute lengths can be shown by plotting (**e**) median values as lines with individual data points, (**f**) as mean values with standard error of the mean (SEM) as error bars, (**g**) as means with standard deviations (SD) in transparent ribbons or (**h**) as individual data points with a fitted smoothed line (method "gam") of the median. Similarly, different options are available to plot growth rates obtained from multiple time points. For example, (**i**) line graphs representing mean values with SEM in ribbons and individual data points, (**j**) means with SD as error bars, (**k**) growth rates of individual roots as half transparent lines and a fitted and smoothed (method "gam") line of mean values including 95% confidence intervals, or (**l**) step-line presentation of mean values with a fitted and smoothed (method "loess") line of the mean

photography is provided by several specialized vendors (e.g., https://irrecams.de/en/). Cameras with a rotatable display have the advantage of easy handling during set up of the experiment (*see* Subheading 3.3).

6. 840 nm is outside of the visual spectrum and outside the excitation spectrum of the known plant photoreceptors (so plants are considered inert to this wavelength). The modified SLR camera is needed to check if the LEDs are properly working.
7. Install R (https://www.r-project.org) and Rstudio (https://posit.co/products/open-source/rstudio/) to use the provided R package "*rootdetectR*." Scripts and packages have been tested with R version 4.3.0 and Rstudio version 2023.06.00 Build 421 on Windows and Linux operating systems but should operate on R versions > version 2.1.
8. Plating of individual seeds can usually be achieved by using a 1 mL Micropipette and corresponding sterile tips. Set the pipette to a medium volume (~ 500 μL) and take up sterilized seeds. Gently dot the tip of the pipette on the surface of the medium without use of the plunger to allow individual seeds to leave the pipette tip. Adjust volume dispersal only by use of the volume dial, not the plunger throughout the plating process. It can be helpful to add agarose to the water in the vial (to 0.5%) to make the liquid more viscous. Alternatively, seeds can be poured from the vial on a piece of sterilized filter paper and the seeds can be easily picked up from there and placed on the agar using a sterilized (autoclaved) wooden toothpick.
9. If medium is to be partially removed, the growth medium should contain an increased amount of gelling agent (1% w/v) to stabilize the medium when cut. Use a sterilized scalpel and blunt forceps and work under a sterile hood. Cut and remove the agar as cleanly as possible and create a small indentation with desired distance into the cut surface by using a micropipette tip. Place a sterilized seed in the indentation by first setting the seed next to it and gently pushing the seed into the indentation (Fig. 1).
10. Using a mold to ensure a fixed position of the individual plates during photography can speed up the process considerably when large batches of plates need to be processed. You can use simple cardboard or have a square (or circle, depending on the plates you use) cut out of a sheet of acrylic, hard plastic, or Plexiglas. Alternatively, a mold can be printed using a 3D printer.
11. Ensuring a strong contrast between the seedling and the background will make the measurement of seedling organs much easier. It may be necessary to use extra lights from the side to

allow good contrast without using the flash. Most SLR cameras can also be connected to a computer and be operated via a corresponding program. This can help to find the optimal settings for a given batch of plates as the test pictures can directly be evaluated on a bigger screen.

12. While RootDetection has originally be designed for automated measurements of root lengths after a transfer, semiautomatic and manual measurements of any kind of organ length is possible and feasible. Automatic detection of roots in a thermomorphogenesis context is in principle possible but it requires a horizontal black line as a starting point for measurements. Otherwise, use the semiautomated or manual measuring mode according to the user manual (www.labutils.de). If you use alternate software (e.g., ImageJ) and want to make use of the provided R package, make sure to adjust your table to the required settings (*see* Subheading 3.4). Please note that RootDetection cannot determine petiole angles, so use alternative software to assess this trait (e.g., ImageJ).
13. If you want to make use of the R package "*rootdetectR*" to process your data, the "Label" entries must not contain "-"" or any other special characters (e.g., $,%,&,/, ",!, etc). Separate the two factors you want to analyze (e.g., genotype and temperature treatment) by a semicolon and avoid any spaces in names or between the separator. If you want to include more information into the label, use a separator that is different from ";"! A good label example entry is shown in the table excerpt in Fig. 2. Also keep in mind that R is case sensitive and will regard "Col;20C" as something different from "Col;20c"!
14. In RootDetection, every measured data point on a picture will receive a number ("Rootnumber"). These numbers are given consecutively to each measurement from left to right. Make sure to always measure the same amount + rough position of seedlings at each timepoint to ensure that data points from different pictures can be associated with the correct seedling. If you use an alternative program, make sure that data points of the same seedling carry the same sample number!
15. Even though the data output generated in RootDetection contains a column named LengthMM, the values are identical to LengthPx due to the fact that automatic conversion is limited to automatic measurements (only possible for roots that originate from a solid horizontal black line). Therefore, the conversion of pixel values to mm has to be performed manually based on the standard measured for the dummy plate picture (3–5 values of 10 mm, labeled as "10 mm").

References

1. Quint M, Delker C, Franklin KA et al (2016) Molecular and genetic control of plant thermomorphogenesis. Nat Plant 2:15190. https://doi.org/10.1038/nplants.2015.190
2. Delker C, Quint M, Wigge PA (2022) Recent advances in understanding thermomorphogenesis signaling. Curr Opin Plant Biol 68:102231. https://doi.org/10.1016/j.pbi.2022.102231
3. Erwin JE, Heins RD, Karlsson MG (1989) Thermomorphogenesis in Lilium longiflorum. Am J Bot 76:47–52. https://doi.org/10.2307/2444772
4. Delker C, Sonntag L, James GV et al (2014) The DET1-COP1-HY5 pathway constitutes a multipurpose signaling module regulating plant photomorphogenesis and thermomorphogenesis. Cell Rep 9:1983–1989. https://doi.org/10.1016/j.celrep.2014.11.043
5. Crawford AJ, McLachlan DH, Hetherington AM, Franklin KA (2012) High temperature exposure increases plant cooling capacity. Curr Biol 22:R396–R397. https://doi.org/10.1016/j.cub.2012.03.044
6. Park Y-J, Lee H-J, Gil K-E et al (2019) Developmental programming of thermonastic leaf movement. Plant Physiol 180:00139. https://doi.org/10.1104/pp.19.00139
7. Ibañez C, Poeschl Y, Peterson T et al (2017) Ambient temperature and genotype differentially affect developmental and phenotypic plasticity in Arabidopsis thaliana. BMC Plant Biol 17:114. https://doi.org/10.1186/s12870-017-1068-5
8. Box MS, Huang BE, Domijan M et al (2015) ELF3 controls thermoresponsive growth in Arabidopsis. Curr Biol 25:194–199. https://doi.org/10.1016/j.cub.2014.10.076
9. van der Woude L, Piotrowski M, Klaasse G et al (2021) The chemical compound 'Heatin' stimulates hypocotyl elongation and interferes with the Arabidopsis NIT1-subfamily of nitrilases. Plant J 106:1523–1540. https://doi.org/10.1111/tpj.15250
10. van der Woude LC, Perrella G, Snoek BL et al (2019) Histone Deacetylase 9 stimulates auxin-dependent thermomorphogenesis in *Arabidopsis thaliana* by mediating H2A.Z depletion. Proc Natl Acad Sci USA 116:25343–25354. https://doi.org/10.1073/pnas.1911694116
11. Zhu Z, Quint M, Anwer MU (2022) Arabidopsis Early Flowering 3 controls temperature responsiveness of the circadian clock independently of the evening complex. J Exp Bot 73:1049–1061. https://doi.org/10.1093/jxb/erab473
12. Ai H, Bellstaedt J, Bartusch KS et al (2023) Auxin-dependent regulation of cell division rates governs root thermomorphogenesis. EMBO J 42:e111926. https://doi.org/10.15252/embj.2022111926
13. Nijland W, de Jong R, de Jong SM et al (2014) Monitoring plant condition and phenology using infrared sensitive consumer grade digital cameras. Agric For Meteorol 184:98–106. https://doi.org/10.1016/j.agrformet.2013.09.007
14. Anwer MU, Davis A, Davis SJ, Quint M (2020) Photoperiod sensing of the circadian clock is controlled by Early Flowering 3 and Gigantea. Plant J 101:1397–1410. https://doi.org/10.1111/tpj.14604
15. Lincoln C, Britton J, Estelle M (1990) Growth and development of the axr1 mutants of Arabidopsis. Plant Cell 2:1071–1080
16. Liu Z, Zhang Y, Liu R et al (2011) Phytochrome interacting factors (PIFs) are essential regulators for sucrose-induced hypocotyl elongation in Arabidopsis. J Plant Physiol 168:1771–1779. https://doi.org/10.1016/j.jplph.2011.04.009
17. Stewart JJ, Demmig-Adams B, Cohu CM et al (2016) Growth temperature impact on leaf form and function in Arabidopsis thaliana ecotypes from northern and southern Europe. Plant Cell Environ 39:1549–1558. https://doi.org/10.1111/pce.12720
18. Zhang Z, Zhu J-Y, Roh J et al (2016) TOR signaling promotes accumulation of BZR1 to balance growth with carbon availability in Arabidopsis. Curr Biol 26:1854–1860. https://doi.org/10.1016/j.cub.2016.05.005

Chapter 2

Dissection of Daytime and Nighttime Thermoresponsive Hypocotyl Elongation in *Arabidopsis*

De Fan and Meng Chen

Abstract

Hypocotyl elongation in *Arabidopsis* is widely utilized as a readout for phytochrome B (phyB) signaling and thermomorphogenesis. Hypocotyl elongation is gated by the circadian clock and, therefore, it occurs at distinct times depending on day length or seasonal cues. In short-day conditions, hypocotyl elongation occurs mainly at the end of nighttime when phyB reverts to the inactive form. In contrast, in long-day conditions, hypocotyl elongation occurs during the daytime when phyB is in the photoactivated form. Warm temperatures can induce hypocotyl growth in both long-day and short-day conditions. However, the corresponding daytime and nighttime temperature responses reflect distinct underpinning mechanisms. Here, we describe assays for dissecting the mechanisms between daytime and nighttime thermoresponsive hypocotyl elongation.

Key words Thermomorphogenesis, Phytochrome B, Hypocotyl measurement, Photoperiod

1 Introduction

Temperature is a crucial environmental factor that profoundly influences all aspects of plant development and growth, referred to as thermomorphogenesis [1–4]. Plants sense changes in ambient temperature primarily via phytochrome B (phyB) [5, 6]. PhyB monitors changes in light quality, quantity, and periodicity through photoconversions between two relatively stable conformers, a red (R) light-absorbing inactive Pr and a far-red (FR) light-absorbing active Pfr form [7, 8]. Warm temperatures attenuate the activity of phyB by accelerating the thermal reversion of the active Pfr back to the inactive Pr [9]. Because warm temperatures often coincide with high light intensities during the daytime—a combined strong-light and high-temperature condition where the majority of phyB molecules remain in the active form [10]—the essence of understanding thermomorphogenesis is to elucidate how warm temperatures engage with phyB signaling.

Meng Chen (ed.), *Thermomorphogenesis: Methods and Protocols*, Methods in Molecular Biology, vol. 2795,
https://doi.org/10.1007/978-1-0716-3814-9_2,

PhyB signaling is best studied in *Arabidopsis thaliana* (*Arabidopsis*) during seedling de-etiolation—a developmental transition that occurs when young seedlings emerge from the ground and first encounter light [11]. *Arabidopsis* seedlings germinated and grown in darkness (mimicking the condition of underground growth) adopt a dark-grown developmental program called skotomorphogenesis or etiolation, which promotes the elongation of the embryonic stem (hypocotyl) and inhibits leaf development and chloroplast biogenesis. Exposing dark-grown seedlings to light initiates de-etiolation to reprogram seedlings to a light-dependent developmental program called photomorphogenesis, which attenuates hypocotyl elongation, stimulates leaf expansion, and promotes chloroplast biogenesis [12, 13]. De-etiolation presents visibly trackable readouts, hypocotyl elongation, and leaf greening, which report two major functions of phyB signaling in controlling plant growth and chloroplast biogenesis, respectively [12, 13].

The effectiveness of phyB in restraining hypocotyl elongation has also been widely used as a physiological assay for phyB-mediated thermomorphogenesis because warm temperatures significantly accelerate hypocotyl elongation [5, 6, 14, 15]. However, often being ignored is the difference between thermal responses in short-day (SD) and long-day (LD) condition. Because hypocotyl elongation is gated by the circadian clock and partitioned to different times in SD and LD conditions [16–18], the hypocotyl-elongation assay performed in different day-length conditions assess two distinct mechanisms in thermomorphogenesis. In SD conditions, hypocotyl elongation occurs mainly at the end of the night or in the dark, when phyB has mostly reverted to the inactive Pr [17]. By contrast, in LD conditions including continuous light, hypocotyl elongation peaks during the daytime, when PHYB is in the active Pfr [16–18]. In accordance with the contrast in daylength-dependent rhythmic growth patterns, hypocotyl elongation is modulated by temperature changes at distinct times in either dark or light under SD and LD conditions, respectively [19, 20]. Therefore, although hypocotyl elongation under both SD and LD conditions can be strongly influenced by temperature changes, the underpinning mechanisms are different [19, 21]. This discrepancy in the timing of hypocotyl growth between nighttime and daytime represents two mechanistic strategies by which temperature cues engage with phyB signaling in different growing seasons. For example, while EARLY FLOWERING 3 (ELF3) mainly participates in nighttime thermoresponsive hypocotyl elongation in SD conditions [19], HEMERA (HMR) primarily contributes to daytime thermoresponsive hypocotyl elongation in LD conditions [21]. CONSTITUTIVE PHOTOMORPHOGENIC 1 (COP1) and PHYTOCHROME-INTERACTING FACTORs (PIFs), such as PIF4, PIF5, and PIF7, are required for the temperature responses in both LD and SD conditions [19, 21, 22]. Here, we describe assays that dissect the daytime and nighttime thermoresponsive hypocotyl elongation. These assays can be used to discern the role of other genes in these two distinct modes of thermomorphogenesis responses.

2 Materials

2.1 Plant Materials and Seedling Growth

1. The *Arabidopsis thaliana* accession Columbia (Col-0) and previously described mutants in Col-0 background, including *cop1-4* [23], *hmr-5* [24], and *elf3-1* [25] (*see* **Note 1**).
2. Half-strength Murashige and Skoog (1/2 MS) media: 2.2 g Murashige and Skoog with Gamborg's vitamins (Caisson Labs, Smithfield, UT), 0.5 g MES, 0.8% (w/v) agar, adjust pH to 5.7, add water to a total volume of 1 L, and autoclave.
3. Seed sterilization solution: 0.02% (v/v) Triton X-100 in 50% (v/v) bleach.
4. Percival LED chamber (Percival Scientific, Perry, IA) for seeding growth under monochromatic R light.
5. Apogee spectroradiometer model PS-200 (Apogee Instruments, Logan, UT) for measuring the fluence rate of light.

2.2 Hypocotyl Measurement

1. Black agarose plate: 2% agarose containing 1% (v/v) of black ink (Hello Hobby).
2. Leica M165 FC stereo microscope equipped with a Leica FLEXACAM C1 12 MP stand-alone microscope camera (Leica Microsystems, Deerfield, IL) (*see* **Note 2**).
3. Adobe Illustrator (Adobe, San Jose, CA).
4. Epson Perfection V700 photo scanner (Epson America, Long Beach, CA).
5. NIH ImageJ software (http://rsb.info.nih.gov/nih-image/).
6. Prism 8 software (GraphPad Software, San Diego, CA).

3 Methods

3.1 Temperature Treatments Under LD and SD Conditions

1. Sterilize Col-0, *cop1-4*, *hmr-5*, and *elf3-1* seeds in seed sterilization solution for 10 min and wash four times in sterile ddH_2O.
2. Pipette seeds individually onto four 1/2 MS plates (*see* **Note 3**).
3. Stratify seeds in the dark at 4 °C for 5 days.
4. For LD experiments, place the two plates containing stratified seeds into a Percival LED chamber in LD conditions (16 h 50 μmol m^{-2} s^{-1} monochromatic R light/8 h darkness, or

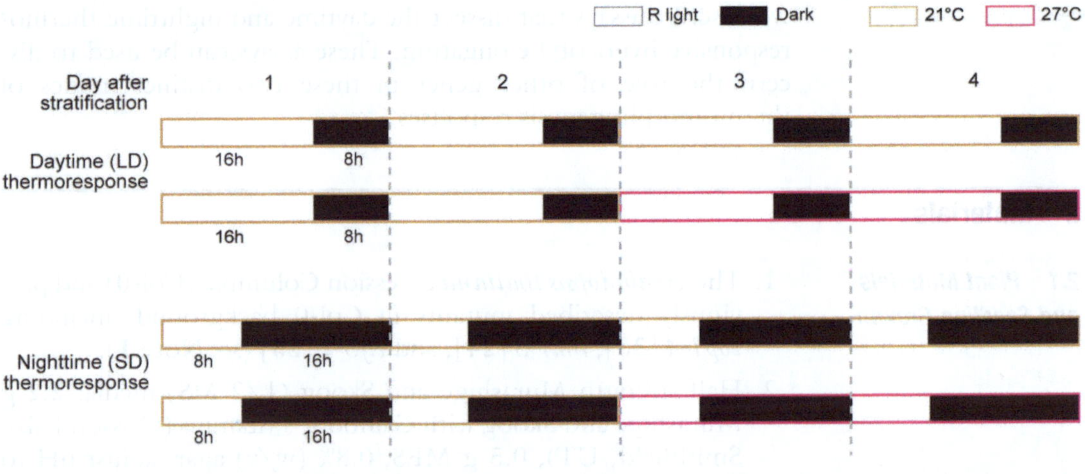

Fig. 1 Schematic diagram showing the seedling growth conditions for assessing daytime (top) and nighttime (bottom) thermoresponsive hypocotyl elongation. For measuring daytime thermomorphogenesis, a 16 h 50 μmol m^{-2} s^{-1} R light and 8 h of darkness diurnal cycle is used. For measuring nighttime thermomorphogenesis, an 8 h 50 μmol m^{-2} s^{-1} R light and 16 h of darkness diurnal cycle is used. Seedlings are incubated in the first 2 days (48 h) at 21 °C to induce germination. During days 3 and 4, seedlings are grown at either 21 °C or 27 °C to assess the temperature effects

continuous R light). For SD experiments, place two plates under 50 μmol m^{-2} s^{-1} monochromatic R light in SD conditions (8 h 50 μmol m^{-2} s^{-1} monochromatic R light/16 h darkness), set both chambers at 21 °C, and incubate the plates for 2 days (48 h) to induce seed germination (Fig. 1) (*see* **Note 4**).

5. For warm temperature treatments, keep one plate at 21 °C and transfer the other plate to 27 °C under the corresponding daylength conditions, and grow both groups of seedlings for two additional days (48 h) (Fig. 1) (*see* **Note 5**).
6. Collect 4-day-old seedling samples at 96 h after stratification.

3.2 Hypocotyl Measurement

1. For taking seedling images, place representative seedlings of each genotype grown under various conditions on a 2% agarose black plate.
2. Take seedling images using a Leica M165 FC stereo microscope and save the images in TIFF format. Combine the images of all genotypes using Adobe Illustrator (Fig. 2a, c).
3. For measuring hypocotyl length, transfer at least 60 seedlings of each genotype under each growth condition on a transparency film covered with wet Kimwipes.
4. Scan seedlings using an Epson Perfection V700 photo scanner. Set the Epson Perfection V700 scanner to "Professional Mode." In the scanning software, select the following settings: Document Type: Film (with Film Area Guide); Film Type:

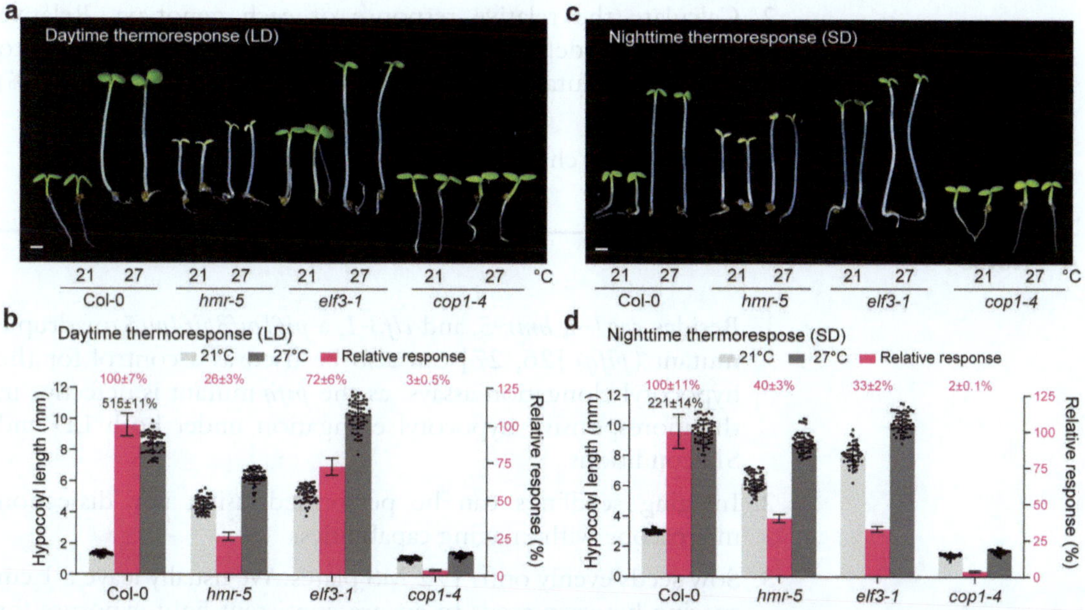

Fig. 2 Daytime and nighttime thermoresponsive hypocotyl elongation. (**a**) Images of representative 4-d-old Col-0, *hmr-5*, *elf3–1*, and *cop1–4* seedlings grown in LD conditions. The scale bar equals 1 mm. (**b**) Hypocotyl length measurements of LD-grown seedlings described in (**a**). (**c**) Images of representative 4-d-old Col-0, *hmr-5*, *elf3–1*, and *cop1–4* seedlings grown in SD conditions. The scale bar equals 1 mm. (**d**) Hypocotyl length measurements of SD seedlings described in (**c**). For (**b**) and (**d**), the light- and dark-gray bars represent hypocotyl length measurements at 21 °C and 27 °C, respectively. The percent increase in hypocotyl length (mean ± SD, $n =$ at least 3 biological replicates) of Col-0 at 27 °C is shown in black above its columns. The magenta bars show the relative response, which is defined as the relative hypocotyl response to 27 °C of a mutant compared with that of Col-0 (set at 100%). Error bars for the hypocotyl measurements represent SD ($n > 30$ seedlings); error bars for the relative responses represent SD of at least three biological replicates. The centers of the error bars represent the mean values

Positive film; Image Type: 16-bit Grayscale, Resolution: 300 dpi. Save images in TIFF format.

5. Measure hypocotyl length by using NIH ImageJ software. Import the scanned seedling images. Calibrate the scale using the image of a ruler. Measure hypocotyl length using the "Segmented Line" function. Save the results in a spreadsheet.

3.3 Data Analysis

1. Assess daytime and nighttime thermoresponsive hypocotyl elongation by calculating the percent increase in hypocotyl length of each genotype at 27 °C compared to 21 °C under either LD or SD conditions, respectively. For example, Col-0 seedlings exhibit 221 ± 14% and 515 ± 11% increases in hypocotyl length for daytime (LD) and nighttime (SD) thermoresponses, respectively (Fig. 2b, d).

2. Calculate the relative response of each genotype. Relative response is defined as the relative hypocotyl response to 27 °C of a mutant compared with that of Col-0 (set at 100%) (Fig. 2b, d).
3. Generate bar charts using GraphPad Prism 8 (Fig. 2b, d).

4 Notes

1. Besides *cop1-4*, *hmr-5*, and *elf3-1*, a *pif1pif3pif4pif5* quadruple mutant (*pifq*) [26, 27] can also be used as a control for the hypocotyl elongation assays, as the *pifq* mutant is defective in thermoresponsive hypocotyl elongation under both LD and SD conditions.
2. Imaging seedlings can be performed using any dissection microscope with imaging capabilities.
3. Sow seeds evenly onto 1/2 MS plates. We usually leave a 1 cm spacing between seeds to ensure consistent light exposure for all seedlings.
4. Monochromatic R light is used for the assays to avoid the repression of thermoresponsive hypocotyl elongation by the blue light photoreceptor CRYPTOCHROME 1 during the first four days of seedling development [28]. The assay could also be performed under white light. However, instead of using 4-day-old seedlings, the temperature-dependent hypocotyl elongation response is best observed using 8-day-old seedlings [21].
5. Thermoresponsive hypocotyl elongation is monitored during the third and fourth days (48–96 h) after stratification because hypocotyl elongation occurs mainly during this seedling developmental period [29, 30].

Acknowledgments

This work was supported by the National Institute of Health grant R01GM087388 to M.C.

References

1. Quint M, Delker C, Franklin KA et al (2016) Molecular and genetic control of plant thermomorphogenesis. Nat Plants 2:15190
2. Casal JJ, Balasubramanian S (2019) Thermomorphogenesis. Annu Rev Plant Biol 70:321–346
3. Delker C, Quint M, Wigge PA (2022) Recent advances in understanding thermomorphogenesis signaling. Curr Opin Plant Biol 68:102231
4. Quint M, Delker C, Balasubramanian S et al (2023) 25 years of thermomorphogenesis research: milestones and perspectives. Trends Plant Sci 28:1098–1100

5. Legris M, Klose C, Burgie ES et al (2016) Phytochrome B integrates light and temperature signals in *Arabidopsis*. Science 354:897–900
6. Jung J-H, Domijan M, Klose C et al (2016) Phytochromes function as thermosensors in *Arabidopsis*. Science 354:886–889
7. Burgie ES, Vierstra RD (2014) Phytochromes: an atomic perspective on photoactivation and signaling. Plant Cell 26:4568–4583
8. Rockwell NC, Lagarias JC (2020) Phytochrome evolution in 3D: deletion, duplication, and diversification. New Phytol 225:2283–2300
9. Klose C, Nagy F, Schäfer E (2019) Thermal reversion of plant phytochromes. Mol Plant 13:386–397
10. Hahm J, Kim K, Qiu Y et al (2020) Increasing ambient temperature progressively disassemble *Arabidopsis* phytochrome B from individual photobodies with distinct thermostabilities. Nat Commun 11:1660
11. Chen M, Chory J (2011) Phytochrome signaling mechanisms and the control of plant development. Trends Cell Biol 21:664–671
12. Chen M, Chory J, Fankhauser C (2004) Light signal transduction in higher plants. Annu Rev Genet 38:87–117
13. Yoo CY, Han S, Chen M (2020) Nucleus-to-plastid phytochrome signaling in controlling chloroplast biogenesis. Annu Plant Rev 3: 251–280
14. Gray WM, Ostin A, Sandberg G et al (1998) High temperature promotes auxin-mediated hypocotyl elongation in *Arabidopsis*. Proc Natl Acad Sci USA 95:7197–7202
15. Foreman J, Johansson H, Hornitschek P et al (2011) Light receptor action is critical for maintaining plant biomass at warm ambient temperatures. Plant J 65:441–452
16. Dowson-Day MJ, Millar AJ (1999) Circadian dysfunction causes aberrant hypocotyl elongation patterns in *Arabidopsis*. Plant J 17:63–71
17. Nozue K, Covington MF, Duek PD et al (2007) Rhythmic growth explained by coincidence between internal and external cues. Nature 448:358–361
18. Michael TP, Breton G, Hazen SP et al (2008) A morning-specific phytohormone gene expression program underlying rhythmic plant growth. PLoS Biol 6:e225
19. Park Y-J, Lee H-J, Ha J-H et al (2017) COP1 conveys warm temperature information to hypocotyl thermomorphogenesis. New Phytol 215:269–280
20. Zhu J-Y, Oh E, Wang T et al (2016) TOC1-PIF4 interaction mediates the circadian gating of thermoresponsive growth in *Arabidopsis*. Nat Commun 7:13692
21. Qiu Y, Li M, Kim RJ-A, et al (2019) Daytime temperature is sensed by phytochrome B in *Arabidopsis* through a transcriptional activator HEMERA. Nat Commun 10:140
22. Fiorucci A-S, Galvão VC, Ince YÇ et al (2020) Phytochrome Interacting Factor 7 is important for early responses to elevated temperature in *Arabidopsis* seedlings. New Phytol 226:50–58
23. McNellis TW, von Arnim AG, Araki T et al (1994) Genetic and molecular analysis of an allelic series of *cop1* mutants suggests functional roles for the multiple protein domains. Plant Cell 6:487–500
24. Galvão RM, Li M, Kothadia SM et al (2012) Photoactivated phytochromes interact with HEMERA and promote its accumulation to establish photomorphogenesis in *Arabidopsis*. Genes Dev 26:1851–1863
25. Hicks KA, Millar AJ, Carré IA et al (1996) Conditional circadian dysfunction of the *Arabidopsis* early-flowering 3 mutant. Science 274: 790–792
26. Leivar P, Monte E, Oka Y et al (2008) Multiple phytochrome-interacting bHLH transcription factors repress premature seedling photomorphogenesis in darkness. Curr Biol 18:1815–1823
27. Shin J, Kim K, Kang H et al (2009) Phytochromes promote seedling light responses by inhibiting four negatively-acting phytochrome-interacting factors. Proc Natl Acad Sci USA 106:7660–7665
28. Ma D, Li X, Guo Y et al (2016) Cryptochrome 1 interacts with PIF4 to regulate high temperature-mediated hypocotyl elongation in response to blue light. Proc Natl Acad Sci USA 113:224–229
29. Van Buskirk EK, Reddy AK, Nagatani A et al (2014) Photobody localization of phytochrome B is tightly correlated with prolonged and light-dependent inhibition of hypocotyl elongation in the dark. Plant Physiol 165: 595–607
30. Gendreau E, Traas J, Desnos T et al (1997) Cellular basis of hypocotyl growth in *Arabidopsis thaliana*. Plant Physiol 114:295–305

Chapter 3

Light-Dependent High Ambient Temperature-Induced Senescence Assay Using Whole Seedlings

Chanhee Kim and Giltsu Choi

Abstract

High ambient temperature affects various plant developmental and physiological processes, including senescence. Here, we present a protocol for assaying light-dependent high ambient temperature-induced senescence using whole seedlings. The protocol covers all steps, from inducing senescence by darkness at high ambient temperature to determining the degree of senescence, and includes experimental tips and notes. The onset of senescence is established by quantifying the increased expression of senescence marker genes by quantitative real-time PCR (RT-qPCR). The degree of senescence is determined by measuring the loss of chlorophyll and the increase of ion leakage. This protocol can be adapted to study light-dependent high ambient temperature-induced senescence in other plant species by adjusting the temperature and duration of darkness.

Key words Leaf senescence, High ambient temperature, Senescence marker gene, Chlorophyll, RT-qPCR, *Arabidopsis thaliana*

1 Introduction

Leaf senescence is a plant developmental process accompanied with the disassembly of cellular organelles, the degradation of macromolecules, and the remobilization of nutrients to developing organs in plants [1]. This natural process occurs as a leaf ages or experiences stressful conditions that hinders photosynthesis such as prolonged darkness, abiotic stresses, and pathogen attacks. Due to the irreversibility of leaf senescence, plants must integrate various internal and external signals to initiate and regulate the senescence process. These distinctive characteristics of leaf senescence make it an excellent model system for studying how different external and internal factors interact to end the life of an organ and recycle biological components.

Meng Chen (ed.), *Thermomorphogenesis: Methods and Protocols*, Methods in Molecular Biology, vol. 2795,
https://doi.org/10.1007/978-1-0716-3814-9_3,

High ambient temperature affects various developmental and physiological processes throughout the plant life cycle. High ambient temperature inhibits seed germination at the beginning of the life cycle, promotes the elongation of hypocotyl and petiole, and results in the production of small-sized hyponastic leaves during the vegetative stage [2–5]. These adaptations allow plants to dissipate heat more efficiently. Furthermore, high ambient temperature stimulates flowering during the final stage of the life cycle [5–7]. Notably, high ambient temperature also accelerates leaf senescence, presumably as a strategy to shed off photosynthetically less efficient old leaves and mobilize nutrients to support the growth of developing young organs, enabling survival under a less optimal temperature conditions [5, 8]. This promotion of leaf senescence is partly achieved through the inactivation of phytochrome B (phyB), which suppresses senescence by inhibiting its interacting transcription factors, PIF4 and PIF5 [9–11]. A method to assay light-dependent high ambient temperature-induced leaf senescence would help to investigate the role of light and high ambient temperature in this crucial process.

The onset and degree of leaf senescence can be estimated by measuring the loss of chlorophyll molecules, ion leakage, and the expression of senescence marker genes, as these are closely associated with leaf senescence [12]. Here, we describe a detailed method for assaying light-dependent high ambient temperature-induced leaf senescence using whole seedlings of *Arabidopsis*, with experimental tips provided for each step.

2 Materials

MS agar media is prepared using distilled water, while other solutions are made with Milli-Q quality ultrapure water (18 MΩ/cm at 25 °C). Ensure the use of analytic grade reagents during the preparation process.

2.1 Plant Materials

1. *Arabidopsis thaliana* seeds.
2. *Nicotiana benthamiana* seeds.
3. *Solanum lycopersicum* seeds.

2.2 Reagents

1. Seed sterilization solution: 0.015% (v/v) Triton-X 100, 10% (v/v) bleach in distilled water.
2. Half MS-agar plate: 1/2 Murashige and Skoog (MS) medium (2.2 g/L), 0.8% phytoagar, 0.05% MES, KOH-titrated pH 5.7 (*see* **Note 1**).

3. RT-qPCR reagents.
   (a) RNase-free ultrapure water.
   (b) 10 mM dNTP.
   (c) RNase Inhibitor such as Solg™ RNase Inhibitor (SOR51-E20h, Solgent).
   (d) Reverse transcriptase such as M-MLV reverse transcriptase and 5X reaction buffer (M1701, Promega).
   (e) Synthetic single-stranded 20-mer oligonucleotide Oligo (dT)20 primer: 5′-TTT TTT TTT TTT TTT TTT TT-3′.
   (f) SYBR green qPCR premix such as TOPreal SYBR Green qPCR preMIX (RT500M, Enzynomics).
4. Absolute Ethanol: 95%.
5. Ultrapure water: Milli-Q quality ultrapure water (18 MΩ/cm at 25 °C).
6. Senescence marker gene primers such as:
   (a) *PP2AA3* (AT1G13320):
   Forward: 5'-CAT TGT AGA ACT TGC TGA AGA CAG G-3'.
   Reverse: 5'-CAA GAC GCT TCA GAT TGT TTG CAG C-3'.
   (b) *ORESARA1* (*ORE1*, AT5G39610):
   Forward: 5'-GGA ACC AGC CGG TTT ACC TTC GT-3'.
   Reverse: 5'-ACC GTC GAA AGG TTT GCC GGA G-3'.
   (c) *NONYELLOWING1* (*NYE1*, AT4G22920):
   Forward: 5'-TTG TGC ATG GAG ATG GGA ACT TG-3'.
   Reverse: 5'-CCT CGG GAA GAG TCT CAG TCT T-3'.
7. Silica-type column-based RNA purification kit such as TaKaRa MiniBEST Plant RNA extraction kit (9769; TaKaRa).
8. Tungsten beads: 6 mm.

2.3 Plasticware and Others

1. Petri dishes: 10 mm height for *Arabidopsis thaliana*, 40 mm height for taller seedlings such as *Nicotiana benthamiana* and *Solanum lycopersicum*.
2. Cell culture plates: 12 well.
3. Falcon tubes: 15 mL, 50 mL.
4. Microfuge tubes: 1.7 mL, round bottom 2 mL.
5. 8-strip PCR tube.

6. Sterilized Pasteur pipettes with rubber bulbs.
7. Green LED (500–510 nm) or Rosco #389 Chroma Green Filter.
8. Quartz cuvette.

2.4 Equipment

1. Microbalance.
2. Tube rotator.
3. Tabletop microcentrifuge.
4. Electric cooking pot.
5. UV Spectrophotometer.
6. Conductivity meter.
7. Tissue lyser.
8. Microvolume spectrophotometer.
9. Thermal cycler.
10. Quantitative real-time PCR machine.
11. Growth chambers with programmable red (670 nm) and far-red (730 nm) lighting system set at 20 °C and 28 °C.

3 Methods

3.1 Preparation of Light-Grown Seedlings for Senescence Assay

1. Sterilize seeds for 5 min with 1 mL of the seed sterilization solution in a 1.7 mL microtube by gentle shaking.
2. Centrifuge briefly with a tabletop mini centrifuge and carefully remove the supernatant.
3. Wash seeds five times with 1 mL of ultrapure water each time, shaking vigorously during each wash.
4. Spot seeds 10 mm apart on half-strength MS-agar plates using sterile Pasteur pipettes with rubber bulbs. If comparing genotypes, sow them on the same plate (*see* **Note 2**).
5. Stratify seeds by wrapping the plates with aluminum foil and storing them at 4 °C for 3 days.
6. Remove the aluminum foil cover and incubate the plates under continuous white light (100 μmol m^{-2} s^{-1}) in a plant growth chamber set at 20 °C for 7 days.
   (a) Mark an “X” on seedlings that did not grow to a similar size as the other seedlings; do not use them for quantifying chlorophyll level, ion leakage, and marker gene expression after the induction of senescence.
   (b) Note that the incubation duration under white light may vary for different plant species depending on their growth characteristics (*see* **Note 3**).

3.2 Induction of Whole Seedling Senescence

1. To ensure uniform starting conditions of phytochromes for senescence induction, reset phytochrome to Pr or Pfr. (If the senescence is induced by darkness with intermittent light pulses, proceed to step 4.)
   (a) Install growth chambers equipped with both red LEDs (670 nm) and far-red LEDs (730 nm) in a light-tight dark room with the safety green light (*see* **Notes 4** and **5**). Set the growth chamber at 20 °C.
   (b) To reset to Pr, irradiate seedlings with a far-red light pulse (2.5 μmol/m^2/s, 10 min) without opening the plate cover. Ensure there are no water droplets on the cover.
   (c) To reset to Pfr, irradiate seedlings with a red light pulse (25 μmol/m^2/s, 10 min) without opening the plate cover.
   (d) Wrap the plates with aluminum foil under the safety green light immediately after irradiating a red light or a far-red light pulse.
2. Transfer the wrapped plates to growth chambers with temperatures set at 20 °C or 28 °C. Set aside a set of plates for 0-day sample.
3. Take out the plates from the growth chamber at 5, 7, 9 days after the transfer (Fig. 1) (*see* **Note 6**). Use seedlings to quantify chlorophyll levels, electrolyte leakage, and the expression of senescence marker gene mRNAs.
4. This step is for the induction of whole seedling senescence by darkness with intermittent light pulses.
   (a) Install four growth chambers equipped with programmable red and far-red LED system in a light-tight dark room with the safety green light. Set two growth chambers at 20 °C and the other two growth chambers at 28 °C (see **Note** 7).
   (b) Set one growth chamber to emit a red light pulse (25 μmol/m^2/s, 10 min) at regular intervals (e.g., every 4 h) and set the other chamber to emit a red light pulse immediately followed by a far-red light pulse (2.5 μmol/m^2/s, 10 min) at the same regular interval.
   (c) Divide the light-grown seedling plates into two groups, transferring one group to a growth chamber with red light pulse and the other group to the other growth chamber with both red and far-red light pulses. Reserve a set of plates for 0-day sample.
   (d) Take out a set of plates from the growth chambers at 5, 7, 9 days after the transfer and quantify the degree of senescence.

Fig. 1 Induction of whole seedling senescence of three plant species at different temperatures. Light-grown 7-day-old *Arabidopsis thaliana* (**a**), 12-day-old *Solanum lycopersicum* (**b**), and 9-day-old *Nicotiana benthamiana* (**c**) seedlings were irradiated with a far-red pulse (2.5 μmol/m^2/s, 10 min), transferred to darkness, and incubated for various days at different temperatures. DDI refers to days of dark incubation

3.3 Quantification of Chlorophyll Levels

1. Collect cotyledons from 5 seedlings into a pre-weighed microfuge tube. Cotyledons can be excised by holding the lower hypocotyl of a seedling with forceps and cutting at the base of the cotyledon using sharp scissors. Avoid using seedlings whose cotyledons touch the surface of the half-strength MS agar medium. Additionally, exclude seedlings marked with an X in Subheading 3.1, **step 6-a**.
2. Weigh 10 cotyledons by measuring the total weight of a microfuge tube with cotyledons and subtracting the weight of the pre-weighed empty microfuge tube.
3. Add 1 mL of absolute ethanol to each microfuge tube and wrap the microfuge tube tightly with aluminum foil to protect from light exposure.

4. Extract chlorophylls by rotating the microfuge tube at 4 °C for 24 h at 15 rpm.
5. Carefully transfer 500 μL of the supernatant to a Quartz cuvette (*see* **Note 8**).
6. Measure the absorbance at 648.6 and 664.2 nm using a pre-warmed UV spectrophotometer. Use absolute ethanol for the blank.
7. Thoroughly wash the cuvette with ultrapure water and ethanol before measuring other samples.
8. Calculate chlorophyll concentration (mg/mL) using the following equations [13]:

   $$\text{Total chlorophylls} = 5.24^*A_{664.2} + 22.24^*A_{648.6}$$

9. Calculate the normalized chlorophyll level (mg/g) by dividing the total chlorophyll concentration by the fresh weight of each sample (Fig. 2).

3.4 Electrolyte Leakage Assay

1. Collect cotyledons of 10 seedlings into 6 mL ultrapure water in a 50 mL Falcon tube (*see* **Notes 9** and **10**).
2. Gently shake the sample for 3 h at room temperature.
3. Measure the conductivities of ultrapure water (C0) and the sample (C3) with a conductivity meter (*see* **Note 11**).
4. Boil the sample for 10 min in an electric cooker pot.
5. Spin down briefly to allow droplets to fall to the bottom. Allow the sample to cool at room temperature.
6. Measure the conductivity of the sample (C10) with a conductivity meter.
7. Calculate the electrolyte leakage value using the formula:

   $$\text{Electrolyte leakage} = (C3 - C0)/(C10 - C0).$$

3.5 Marker Gene mRNA Expression Levels

1. Collect approximately 80 seedlings into a 2 mL round bottom microfuge tube with a single 6 mm tungsten bead (*see* **Notes 12** and **13**).
2. Freeze the sample in the microfuge tube by immersing in liquid nitrogen.
3. Grind the sample with the tissue lyser such as Tissue Lyser II (85300, QIAGEN).
   (a) Set the shaking frequency of the tissue lyser to 28 Hz and the duration to 1 min and 30 s.
   (b) Place the frozen sample tube in a tissue lyser adapter set that was chilled by dipping it in liquid nitrogen.

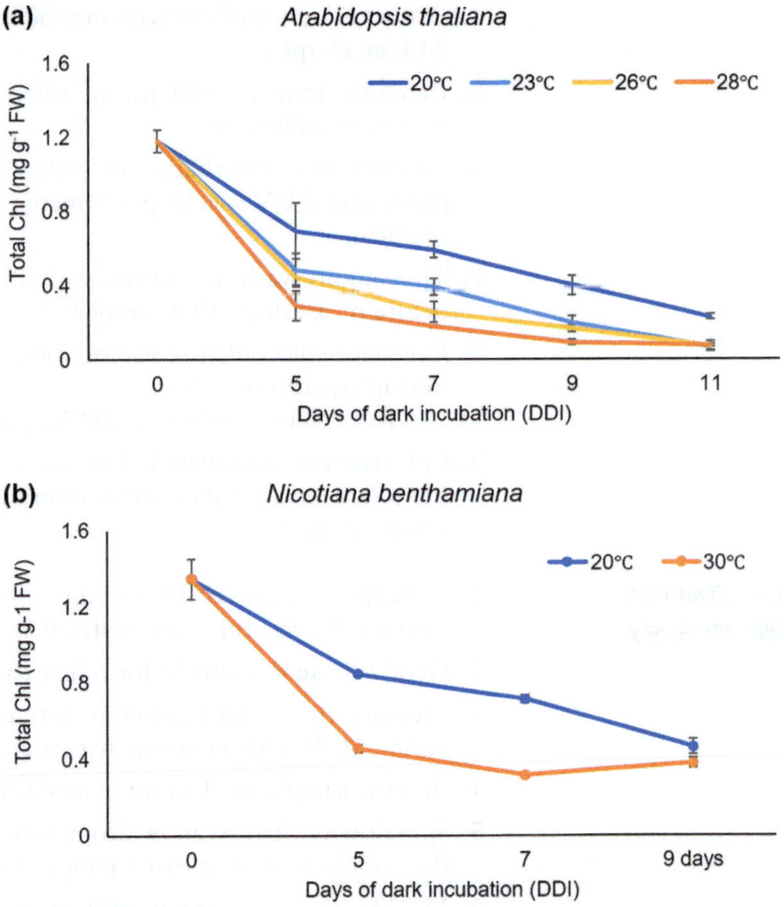

Fig. 2 Quantification of chlorophyll levels after the induction of senescence at different temperatures. Light-grown 7-day-old *Arabidopsis thaliana* and 9-day--old *Nicotiana benthamiana* seedlings were irradiated with a far-red pulse (2.5 μmol/m^2/s, 10 min), transferred to darkness, and incubated for various days at different temperatures. After the induction of senescence, cotyledons were excised and chlorophylls were extracted from excised cotyledons and chlorophyll levels were determined by a spectrophotometer. DDI refers to days of dark incubation. FW refers to fresh weight

(c) Install the adaptor set in the tissue lyser and grind the sample.

(d) Remove the adaptor set and chill it again by dipping it in liquid nitrogen.

(e) Repeat step 3-c and 3-d two more times.

4. Extract the total RNA from each sample using a spin column-based plant RNA purification kit, such as the TaKaRa MiniBEST Plant RNA extraction kit, following the instructions provided in the kit's manual.

5. Determine the concentration of the purified total RNA sample by measuring the absorbance at 230 nm, 260 nm, and 280 nm using a microvolume spectrophotometer such as the NanoDrop spectrophotometer (ND-2000, Thermo Fisher Scientific).
6. Synthesize cDNAs from the total RNA using oligo dT20 primers and the M-MLV reverse transcriptase.
   (a) In an 8-strip PCR tube, combine 2 μg of total RNA and 1 μL of oligo dT20 (100 pmol/μL). Add RNase-free ultrapure water to make the total volume of 5 μL.
   (b) Mix the contents by gently tapping the tube and then briefly spin the tube to ensure proper mixing.
   (c) Place the sample tube in a thermal cycler and run a cycle at 70 °C for 5 min, then cool to 4 °C.
   (d) Prepare a sufficient amount of reverse transcription reaction mixture, considering the number of total samples plus one (1 reverse transcription reaction mixture: 9 μL of RNase-free ultrapure water, 4 μL of M-MLV 5X reaction buffer, 1 μL of 10 mM dNTP, 0.5 μL of RNase Inhibitor, 0.5 μL of M-MLV reverse transcriptase).
   (e) Add 15 μL of the reverse transcription reaction mixture to each sample tube.
   (f) Mix the contents by gently tapping the tube and briefly spin the tube to ensure proper mixing.
   (g) Place the sample tube in a thermal cycler and run a cycle at 42 °C for 60 min, followed by 70 °C for 15 min and finally cool to 4 °C.
7. Quantify senescence marker gene expression levels by qPCR.
   (a) Add 180 μL of ultrapure water to each cDNA sample tube for a ten-fold dilution. Mix by gently tapping and spin briefly. Put it on ice.
   (b) Prepare the qPCR reaction mixture for the number of samples plus one (1 qPCR reaction mixture: 10 μL of 2X SYBR green qPCR premix, 1 μL of forward primer (10 pmol/μL), 1 μL of reverse primer (10 pmol/μL), 5 μL of ultrapure water). Ensure to prepare the qPCR reaction mixture also for a control gene mRNA such as PP2AA3.
   (c) Transfer 3 μL of the diluted cDNA sample into a new 8-strip PCR tube. Add 17 μL of the qPCR reaction mixture to each cDNA sample tube.
   (d) Mix the contents by gently tapping the tube and then briefly spin the tube to ensure proper mixing.

(e) Run the quantitative real-time PCR machine with the specific program (95 °C 10 min and 50 cycles of (95 °C 20 s – 56 °C 20 s – 72 °C 20 s).

(f) Normalize the expression level of the specific marker gene mRNA to that of *PP2AA3* mRNA (*see* **Note 14**).

4 Notes

1. Do not include sucrose in half MS-agar medium as its presence delays senescence.
2. For *Arabidopsis*, use 10 mm height petri dishes, while for taller seedlings such as *Nicotiana benthamiana* and *Solanum lycopersicum*, use 40 mm height petri dishes.
3. Transfer light-grown seedlings to darkness before the emergence of the first true leaves. Typically, transfer 7-day-old seedlings for *Arabidopsis thaliana*, 9-day-old seedlings for *Nicotiana benthamiana*, and 12-day-old seedlings for *Solanum lycopersicum*.
4. Alternatively, two growth chambers can be used: one with red LEDs and the other with far-red LEDs.
5. The safety green light can be made by using LED (500–510 nm) or by wrapping a light lamp with Rosco #389 Chroma Green Filter.
6. The most suitable time intervals for dark-induced and/or high ambient temperature-induced senescence may vary depending on the plant species, genotype, and temperature treatment conditions. It is recommended to conduct experiments with various time intervals to identify the optimal timing for senescence induction.
7. Need at least two growth chambers equipped with programable red and far-red LED system. If only two growth chambers are available, perform the experiment by adjusting the temperature of growth chambers to 20 °C and repeat the experiment by adjusting the temperature of growth chambers to 28 °C.
8. Remove the aluminum foil cover from each sample just before measuring the absorbance.
9. Cut cotyledons by holding the lower hypocotyl with forceps and cutting at the upper region of the hypocotyl with clean scissors. Do not use seedlings whose cotyledons touched the half-strength MS agar medium. Additionally, exclude seedlings marked with an X in step 3.1-6-a.
10. Ensure that cotyledons and scissor blades do not touch the medium during handling.

11. When measuring the conductivity, clean the forceps and the tip of the conductivity meter with ethanol and ultrapure water between each sample measurement.

12. The expression of the senescence marker gene precedes the visible decrease of chlorophyll levels. Thus, it is recommended to determine the marker gene mRNA expression at 2–4 days after the transfer to darkness.

13. It is highly recommended to use a 2 mL round bottom microfuge tube. If you use 1.7 mL microfuge tube during step 3.5-3, both the cap and bottom part of the tube may be broken by the tungsten bead.

14. In previous studies, various senescence marker genes have been reported, such as the leaf senescence regulator gene *ORE1* and the chlorophyll degradation regulator gene *NYE1*. Consider checking the expression levels of several senescence marker genes with your sample to get a comprehensive understanding of the senescence process.

References

1. Lim PO, Kim HJ, Nam HG (2007) Leaf senescence. Annu Rev Plant Biol 58:115–136
2. Gray WM, Ostin A, Sandberg G, Romano CP, Estelle M (1998) High temperature promotes auxin-mediated hypocotyl elongation in Arabidopsis. Proc Natl Acad Sci USA 95(12): 7197–7202
3. Millenaar FF, Cox MCH, van Berkel YEMDJ, RAM W, Pierik R, LAJC V, AJM P (2005) Ethylene-induced differential growth of petioles in Arabidopsis. Analyzing natural variation, response kinetics, and regulation. Plant Physiol 137(3):998–1008
4. Casal JJ, Balasubramanian S (2019) Thermomorphogenesis. Annu Rev Plant Biol 4
5. Lippmann R, Babben S, Menger A, Delker C, Quint M (2019) Development of wild and cultivated plants under global warming conditions, current biology, vol 29. Cell Press, pp R1326–R1338
6. Halliday KJ, Salter MG, Thingnaes E, Whitelam GC (2003) Phytochrome control of flowering is temperature sensitive and correlates with expression of the floral integrator FT. Plant J 33(5):875–885
7. Kumar SV, Lucyshyn D, Jaeger KE, Alós E, Alvey E, Harberd NP, Pa W (2012) Transcription factor PIF4 controls the thermosensory activation of flowering. Nature 484(7393): 242–245
8. Al-Khatib K, Paulsen GM (1984) Mode of high temperature injury to wheat during grain development. Physiol Plant 61(3):363–368
9. Sakuraba Y, Jeong J, Kang MY, Kim J, Paek NC, Choi G (2014) Phytochrome-interacting transcription factors PIF4 and PIF5 induce leaf senescence in Arabidopsis. Nat Commun 5: 4636
10. Song Y, Yang C, Gao S, Zhang W, Li L, Kuai B (2014) Age-triggered and dark-induced leaf senescence require the bHLH transcription factors PIF3, 4 and 5. Mol Plant:ssu109
11. Kim C, Kim SJ, Jeong J, Park E, Oh E, Park YI, Lim PO, Choi G (2020) High ambient temperature accelerates leaf senescence via phytochrome-interacting factor 4 and 5 in Arabidopsis. Mol Cells 43(7):645–661
12. Bresson J, Bieker S, Riester L, Doll J, Zentgraf U (2017) A guideline for leaf senescence analyses: from quantification to physiological and molecular investigations. J Exp Bot 69(4): 769–786
13. Lichtenthaler HK (1987) [34] chlorophylls and carotenoids: pigments of photosynthetic biomembranes. Methods Enzymol 148:350–382

Chapter 4

Assessing Temperature Responses in Roots

Sanghwa Lee and Wolfgang Busch

Abstract

Due to global warming, it is important to understand how plants respond to high ambient temperature. Plant growth responses to high ambient temperature are termed thermomophogenesis and have been explored for more than a decade. However, this was mostly focused on the above-ground part of plants, the shoot. In this chapter, we describe a simple method to assess root growth phenotype to high ambient temperatures. In principle, this protocol can be applied for any other treatments to test overall seedling growth.

Key words Root, Thermomorphogenesis, Root phenotyping, *Arabidopsis thaliana*

1 Introduction

Compared to 1850–1900, the global annual temperature has increased by around 1.1 °C [1]. Temperature affects the change in free energy of biochemical reactions according to the basic principles of thermodynamics [2, 3]. Therefore, even small changes in temperature will affect plants [4]. The developmental response of plants to higher temperature is termed thermomorphogenesis. Typical responses that are associated with thermomorphogenesis include the development of longer hypocotyl and petiole, decreased stomatal development, early flowering, and primary root elongation [5–8].

Increased plant height may result in decreased yield with low stress resistance, thereby negatively affecting agronomic goals [9]. Also, temperature-induced changes in wild plants will affect the entire ecosystems [10] and can lead to imbalances and ecosystem collapse. Overall, understanding the effects and their genetic and molecular bases in plants at higher temperature is crucial for understanding changes that will occur due to global warming and for mitigating its effects via genetics- and genomics-informed breeding and restoration approaches.

Meng Chen (ed.), *Thermomorphogenesis: Methods and Protocols*, Methods in Molecular Biology, vol. 2795,
https://doi.org/10.1007/978-1-0716-3814-9_4,

Many studies in thermomorphogenesis have been conducted over more than a decade. So far, the molecular mechanism of how plants perceive high ambient temperature is mainly explained by the thermosensors—PHYTOCHROME B (PHYB), EARLY FLOWERING 3 (ELF3), and PHYTOCHROME INTERACTING FACTOR 7 (PIF7) [11–13]. Furthermore, the main hub of thermomorphogenesis which acts as a downstream target of these thermosensors is PIF4 with minor contribution of other PIFs [14–16]. Therefore, this thermosensor-PIF4 cascade was considered a core mechanism to understand thermomorphogenesis. However, most of the discoveries were made in the shoot, which is the above-ground part of the plant. The responses of plant's underground parts, i.e., the root system, are relatively less explored at high ambient temperature since the root grows in the soil. Recently, primary root elongation was identified as a thermomorphic response at high ambient temperature which is mainly regulated by ELONGATED HYPOCOTYL 5 (HY5) [5, 6].

As described above, it is critical to develop a method to observe root phenotype since the root is covered in the soil in nature. In the laboratory, growing early seedlings on agar medium enables the assessment of phenotypes under highly controlled conditions and to perform cellular and molecular work. Here, we describe a simple procedure to test temperature responses in the root. The critical step would be transferring the seedling plates after the germination and proper early seedling growth since temperature affects germination [17]. This method has an advantage of measuring the primary root length in a convenient way and can be utilized for diverse applications.

2 Materials

2.1 Media Preparation

1. 1/2x MS media: Murashige and Skoog (MS) medium (2.16 g/L), 0.9% phytoagar, adjust pH to 5.7 using 1 M KOH (*see* **Note 1**).
2. 1 M KOH (1X): 56 g of KOH in 1 L H_2O.

2.2 Seed Sterilization

1. 5 or 2 mL microtube.
2. *Arabidopsis* seeds accession Columbia-0 (Col-0).
3. 70% EtOH (1X): 70 mL of 100% EtOH and 30 mL of H_2O.
4. Square plate (120 mm × 120 mm × 17 mm).
5. 3 M Surgical tape.
6. Aluminum foil.

2.3 Root Growth Analysis

1. Growth chamber (PGC FLEX-2TIER, Conviron).
2. Scanner (Perfection v600, Epson).

3 Methods

3.1 Seed Sterilization and Plating

1. Aliquot the seeds into 1.5 mL microtube.
2. Put 1 mL of 70% EtOH into the seed containing microtube.
3. Close the cap and vortex for 10 s and leave for 30 s.
4. Remove the liquid with pipette.
5. Put 1 mL of DW into the seed-containing microtube.
6. Repeat steps 3–5 for 3 times.
7. Bring the microtube to the hood.
8. Cut the edge of 200 μL tip using blade for seed plating.
9. Put 10–15 seeds to the plate for each condition (*see* **Note 2**).
10. Tape the plates using surgical tape to prevent excess evaporation.
11. Seal the plates with aluminum foil and place the plates at 4 °C for 3 days.

3.2 Analysis

1. Uncover the plates and place them into 21 °C growth chamber (100 μmol m^{-2} s^{-1} light intensity using LED light with long day condition—16 hr. light:8 h dark) (*see* **Note 3**).
2. After 4 days of germination and early seedling growth, transfer the plates to either 21 °C or 28 °C chamber for additional 5 days (*see* **Note 4**).
3. Scan the plates (6608 × 6614 pixels, 400 pixels/inch) using Multiscan GUI described in https://busch.salk.edu/tools/ [18]. An example result is shown in Fig. 1A (*see* **Note 5**).
4. Analyze the primary root length using Fiji [19]. Using the "Segmented Line" mode, trace the primary root length. After following each primary root, please follow "Analyze-Measure" to get the measurements. A brief example is shown in Fig. 2.
5. Copy the values from the measurements and plot the data with appropriate program. Figure 1b is shown using "Prism 9" with scatter plot mode.

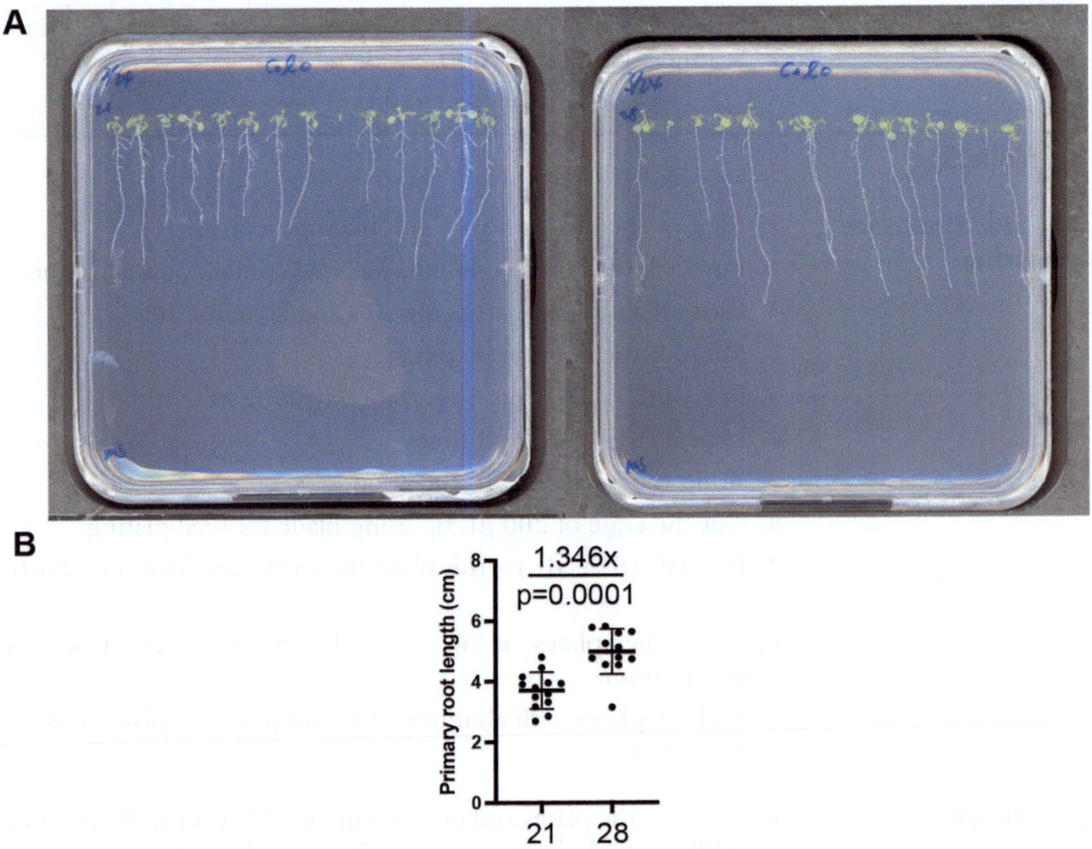

Fig. 1 Root phenotyping for temperature responses. (**a**) Raw images showing 1/2 ms plate grown Col-0 seedlings. Plates were placed at 21 °C for 4 days and transferred to either 21 °C (left panel) or 28 °C (right panel) for additional 5 days. (**b**) Scatter plot of primary root length from (**a**). Average difference of primary root length and *p* value are stated as 1.346× and 0.0001, respectively. *P* value was assessed by two-sided Student's *t*-test

4 Notes

1. Media is not limited to MS, any kind of medium or adding sucrose would work. Also, we use 0.9% phytoagar; however, this will depend on the user's preference.
2. As long as the seedlings do not overlap, it will not affect the result. Also, if you want to mimic the natural condition so that root is in the dark, please follow the protocol described in [20].
3. We confirmed that root thermomorphogenesis phenotype is conserved in three different conditions: 30 μmol m^{-2} s^{-1} of light intensity using white light combined with either constant light or long day condition (16 h light: 8 h dark), and 100 μmol m^{-2} s^{-1} light intensity using LED light combined with long day condition [5, 6].

A

B

C

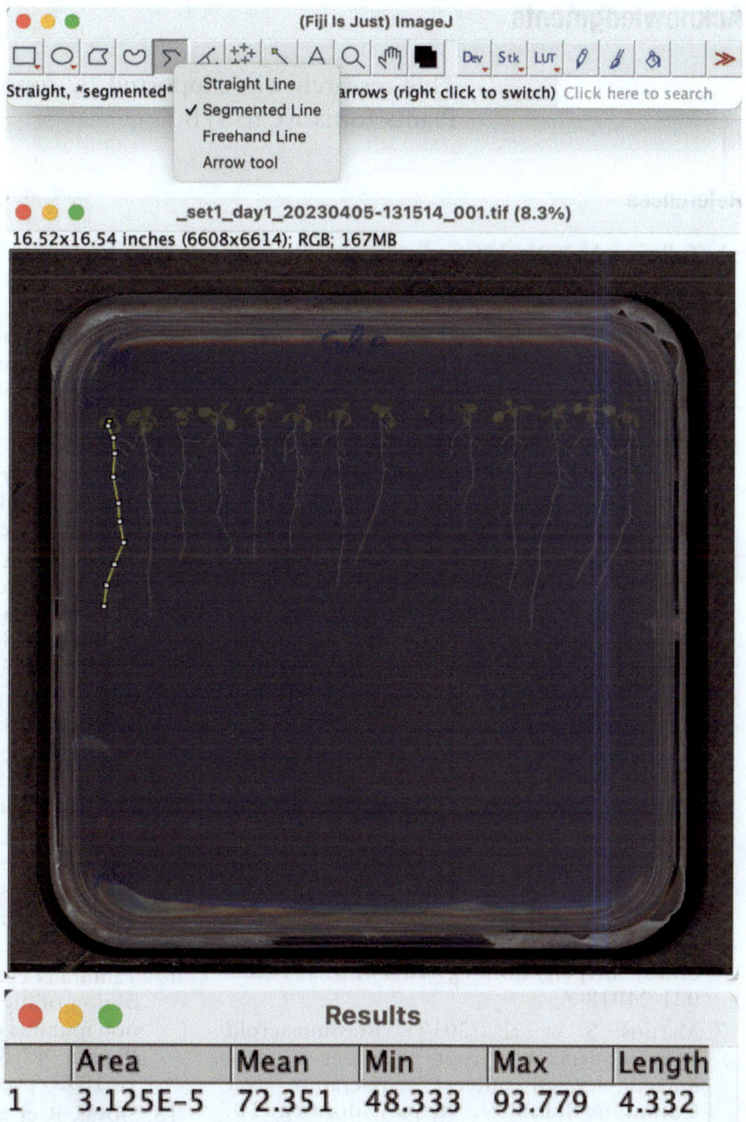

Fig. 2 Root phenotype analysis using Fiji. (**a**) "Segmented Line" mode is needed to measure the root length properly. (**b**) Example of root length measurement from the plate image using "Segmented Line" mode. (**c**) Example of the extracted values from (**b**). Using the original plate scanned image, the data from the length column is the actual root length (cm)

4. For even growth and germination, many studies were performed in the range of 3–4 days in the 21 °C growth chamber [5, 6, 21]. Also, 28 °C growth chamber must have same light intensity and diurnal condition compared with 21 °C growth chamber.
5. Here we used Multiscan GUI; however, images from the camera are still sufficient for this analysis.

Acknowledgments

This research was supported by funds from the Salk Harnessing Plants Initiative to WB.

References

1. Tollefson J (2021) IPCC climate report: earth is warmer than it's been in 125,000 years. Nature 596(7871):171–172. https://doi.org/10.1038/d41586-021-02179-1
2. Held C, Sadowski G (2016) Thermodynamics of bioreactions. Annu Rev Chem Biomol Eng 7:395–414. https://doi.org/10.1146/annurev-chembioeng-080615-034704
3. Ibanez C et al (2017) Ambient temperature and genotype differentially affect developmental and phenotypic plasticity in Arabidopsis thaliana. BMC Plant Biol 17(1):114. https://doi.org/10.1186/s12870-017-1068-5
4. Schleuning M et al (2016) Ecological networks are more sensitive to plant than to animal extinction under climate change. Nat Commun 7:13965. https://doi.org/10.1038/ncomms13965
5. Gaillochet C et al (2020) HY5 and phytochrome activity modulate shoot-to-root coordination during thermomorphogenesis in Arabidopsis. Development 147(24). https://doi.org/10.1242/dev.192625
6. Lee S, Wang W, Huq E (2021) Spatial regulation of thermomorphogenesis by HY5 and PIF4 in Arabidopsis. Nat Commun 12(1):3656. https://doi.org/10.1038/s41467-021-24018-7
7. Martins S et al (2017) Brassinosteroid signaling-dependent root responses to prolonged elevated ambient temperature. Nat Commun 8(1):309. https://doi.org/10.1038/s41467-017-00355-4
8. Vu LD, Xu X, Gevaert K, De Smet I (2019) Developmental plasticity at high temperature. Plant Physiol 181(2):399–411. https://doi.org/10.1104/pp.19.00652
9. Challinor A, Watson J, Lobell D (2014) A meta-analysis of crop yield under climate change and adaptation. Nat Clim Chang 4:287–291. https://doi.org/10.1038/nclimate2153
10. Traill LW, Lim ML, Sodhi NS, Bradshaw CJ (2010) Mechanisms driving change: altered species interactions and ecosystem function through global warming. J Anim Ecol 79(5):937–947. https://doi.org/10.1111/j.1365-2656.2010.01695.x
11. Chung BYW et al (2020) An RNA thermoswitch regulates daytime growth in Arabidopsis. Nat Plants 6(5):522–532. https://doi.org/10.1038/s41477-020-0633-3
12. Jung JH et al (2020) A prion-like domain in ELF3 functions as a thermosensor in Arabidopsis. Nature 585(7824):256–260. https://doi.org/10.1038/s41586-020-2644-7
13. Jung JH et al (2016) Phytochromes function as thermosensors in Arabidopsis. Science 354(6314):886–889. https://doi.org/10.1126/science.aaf6005
14. Franklin KA et al (2011) Phytochrome-interacting factor 4 (PIF4) regulates auxin biosynthesis at high temperature. Proc Natl Acad Sci USA 108(50):20231–20235. https://doi.org/10.1073/pnas.1110682108
15. Koini MA et al (2009) High temperature-mediated adaptations in plant architecture require the bHLH transcription factor PIF4. Curr Biol 19(5):408–413. https://doi.org/10.1016/j.cub.2009.01.046
16. Lee S, Zhu L, Huq E (2021) An autoregulatory negative feedback loop controls thermomorphogenesis in Arabidopsis. PLoS Genet 17(6):e1009595. https://doi.org/10.1371/journal.pgen.1009595
17. Tamura N et al (2006) Isolation and characterization of high temperature-resistant germination mutants of Arabidopsis thaliana. Plant Cell Physiol 47(8):1081–1094. https://doi.org/10.1093/pcp/pcj078
18. Slovak R et al (2014) A scalable open-source pipeline for large-scale root phenotyping of Arabidopsis. Plant Cell 26(6):2390–2403. https://doi.org/10.1105/tpc.114.124032
19. Schindelin J et al (2012) Fiji: an open-source platform for biological-image analysis. Nat Methods 9(7):676–682. https://doi.org/10.1038/nmeth.2019
20. Burko Y et al (2020) Local HY5 activity mediates hypocotyl growth and shoot-to-root communication. Plant Commun 1(5). https://doi.org/10.1016/j.xplc.2020.100078
21. Ding L et al (2018) Two B-box domain proteins, BBX18 and BBX23, interact with ELF3 and regulate Thermomorphogenesis in Arabidopsis. Cell Rep 25(7):1718–1728 e1714. https://doi.org/10.1016/j.celrep.2018.10.060

Chapter 5

A Luciferase Imaging-Based Assay for Studying Temperature Compensation of the Circadian Clock

Hongtao Zhang and Stacey L. Harmer

Abstract

The pace of circadian rhythms remains relatively unchanged across a physiologically relevant range of temperatures, a phenomenon known as temperature compensation. Temperature compensation is a defining characteristic of circadian rhythms, ensuring that clock-regulated processes occur at approximately the same time of day across a wide range of conditions. Despite the identification of several genes involved in the regulation of temperature compensation, the molecular mechanisms underlying this process are still not well understood. High-throughput assays of circadian period are essential for the investigation of temperature compensation. In this chapter, we present a luciferase imaging-based method that enables robust and accurate examination of temperature compensation in the plant circadian clock.

Key words Temperature buffering, Circadian oscillator, Period analysis, Bioluminescence, Low-light imaging, Arabidopsis

1 Introduction

Most organisms exhibit daily rhythms in development and physiology. Many of these rhythms are driven by an internal timekeeping machinery called the circadian clock [1]. The circadian clock provides a fitness advantage when its periodicity aligns with that of diurnal environmental cycles [2]. Environmental cues, such as light and temperature, play important roles in the clock input pathways. For example, both photocycles and thermocycles are sufficient to entrain the clock and reset its phase [1]. Nonetheless, circadian clock pace remains largely invariant across a physiologically relevant range of temperatures [3]. This phenomenon, termed temperature compensation, ensures that the timing of clock-regulated output pathways is relatively insensitive to temperature changes [4]. Therefore, temperature compensation allows plants to schedule circadian-regulated processes to the most appropriate time of day across a range of ambient temperatures.

Meng Chen (ed.), *Thermomorphogenesis: Methods and Protocols*, Methods in Molecular Biology, vol. 2795, https://doi.org/10.1007/978-1-0716-3814-9_5,

Although temperature compensation is a defining property of circadian rhythms, the underlying molecular mechanisms are not conserved across higher taxa [3]. In plants, the circadian oscillator is mainly composed of multiple interlocking transcriptional-translational feedback loops [1]. Studies in the model species *Arabidopsis thaliana* have revealed that components of these feedback loops, often referred to as core clock genes, are essential for temperature compensation. For instance, PSEUDO RESPONSE REGULATOR7 (PRR7) and PRR9 regulate proper temperature compensation by preferentially inhibiting the expression of *CIRCADIAN CLOCK ASSOCIATED1* (*CCA1*) and *LATE ELONGATED HYPOCOTYL* (*LHY*) at higher temperatures [5]. Importantly, accumulating evidence suggests that alternative splicing may also play a crucial role in temperature compensation. Several core clock genes undergo significant alternative splicing in response to temperature changes [6, 7]. Furthermore, mutations in splicing factors, including *GEMIN2*, *SICKLE*, and *SKIP*, result in temperature compensation defects [8–10]. Interestingly, recent studies indicate that pre-mRNA splicing itself is subject to extensive circadian regulation [11–13]. Therefore, both core clock genes and splicing factors play important roles in the responses of circadian clock to ambient temperature. However, the cooperative contributions of these individual regulators to the molecular basis of temperature compensation remain unclear.

Investigations into temperature compensation require robust and accurate assays of circadian rhythms under different temperature conditions. In early studies, researchers successfully identified mutants and quantitative trait loci (QTL) responsible for circadian period length by monitoring leaf movement and stomatal conductance rhythms [14–16]. At the molecular level, rhythmic expression of transcripts can be examined using genome-wide microarray, quantitative reverse-transcriptase polymerase chain reaction (qRT-PCR), and RNA-sequencing experiments [11, 17, 18]. However, those methods are time- and labor-intensive, particularly when multiple genotypes and temperature conditions need to be tested. Alternatively, a rhythmic assay based on the monitoring of delayed chlorophyll fluorescence (DF) signals offers a high throughput approach but may generate noisier data compared to other methods [19].

In this chapter, we describe a luciferase imaging-based method for studying temperature compensation of plant circadian clocks. The inherent instability of the luciferase (LUC) enzyme in the presence of its substrate luciferin makes it a powerful tool for real-time gene expression studies [20]. By utilizing a *LUC* reporter gene under the control of a circadian clock-regulated promoter, circadian parameters can be easily determined in live plants [21]. This method is both high throughput and accurate, making it widely used in circadian clock assays including large-scale projects such as forward-genetic screens [22, 23]. Moreover, recent studies have demonstrated the effectiveness of this approach in identifying novel regulators involved in temperature compensation [5, 24–26]. Here we present a detailed protocol describing the experimental procedures and data analysis methods needed for assessment of temperature compensation in the circadian clock.

2 Materials

2.1 Plant Materials

1. Wild-type and mutant *Arabidopsis* seeds carrying a *circadian-clock-regulated promoter::LUC* fusion reporter (*see* **Note 1**).
2. *Arabidopsis* mutants with known defects in temperature compensation can serve as positive controls (*see* **Note 2**).

2.2 Reagents

1. 1× Murashige and Skoog (MS) growth medium: Dissolve 1× strength of MS salt powder in Milli-Q or deionized water. Optionally, add 3% (w/v) of D(+)-sucrose and mix well (*see* **Note 3**). Adjust the pH to 5.7 using 1 M KOH. Then, add 0.7% (w/v) of agar powder. Sterilize the growth medium by autoclaving at 121 °C and 1.05 kg/cm^2 (15 psi) for 30 min. Let cool on the bench or in a 55 °C water bath until the bottle can be handled comfortably. Pour approximately 40 mL of warm medium into each square petri dish (100 mm × 100 mm) in a laminar flow hood. Let the agar solidify in the hood with lid partially off for at least an hour (*see* **Note 4**). If not using immediately, store the agar plates in a well-sealed plastic sleeve at 4 °C.
2. Seed sterilization solution: Prepare 70% (v/v) ethanol by mixing 7 parts of absolute ethanol with 3 parts of 0.1% (v/v) Triton X-100.
3. 3 mM D-luciferin potassium salt (GoldBio) working solution: Dilute from a 100 mM D-luciferin stock solution using 0.01% (v/v) Triton X-100 (*see* **Note 5**).

2.3 Equipment and Other Supplies

1. pH meter.
2. 100 mm × 100 mm square petri dishes.
3. A sealable desiccator.
4. Plastic drinking straws (about 0.6 mm in diameter).
5. 1.5 mL microcentrifuge tubes.
6. Micropore surgical tapes (3 M).
7. 0.22 μm syringe-driven filter units (Millipore).
8. 10 mL needless syringes.
9. 15 mL conical centrifuge tubes.

10. A plastic pump sprayer (*see* **Note 10**).
11. Light meter (e.g., LI-COR).
12. Temperature data logger (e.g., HOBO).
13. A plant growth chamber suitable for entrainment (must have light and temperature controls).
14. A plant imaging chamber equipped with LED lights (e.g., Snap-Lites by Quantum Devices) and a sensitive, cooled CCD camera (e.g., an ORCA-R2 [Hamamatsu] or a DU434-BV [Andor Technology]).

2.4 Software

1. HOBOware software (Onset Brands).
2. Image capture software compatible with CCD camera (e.g., Andor Solis, Micro-Manager or equivalent).
3. MetaMorph (Molecular Devices) or ImageJ/FIJI software (https://imagej.nih.gov/ij/).
4. Microsoft Excel (or equivalent).
5. BioDare2 account (https://biodare2.ed.ac.uk).

3 Methods

3.1 Prepare *Arabidopsis* Seeds

1. Aliquot *Arabidopsis* seeds into 1.5 mL microcentrifuge tubes. Surface-sterilize the seeds using one of the following methods:
   (a) Chlorine gas: Sterilize the seeds with chlorine gas for 3 h in a well-sealed desiccator placed in a chemical fume hood. Place the tubes of seeds within the desiccator. To generate chlorine gas, pour 50 mL of 100% bleach into a beaker and place it in the desiccator. Add 3 mL of 1 M HCl to the bleach and immediately seal the desiccator. After sterilization, add 1 mL of autoclaved water to each tube (*see* **Note 6**).
   (b) Ethanol: Alternatively, sterilize the seeds by vortexing them intermittently in 1 mL of seed sterilization solution for 5 min, followed by a 20 min incubation in 100% ethanol. Wash the seeds three times with autoclaved water to remove any residual ethanol. Add 1 mL of autoclaved water to each tube (*see* **Note 6**).
2. In a laminar flow hood, plant the seeds onto 1× MS medium using a clean pipette and autoclaved tips (*see* **Note 7**). Ensure that there is sufficient space between the seeds (Fig. 1a). For better capture of bioluminescence signals, we recommend placing an autoclaved straw around each seed immediately after plating to restrict seedling growth (*see* **Note 8**). When imaging plants with dim luciferase signals, 5–10 seeds can be clustered within one straw.

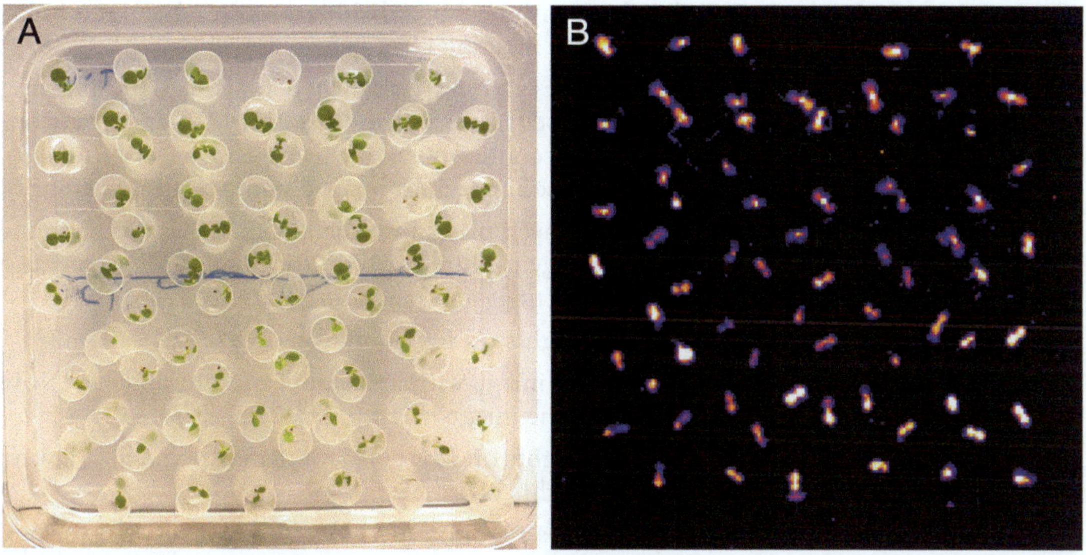

Fig. 1 An example experimental setup illustrating proper spacing between plants and the use of straw fragments. Approximately 10-day-old *Arabidopsis* seedlings carrying a *PRR9::LUC2* promoter were imaged in the bright field (**a**) and in the imaging chamber by a CCD camera (**b**). The petri dish is 100 mm × 100 mm. The colormap in (b) is set to "pseudocolor" for better visualization

3. Cover the plates and leave them in the hood until there is no visible water surrounding the seeds. Seal the plates with breathable paper tape (e.g., 3 M Micropore surgical tape).
4. Wrap the plates with aluminum foil and stratify at 4 °C for 3 days to synchronize germination of seeds.

3.2 Germinate Seeds in the Entraining Chamber

1. Set appropriate light and temperature conditions in the entraining chamber. We typically entrain *Arabidopsis* seedlings in 12-hour light (cool-white fluorescent bulbs, 55 μmol m^{-2} s^{-1})/12-hour dark photocycles at a constant temperature of 22 °C. Alternatively, seedlings can also be entrained in thermal cycles such as 12-hour at 12 °C/12-hour at 22 °C while being maintained in constant light (*see* **Note 9**).
2. After stratification, transfer the plates into the entraining chamber. Allow the seedlings to grow under the entraining conditions for 6 days before transferring them to the imaging chamber.

3.3 Set Up Luciferase Imaging

1. In a laminar flow hood, sterilize freshly prepared 3 mM D-luciferin solution by passing it through a 0.22 μm filter paired with a needleless syringe. Use a sterile sprayer (*see* **Note 10**) to evenly spray approximately 1 mL of the 3 mM D-luciferin solution onto each plate. Leave the plates in the hood with lids off for another 6–8 min to allow extra moisture to evaporate (*see* **Note 11**).

2. Seal the plates with new breathable paper tape before transferring them to the imaging chamber.
3. To measure free-running circadian periods, set the imaging chamber to constant temperature (*see* **Note 12**) and light conditions. Provide continuous light using LEDs (*see* **Note 13**). Alternatively, seedlings can be imaged under constant darkness.
4. Transfer the sprayed plates into the imaging chamber. Use the appropriate image capture software to set the CCD camera to the following settings and begin imaging:
   (a) Turn the lights off for 10 min (*see* **Note 14**).
   (b) Capture an image with a 20-min exposure time (*see* **Note 15**). Save the image in .tif format.
   (c) Turn the lights on (for constant dark experiments, always keep the lights off).
   (d) Repeat steps a–c and capture one image every 2 h for 6–7 consecutive days.

3.4 Analyze Temperature Compensation of Circadian Period

1. Quantify bioluminescence signals from the images using one of the two approaches below:
   (a) MetaMorph software:
      (i) Import the raw image series through "File > Open Special > Build Stack > User Defined." Adjust the scale slider to manually change the displayed range of gray levels for optimal visualization.
      (ii) Use the "Region > Region tools" function to generate a circle; place one circle around each seedling (Fig. 1b) or the tissue of interest. Use the same size circle for each region of interest.
      (iii) Open a Microsoft Excel workbook for data collection by clicking "Log > Open Data Log" and checking "Dynamic Data Exchange (DDE)."
      (iv) Measure the intensities within the regions using the "Apps > Graph Intensities" function (*see* **Note 16**). Select "Stack" for "Measure From" and "Plane Number" for "Measure Regions Over." The reads will be automatically recorded in the Excel Workbook.
   (b) ImageJ/FIJI:
      (i) Import raw images through "File > Import > Image Sequence." Adjust the brightness/contrast of the images.
      (ii) Set up measurement by clicking "Analyze > Set Measurements" and selecting "Integrated Density."

(iii) To measure signal density from multiple seedlings, go to "Analyze > Tools > ROI Manager." Click "Show all" to view all the regions that you have selected.

(iv) Select plants using the "Circle Tool" (a blue circular icon in the toolbar) (*see* **Note 17**). Press "t" on the keyboard (or click "Add" in ROI Manager) to proceed to selecting the next plant.

(v) After selecting all the plants, measure luciferase signal density by clicking "More > Multiple Measure" in ROI Manager. Copy and paste the data into an Excel workbook.

2. Estimate the circadian period length for each temperature and genotype. First, import the bioluminescence data into BioDare2 (https://biodare2.ed.ac.uk). In the "period analysis" module, select the appropriate data window (*see* **Note 18**). For detrending and period analysis methods, we recommend choosing the "linear dtr" (linear detrending) and "FFT-NLLS" (Fourier Fast Transform-Non-Linear Least Squares) options, which are commonly used in circadian period studies (*see* **Note 19**). Identify and exclude individuals with poor rhythmicity (*see* **Note 20**).

3. Plot the estimated circadian period lengths against temperature for each genotype (Fig. 2). The slopes illustrate the variation in period lengths in response to temperature changes, thereby

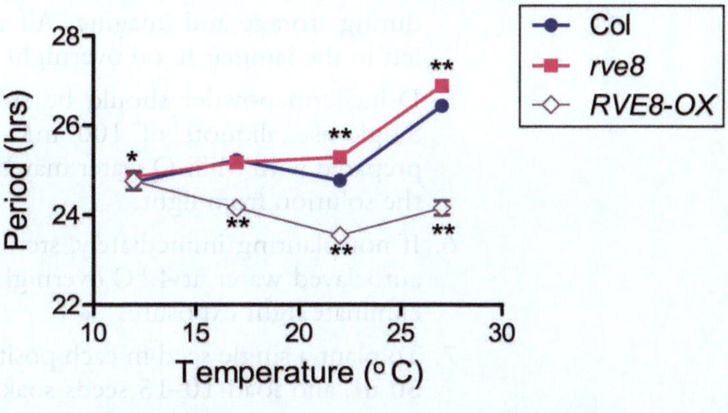

Fig. 2 Loss of function and overexpression of *RVE8* significantly disturb temperature compensation of the *Arabidopsis* circadian clock. Plants were entrained in 12 h light/12 h dark for 6 days at 22 °C before released to constant red light at the indicated temperatures. Data represent average free-running period lengths ± SEM from at least two independent experiments. Statistical analysis is performed by student's two-tail heteroscedastic t test (*$p < 0.05$; **$p < 0.001$). This figure is reprinted from Fig. 7E of [18] with copyright permitted

reflecting the temperature compensation phenotype. In wild-type plants, the period length typically remains relatively unchanged within a physiologically relevant range of temperatures [14]. However, mutation or overexpression of genes involved in temperature compensation may result in significantly different slopes compared to the wild type [5, 15, 18].

4 Notes

1. Previous studies have demonstrated the successful application of *CCR2::LUC*, *CCA1::LUC*, and *CAB2::LUC* reporters in temperature compensation studies [5, 18, 24]. Other luciferase reporters that show robust rhythmicity at investigated temperatures can also be used.
2. *lhy-21* [15], *rve8-1* [18], (Fig. 2) and *prr7-3 prr9-1* mutants [5] show defects in temperature compensation of circadian period at higher temperatures, while *gi-11* and *cca1-11* have more impaired temperature compensation at lower temperatures [15].
3. Addition of sucrose to the growth medium can enhance the amplitude of circadian rhythms, especially under constant dark conditions [27]. However, it is worth noting that sucrose also significantly shortens the circadian period [28, 29]. Therefore, it is important to know whether the gene of interest is involved in the response of the plant circadian clock to sucrose.
4. This is to avoid any condensation that may form in the plate during storage and imaging. Alternatively, agar plates can be left in the laminar hood overnight with lid closed to solidify.
5. D-luciferin powder should be stored desiccated at −20 °C. Single-use aliquots of 100 mM D-luciferin stock solution prepared with Milli-Q water may be stored at −20 °C. Protect the solution from light.
6. If not planting immediately, sterilized seeds can be stored in autoclaved water at 4 °C overnight. Wrap the tubes in foil to eliminate light exposure.
7. To plant a single seed in each position, set a 200 uL pipette to 50 uL and load 10–15 seeds soaked in autoclaved water into the pipette tip. Gently pipette one seed out of the tip each time.
8. Cut the drinking straws into 1-cm long pieces. Autoclave the straws and a pair of forceps before each use. Gently insert one end of each straw fragment vertically into the growth medium around each seed using the forceps (Fig. 1a).
9. Note that some *Arabidopsis* mutants, such as *prr7-3 prr9-1*, are insensitive to entrainment by thermal cycles [30].

10. Plastic pump sprayers used for cosmetics may be obtained from a local drug store. To keep the sprayer sterile, store it in 75% ethanol between uses. Make sure to remove as much ethanol as possible from the sprayer tube before placing it in the luciferin solution.
11. During the imaging process, condensation forms on the lid of the plate, especially at lower temperatures. Severe condensation may impair the signal-to-noise ratio. Therefore, try to reduce excess moisture from the sprayed plates as much as possible.
12. A physiologically relevant range of temperatures should be selected [5, 18, 24]. Each temperature should be tested in at least two independent experiments. During the experiment, we recommend using a HOBO Temperature/Relative Humidity Data Logger device to continuously monitor the temperature in the imaging chamber.
13. The highly sensitive CCD cameras required to detect luciferase luminescence from plants will also detect light leaking into the imaging chamber from the exterior or produced within the chamber by fluorescence or phosphorescence. We therefore recommend that the growth cabinet containing the CCD camera be placed within a darkroom and that LED, not fluorescent, lights be mounted within the chamber. We routinely image plants illuminated with 35 μmol m^{-2} s^{-1} each of monochromatic red and blue light, but other light regimes can be used. Since both fluence rate and light quality affect circadian period [31], we recommend calibrating the intensity of the LED lights in the imaging chamber using a light meter (e.g., LI-250A). Make sure that the sensor is placed at the same distance from the light source as the plates to be imaged.
14. Under constant light conditions, a 10-min dark treatment is needed prior to camera exposure. This is to allow the decay of any chlorophyll auto-fluorescence that would interfere with detection of the bioluminescence signals [21] and for any phosphorescence generated by chamber components to subside.
15. For some brighter versions of luciferase reporters such as *LUC2* [32], we recommend reducing the exposure time to as short as 2 min to avoid saturation of signals.
16. For dim sample images, use the "stack arithmetic" function to increase visualization when placing circles around seedlings using the "region tool." Try the "sum," "max," or "average" options to obtain the best processed images.
17. If the "Circle Tool" icon cannot be found in the tool bar, install it by clicking "Plugins > Macros > Install." In the pop-up window, open the file named "Circle_Tool.ijm" under "/

ImageJ.app/macros/Examples/Macro/." Alternatively, the source file can be downloaded from https://imagej.nih.gov/ij/source/macros/Circle_Tool.txt.

18. Luciferase protein is relatively stable in plants until exogenous luciferin substrate is added. Therefore, the bioluminescence level measured immediately after luciferin application reflects the abundance of accumulated luciferase protein, rather than the real-time activity of the promoter [21]. Consequently, data collected from the first 24 h of imaging should be excluded.

19. Check https://biodare2.ed.ac.uk/documents/all and [33] for detailed definitions and comparisons among methods for circadian period analysis.

20. We use relative amplitude error (RAE or ERR) to determine whether an individual data trace is rhythmic. RAE is defined as the ratio of the amplitude error to the amplitude value. In general, we consider a plant as rhythmic if its RAE is less than 0.6.

References

1. Harmer SL (2009) The circadian system in higher plants. Annu Rev Plant Biol 60:357–377
2. Dodd AN, Salathia N, Hall A et al (2005) Plant circadian clocks increase photosynthesis, growth, survival, and competitive advantage. Science 309:630–633
3. Gil K-E, Park C-M (2019) Thermal adaptation and plasticity of the plant circadian clock. New Phytol 221:1215–1229
4. Somers DE (1999) The physiology and molecular bases of the plant circadian clock. Plant Physiol 121:9–20
5. Salomé PA, Weigel D, McClung CR (2010) The role of the Arabidopsis morning loop components CCA1, LHY, PRR7, and PRR9 in temperature compensation. Plant Cell 22:3650–3661
6. James AB, Syed NH, Bordage S et al (2012) Alternative splicing mediates responses of the Arabidopsis circadian clock to temperature changes. Plant Cell 24:961–981
7. Kwon Y-J, Park M-J, Kim S-G et al (2014) Alternative splicing and nonsense-mediated decay of circadian clock genes under environmental stress conditions in Arabidopsis. BMC Plant Biol 14:136
8. Schlaen RG, Mancini E, Sanchez SE et al (2015) The spliceosome assembly factor GEMIN2 attenuates the effects of temperature on alternative splicing and circadian rhythms. Proc Natl Acad Sci USA 112:9382–9387
9. Marshall CM, Tartaglio V, Duarte M et al (2016) The Arabidopsis sickle mutant exhibits altered circadian clock responses to cool temperatures and temperature-dependent alternative splicing. Plant Cell 28:2560–2575
10. Wang X, Wu F, Xie Q et al (2012) SKIP is a component of the spliceosome linking alternative splicing and the circadian clock in Arabidopsis. Plant Cell 24:3278–3295
11. Romanowski A, Schlaen RG, Perez-Santangelo S et al (2020) Global transcriptome analysis reveals circadian control of splicing events in Arabidopsis thaliana. Plant J 103:889–902
12. Yang Y, Li Y, Sancar A et al (2020) The circadian clock shapes the Arabidopsis transcriptome by regulating alternative splicing and alternative polyadenylation. J Biol Chem 295:7608–7619
13. Zhang H, Kumimoto RW, Anver S et al (2023) XAP5 circadian timekeeper regulates RNA splicing and the circadian clock by genetically separable pathways. Plant Physiol kiad 193:2492
14. Somers DE, Webb AA, Pearson M et al (1998) The short-period mutant, toc1-1, alters circadian clock regulation of multiple outputs throughout development in Arabidopsis thaliana. Development 125:485–494
15. Gould PD, Locke JCW, Larue C et al (2006) The molecular basis of temperature compensation in the Arabidopsis circadian clock. Plant Cell 18:1177–1187

16. Edwards KD, Lynn JR, Gyula P et al (2005) Natural allelic variation in the temperature-compensation mechanisms of the Arabidopsis thaliana circadian clock. Genetics 170:387–400
17. Covington MF, Maloof JN, Straume M et al (2008) Global transcriptome analysis reveals circadian regulation of key pathways in plant growth and development. Genome Biol 9: R130
18. Rawat R, Takahashi N, Hsu PY et al (2011) Reveille8 and pseudo-reponse regulator5 form a negative feedback loop within the Arabidopsis circadian clock. PLoS Genet 7: e1001350
19. Rees H, Duncan S, Gould P et al (2019) A high-throughput delayed fluorescence method reveals underlying differences in the control of circadian rhythms in Triticum aestivum and Brassica napus. Plant Methods 15:51
20. Millar AJ, Short SR, Chua NH et al (1992) A novel circadian phenotype based on firefly luciferase expression in transgenic plants. Plant Cell 4:1075–1087
21. Southern MM, Brown PE, Hall A (2006) Luciferases as reporter genes. Methods Mol Biol 323:293–305
22. Millar AJ, Carré IA, Strayer CA et al (1995) Circadian clock mutants in Arabidopsis identified by luciferase imaging. Science 267:1161–1163
23. Martin-Tryon EL, Kreps JA, Harmer SL (2007) GIGANTEA acts in blue light signaling and has biochemically separable roles in circadian clock and flowering time regulation. Plant Physiol 143:473–486
24. Portolés S, Más P (2010) The functional interplay between protein kinase CK2 and CCA1 transcriptional activity is essential for clock temperature compensation in Arabidopsis. PLoS Genet 6:e1001201
25. Gould PD, Ugarte N, Domijan M et al (2013) Network balance via CRY signalling controls the Arabidopsis circadian clock over ambient temperatures. Mol Syst Biol 9:650
26. Kolmos E, Chow BY, Pruneda-Paz JL et al (2014) HsfB2b-mediated repression of PRR7 directs abiotic stress responses of the circadian clock. Proc Natl Acad Sci USA 111:16172–16177
27. Dalchau N, Baek SJ, Briggs HM et al (2011) The circadian oscillator gene GIGANTEA mediates a long-term response of the Arabidopsis thaliana circadian clock to sucrose. Proc Natl Acad Sci USA 108:5104–5109
28. Haydon MJ, Mielczarek O, Robertson FC et al (2013) Photosynthetic entrainment of the Arabidopsis thaliana circadian clock. Nature 502: 689–692
29. Philippou K, Ronald J, Sánchez-Villarreal A et al (2019) Physiological and genetic dissection of sucrose inputs to the Arabidopsis thaliana circadian system. Genes (Basel) 10:334
30. Salomé PA, McClung CR (2005) Pseudo-response regulator 7 and 9 are partially redundant genes essential for the temperature responsiveness of the Arabidopsis circadian clock. Plant Cell 17:791–803
31. Devlin PF, Kay SA (2000) Cryptochromes are required for phytochrome signaling to the circadian clock but not for rhythmicity. Plant Cell 12:2499–2510
32. Hu W, Franklin KA, Sharrock RA et al (2013) Unanticipated regulatory roles for Arabidopsis phytochromes revealed by null mutant analysis. Proc Natl Acad Sci USA 110:1542–1547
33. Zieliński T, Hay J, Millar AJ (2022) Period estimation and rhythm detection in Timeseries data using BioDare2, the free, online, community resource. Methods Mol Biol 2398:15–32

Chapter 6

Investigating the Effects of Temperature on Pathogen Propagation in *Arabidopsis*

Zhan Li, Yang Liu, and Jian Hua

Abstract

Temperature is one of the most prominent environmental factors that influence plant immunity. Depending on the plant-pathogen system, increased temperature may inhibit or enhance disease resistance or immunity in plants. Measuring the effect of temperature on plant immunity is the first step toward revealing climate effects on plant-pathogen interactions and molecular regulators of temperature sensitivity of plant immunity. Quantification of plant disease resistance or susceptibility under different temperatures can be accomplished by assessing pathogen growth over time in infected plants or tissues. Here, we present a protocol for quantifying pathogen growth in the most studied system of *Arabidopsis thaliana* and *Pseudomonas syringae pathovar tomato* (*Pst*) DC3000. We discuss important factors to consider for assaying pathogen growth in plants under different temperatures. This protocol can be used to assess temperature sensitivity of resistance in different plant genotypes and to various pathovars.

Key words Temperature, Plant immunity, *Pst* DC3000, *Arabidopsis*, Pathogen growth

1 Introduction

Temperature, as one of the key environment factors, affects plant growth and development as well as plant immunity or disease resistance. Plants possess multiple branches of immune systems to effectively inhibit the invasion and propagation of most pathogens. The first immune branch recognizes general pathogen patterns, such as microbial- or pathogen-associated molecular patterns (MAMPs or PAMPs), by pattern recognition receptors (PRRs), and the second immune branch recognizes effectors secreted by specific pathogens by nucleotide-binding domain leucine-rich repeat (NLR) proteins [1]. These PAMP-triggered immunity (PTI) and effector-triggered immunity (ETI) could potentiate each other [2]. Different combinations of branches of immune responses are triggered by different pathogen strains in plants [3]. The susceptibility or resistance of a plant to a pathogen is

Meng Chen (ed.), *Thermomorphogenesis: Methods and Protocols*, Methods in Molecular Biology, vol. 2795, https://doi.org/10.1007/978-1-0716-3814-9_6,

dependent on the virulence of pathogen and the immunity of plant, both of which are influenced by temperature [4]. For instance, low temperature enhances salicylic acid (SA)-mediated plant immunity and does not compromise the virulence (secretion of the effectors) of bacterial pathogen [5]. High temperature reduces the SA pathway and inhibits the activity of NLR proteins, resulting in heightened disease in plants [6].

Pseudomonas syringe pathovar tomato (*Pst*) DC3000 is one of the most widely studied virulent bacterial pathogens [7]. Its pathogenicity stems from the secretion of virulence effector proteins to suppress PTI of the host plant *Arabidopsis thaliana* (hereafter *Arabidopsis*), while these effector proteins could trigger robust ETI responses when plants harbor the corresponding NLR receptor proteins. Quantification of pathogen growth is a direct way to assess overall plant disease resistance or susceptibility. Other methods can be used to assay specific aspect or process of immune responses, such as expression of defense response genes [8], production of ROS [9], stomatal closure [10], and programmed cell death [11].

Here we describe a protocol of measuring the growth/propagation of *Pst* DC3000 in *Arabidopsis* under different temperatures. This can be applied to most *Arabidopsis* mutants or natural accessions to identify genes important for temperature sensitivity of plants. It can also be applied to various pathogens to assess the pathogen factors contributing to temperature sensitivity of plant immunity. To compare resistance under different temperatures, plants grown to a similar developmental stage should be used because plant immunity is modulated by plant age and tissue maturity [12]. Pathogen inoculation is commonly done by either dipping (leaf is dipped into pathogen solution) or infiltration (pathogen solution is infiltrated into leaves) [13]. Spraying is sometimes used in place of dipping for leaf surface inoculation. Different methods are used to probe different layers of immune responses. The infiltration method assays apoplastic defense, as the pathogen is delivered into apoplast space directly and therefore bypassing stomatal defense [10]. Dipping method assays the combined stomatal defense and apoplastic defense in plants. Since a detailed protocol of infiltration method is already available [14], we describe here the dipping method used for examining the growth of bacterial pathogen *Pst* DC3000 in *Arabidopsis* under different temperatures.

2 Materials

2.1 Materials

1. Arabidopsis seeds.
2. Soil for general *Arabidopsis* (such as LM-111 from Lambert).
3. Window/door screen mesh (polyester) with a grid size of about 2 mm. The screen mesh can be purchased as a cut

sheet from garden or home stores. One example is SAINT-GOBAIN ADFORS Pet-defence Window Screen, charcoal, polyester.

4. Rubber bands.
5. Bacterial pathogen. Strains of *Pst* DC3000 (*see* **Note 1**).

2.2 Chemicals and Buffers

1. 10 mM $MgCl_2$ buffer: Dissolve 2.03 g $MgCl_2{\cdot}6H_2O$ in 1 L ddH_2O. Autoclave the solution at 120 °C for 20 min. Store the autoclaved solution at room temperature.
2. Silwet L-77.
3. King's B (KB) medium plate: Add 10 g peptone #2 (or 20 g peptone #3), 1.5 g K_2HPO_4, and 15 g agar to deionized H_2O for a final volume of 975 mL. Autoclave the medium at 121 °C for 20 min. Add 20 ml sterile 50% glycerol and 5 mL sterile 1 M $MgSO_4$ to the cooled medium. Add rifampicin to the cooled media to a final concentration of 25 μg mL^{-1}, given that *Pst* DC3000 is rifampicin resistance (*see* **Note 2**). Pour the medium to 10 cm petri dishes to make KB plate.
4. Luria-Bertani (LB) medium: Add 25 g LB powder and 15 g agar to deionized H_2O for a final volume of 1 L. Autoclave the medium at 121 °C for 20 min. Add rifampicin into the cooled media (around 50 °C) to a final concentration of 25 μg mL^{-1}. Pour the medium to the 10 cm petri dishes (*see* **Note 3**).

3 Methods

3.1 Plant Growth

1. Moisten the soil thoroughly with water.
2. Fill moistened soil into pots (8 cm square or similar size) to form a dome on top (Fig. 1a) (*see* **Note 4**).
3. Cut mesh screen into squares of 14 cm (larger than the pot surface). Drape the mesh screen square over the pot and fold it over at the corner of the pot. Fasten the screen to the pot using one or two rubber bands (Fig. 1b). Make it tight to minimize the gap between the soil and the screen (*see* **Note 5**).
4. Sow the *Arabidopsis* seeds on the soil through the holes of the mesh screen (*see* **Note 6**).
5. Move pots to 4 °C for a 3-day cold stratification. Transfer pots to chamber for germination (*see* **Note 7**) and growth under 12 h light/12 h dark with light intensity of 100 μmol m^{-2} s^{-1} and relative humidity at 50–70%.
6. Grow plants at different temperatures to comparable developmental stages for pathogen inoculation (*see* **Note 8**). For instance, 21 days at 16 °C, 14 days at 22 °C, and 12 days at 28 °C.

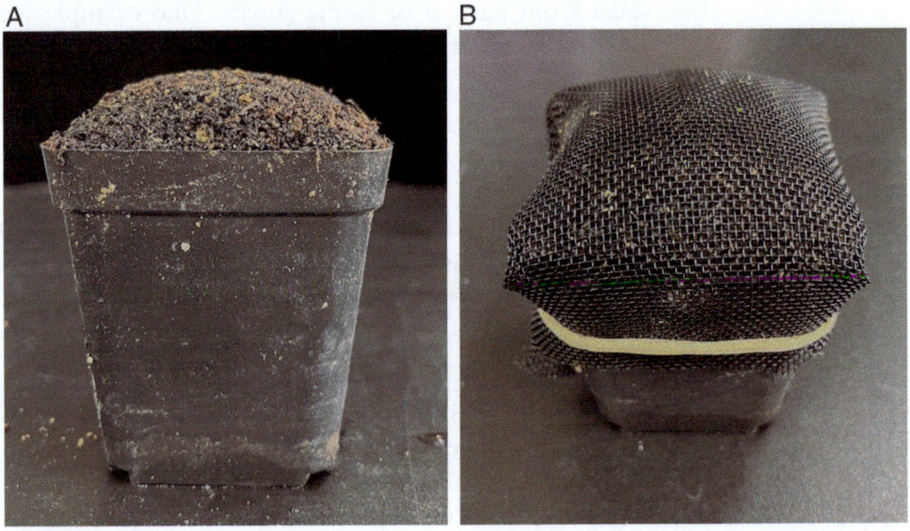

Fig. 1 Soil preparation for sowing *Arabidopsis* seeds. (**a**) A pot overfilled with moistened soil. (**b**) A pot covered with a mesh screen fastened by rubber bands

3.2 Pathogen Inoculation

1. Day 3 before inoculation.

   Revive *Pst* DC3000 strains on KB medium plate by streaking them from −80 °C stock. It usually takes 36–48 h for strains to grow into visible colonies at 28 °C (Fig. 2a).

2. Day 1 before inoculation.
   1. Streak colonies from the above plate to a new KB medium plate. Streak enough for the production of a lawn of bacteria (Fig. 2b) (*see* **Note 9**).
   2. Water plants well (*see* **Note 10**).
3. Inoculation.
   1. Add 10 ml of 10 mM $MgCl_2$ to the lawn of *Pst* DC3000 strain on the KB medium plate. Shake the plate gently on a shaker at 60 rpm for 5–10 min until all bacteria are dislodged off the plate (Fig. 2c).
   2. Transfer the bacteria solution from plate to a 50 mL beaker and add 10 mM $MgCl_2$ to bring the volume to 50 mL.
   3. Measure the optical density at 600 nm (OD_{600}) of the bacteria solution. Use a larger beaker to adjust the solution to $OD_{600} = 0.05$ by adding more 10 mM $MgCl_2$ as needed. Add Silwet L-77 to the solution for a concentration of 200 μL/L. Mix the bacteria inoculation solution thoroughly in the beaker by stirring.

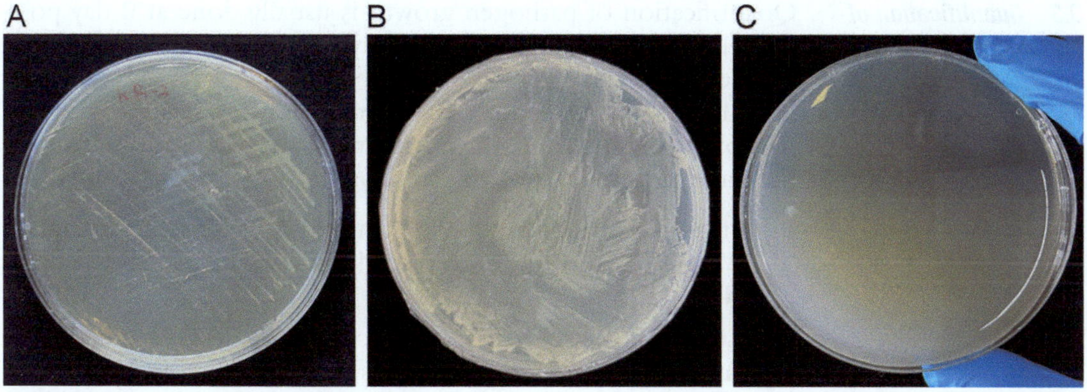

Fig. 2 Preparation of the *Pst* DC3000 strains. (**a**) *Pst* DC3000 strains grown for 48 hours at 28 °C after streaked from −80 °C. (**b**) A lawn of *Pst* DC3000 strain streaked out from plate in (a) and cultivated at 28 °C for 16 h. (**c**) The KB plate in (b) after the bacteria were wash off

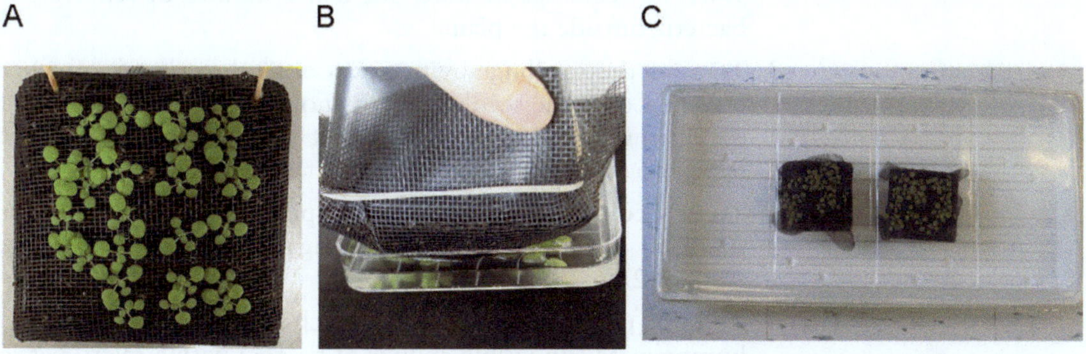

Fig. 3 Dipping inoculation for *Arabidopsis* seedling. (**a**) A pot of *Arabidopsis* seedlings grown at 22 °C for 14 days under a 12 h light/12 h dark photoperiod. (**b**) A pot being inverted and seedlings dipped into *Pst* DC3000 solutions placed in a petri dish. (**c**) Pots placed upright in a tray and covered with a propagation dome

4. Pour the well-mixed solution into a square dish (9 cm × 9 cm × 1.5 cm). The volume is about 40 mL to cover more than 3/4 of the dish.
5. Invert pots of plants (Fig. 3a) to dip plants into the solution for about 30 s. All leaves are submerged into the solution but do not press the bottom of dish to prevent damage to the plants (Fig. 3b).
6. Place pots upright on a tray and cover the pots with a propagation dome to keep humidity for infection (Fig. 3c). Place the flat with dome back into the original growth chamber. Remove the dome 1 h later.
7. Plants will be grown in chambers at respective temperatures for up to 5 days. Water plants as needed.

3.3 Quantification of Pathogen Growth

Quantification of pathogen growth is usually done at 0 day post-inoculation (0 dpi), or more precisely 1 h post-inoculation (1 hpi), to assess the amount of pathogens that have entered the plant after inoculation. It is then done at 2, 3, or 4 dpi to assess the propagation of pathogens inside the plants (*see* **Note 11**). The procedure is the same for any day post-inoculation.

1. Prepare 1.5 mL tubes for plant fresh weight measurement. Weight each tube and record its weight (weight 1) (*see* **Note 12**).
2. Cut aerial part of seedlings by scissors. Collect 3 seedlings as a sample replica, e.g., 9 seedlings for three replica. Wash seedlings to remove bacteria outside the plants by submerging them in sterilized water in a container (such as a 250 ml beaker) and gently shaking the container to dislodge bacteria from the leaf surfaces. Alternatively, gently gather seedlings by tweezers and swirl the seedlings in water for half a minute to remove the bacteria outside the plants.
3. Transfer the seedlings by tweezers onto three layers of paper towels. Cover the seedlings with another set of three layers of paper towels and gently press (not squeeze) to allow paper towels to absorb water from the surface of the seedlings.
4. Three seedlings (one replica) will be placed in one tube. Weight the tube that now contains seedlings and record the weight (weight 2).
5. Add 200 μL of 10 mM $MgCl_2$ to each tube. Grind seedlings to homogenate (no visible chunks of tissues) with a micro pestle (Fig. 4a). Place tubes on ice before and after grinding when processing other samples.

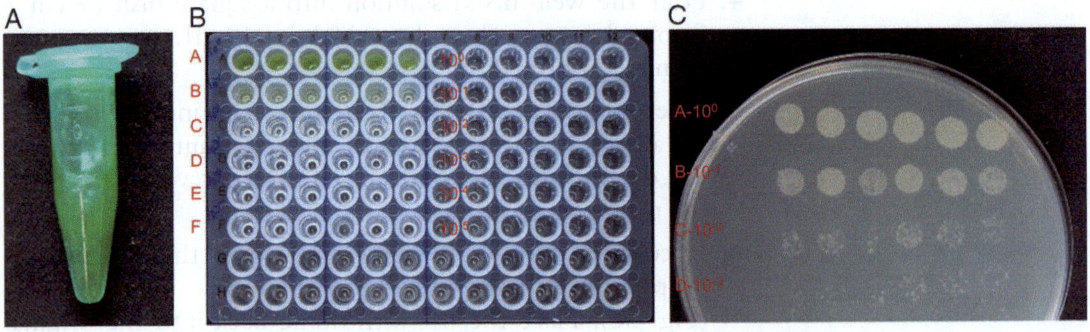

Fig. 4 Quantification of *Pst* DC3000 propagation. (**a**) A thoroughly homogenized tissue solution in a 1.5 mL tube. (**b**) Dilution of 6 homogenized samples in a 96-well plate. Six samples (in columns 1 to 6) are diluted from 10^{-1} to 10^{-5} (in rows A to F). Letters on the left are the row names, and numbers on the right are dilution factors. (**c**) Growth of serial dilutions of *Pst* DC3000 spotted on LB plate after 42 h of incubation at 28 °C

6. When all samples are ground, add 800 μL of 10 mM $MgCl_2$ to each tube. Shake tubes for 1 h on a shaker at 50–60 rpm to further release bacteria from plant tissues. This 1 mL tissue homogenate is the original bacteria solution.
7. Prepare tittering plates by dispensing 180 μL of 10 mM $MgCl_2$ to wells of the 96 well plates except for the first row A that is reserved for the original solution. Add 100 μL of the original bacteria solution to row A. Transfer 20 μL from row A to row B to make a 1:10 dilution. Mix the stock solution by pipetting at least three times before 20 μL is taken out. Mix the dilution solution by pipetting at least three times after stock solution is transferred to the dilution solution. Continue the dilution by transferring 20 μL from row B to row C, C to D, and so on (Fig. 4b) (*see* **Note 13**).
8. Repeat the above step for additional samples. This can be done with multiple channel pipette for processing multiple samples. Change tips for each dilution. For 0 dpi solution, the original solution usually needs to be diluted to 10^{-3}; for 4 dpi (16 °C), 3 dpi (22 °C), and 2 dpi (28 °C), the original solution usually needs to be diluted to 10^{-5}.
9. Spot serial dilutions on pre-dried LB plates (*see* **Note 14**). Label the sample name and dilutions on the bottom of the plate. Dispense 10 μL of dilution gently (avoid splashing) on the plate. Leave enough space between spots. The same tip can be used for the same dilution series by working from the highest dilution to the lowest dilution.
10. Dry spotted dilutions on plate in aseptic hood. Once no liquid can be seen on the plate surface, close the lid, invert the dish, and incubate it at 28 °C incubator for 40–48 h until colonies become visible (Fig. 4c) (*see* **Note 15**).
11. Count the number of colonies on day 2 after spotting. Use the dilution that yields between 10 and 50 colonies for counting.
12. Calculate the bacterial growth. Pathogen growth is usually expressed as Log_{10} (cfu/fresh weight) (Fig. 5a). The total number of colony-forming unit is the bacterial count number x dilution factor x spotting factor (1000 μL/10 μL). For instance, if fresh weight of sample plant is 30 mg and colony number of 10^{-3} dilution is 30, bacterial growth is calculated as Log_{10} (30*1000*100/30) = 5.
13. The above procedure is done shortly after inoculation (0 dpi) and end of pathogen growth assay (such as 3 dpi at 22 °C).
14. Conduct at least two independent sets of experiments (each with three biological replica) on plants grown on separate days and space to minimize the influence from uncontrolled factors (*see* **Note 16**).

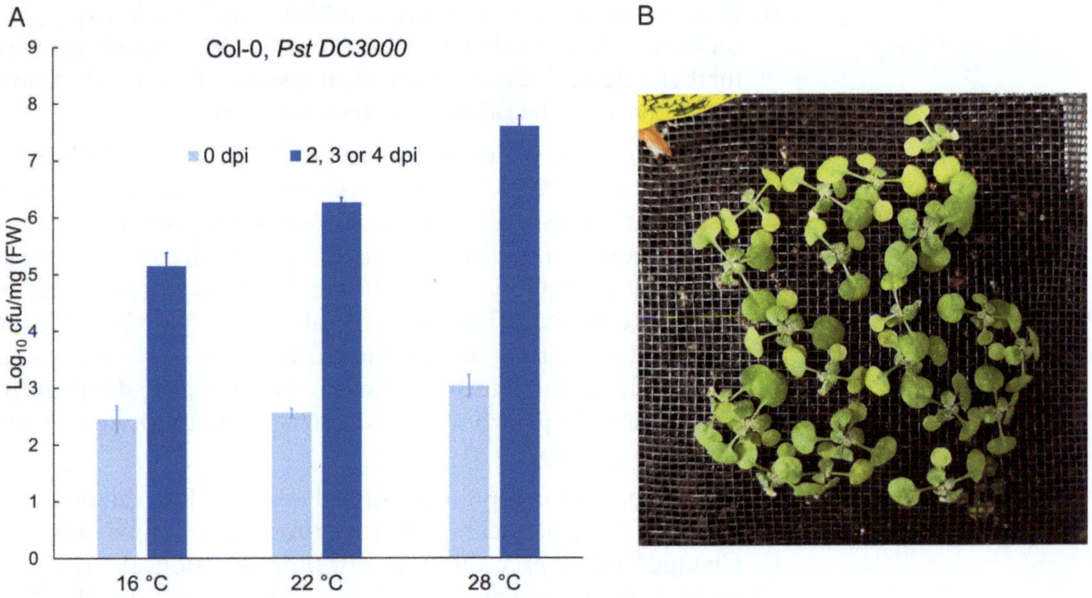

Fig. 5 Assessment of pathogen propagation and disease symptom in *Arabidopsis*. (**a**) Growth of the virulent pathogen *Pst* DC3000 in wild-type Col-0 *Arabidopsis* plants at 0, 2 days post-inoculation (dpi) at 16 °C, 3 dpi at 22 °C, or 2 dpi at 28 °C. Values represent means ± SE ($n = 3$) for three biological repeats. cfu, colony-forming unit. Modified from a previous study [5]. (**b**) Symptoms of *Arabidopsis* Col-0 plants at 3 dpi after infection with *Pst* DC3000

15. Observe disease symptoms and make measurements as needed. These include taking photos (Fig. 5b), measuring chlorophyll content [15], and staining for cell death [16] (*see* **Note 17**).

4 Notes

1. The selection of the pathogenic bacterial strain depends on the specific purpose of the experiment. *Pst* DC3000 strain is used for evaluating basal defense (consisting of mostly PTI and ETI). The *Pst* DC3000 *hrcU*⁻ strain that is defective in secreting effectors can be employed to investigate PTI immunity [17]. Strains containing specific effectors, such as *Pst* DC3000 *AvrRps4*, can be chosen for assaying ETI immunity induced by the NLR gene such as *RPS4* [18].
2. 2Use appropriate antibiotics (which may not be rifamycin) to select bacterial strains.
3. Store KB and LB plates at 4 °C. KB plates are for bacteria reviving, and LB plates are for bacteria counting.
4. Pot size can vary as long as it is suitable. Overfill soil facilitates a tighter and more secure fit with the mesh screen and minimizes gaps between the screen and soil.

5. Prepare the mesh screen by cutting it to the squares that are larger than the upper surface of the pot. In general, an extra of 3–4 cm on each side of the square allows an easy secure by rubber bands. The mesh screen can be washed and sterilized for reuse.
6. Avoid sowing seeds in areas where there are gaps between the soil and the mesh which will lead to the growth of the seedling underneath the mesh screen and poor inoculation by dipping.
7. To mitigate potential challenge of *Arabidopsis* germination at 28 °C, plants can be germinated at 22 °C for a period of 2 days after stratification at 4 °C. They can then be transferred to 28 ° C for subsequent growth.
8. Flowering time can be used to normalize the vegetative growth period and to select a comparable stage of development for pathogen inoculation at different temperatures.
9. Freshly cultivated *Pst* DC3000 is needed for its activity to infect plants. It is preferrable to streak out *Pst* DC3000 from a − 80 ° C stock with an initial growth of 48 h followed by a second streak and growth of 16–24 h.
10. The purpose of this step is to ensure that all plants are free of drought stress and their stomata are in good condition to facilitate the equal infection of *Pst* DC3000.
11. To achieve more detailed assessment of plant immune response, samplings for pathogen growth can be taken at multiple days (from 2 to 4 days). As pathogen grows slower at a lower temperature, a longer propagation time is given to plants before assessing pathogen growth at low temperatures.
12. Utilize a precision balance with an accuracy of 0.001 g to ensure the accuracy of mass measurements.
13. The 96-well plates can be washed and autoclaved for reuse.
14. Dry the LB medium plates in advance in the aseptic hood until no liquid on medium surface before spotting serial dilutions. This procedure makes sure the transferred droplet stays in place until absorbed.
15. Take plates out for colony counting as soon as the colonies become visible. Longer incubation will lead to colony overgrowth and fusion.
16. Plant immune responses are influenced by the circadian rhythm [19]. It is preferrable to perform independent experiments or batch of experiments at same time of the day. For instance, pathogen inoculation can start at the same time of 9:00 am for plants in a set of experiment but are assayed on different days.
17. Visual inspection of disease symptoms can be used as a preliminary assessment of plant immunity. Excessive proliferation of *Pst* DC3000 usually causes chlorosis and wilting of leaves.

Acknowledgments

Research in J. Hua's lab is supported by National Science Foundation USA IOS-1946174.

References

1. Jones JD, Dangl JL (2006) The plant immune system. Nature 444:323–329
2. Yuan M, Ngou BPM, Ding P et al (2021) PTI-ETI crosstalk: an integrative view of plant immunity. Curr Opin Plant Biol 62:102030
3. Grant SR, Fisher EJ, Chang JH et al (2006) Subterfuge and manipulation: type III effector proteins of phytopathogenic bacteria. Ann Rev Microbiol 60:425–449
4. Hua J (2013) Temperature and plant immunity. In: Temperature and plant development. Wiley-Blackwell, Hoboken, pp 163–180
5. Li Z, Liu H, Ding Z et al (2020) Low temperature enhances plant immunity via salicylic acid pathway genes that are repressed by ethylene. Plant Physiol 182:626–639
6. Hua J (2013) Modulation of plant immunity by light, circadian rhythm, and temperature. Curr Opin Plant Biol 16:406–413
7. Xin XF, He SY (2013) *Pseudomonas syringae pv. Tomato* DC3000: a model pathogen for probing disease susceptibility and hormone signaling in plants. Annu Rev Phytopathol 51: 473–498
8. Glazebrook J (2001) Genes controlling expression of defense responses in Arabidopsis – 2001 status. Curr Opin Plant Biol 4:301–308
9. Lamb C, Dixon RA (1997) The oxidative burst in plant disease resistance. Annu Rev Plant Biol 48:251–275
10. Melotto M, Underwood W, Koczan J et al (2006) Plant stomata function in innate immunity against bacterial invasion. Cell 126:969–980
11. Coll NS, Epple P, Dangl JL (2011) Programmed cell death in the plant immune system. Cell Death Differ 18:1247–1256
12. Develey-Riviere MP, Galiana E (2007) Resistance to pathogens and host developmental stage: a multifaceted relationship within the plant kingdom. New Phytol 175:405–416
13. Gou M, Zhang Z, Zhang N et al (2015) Opposing effects on two phases of defense responses from concerted actions of heat shock Cognate70 and Bonzai1 in Arabidopsis. Plant Physiol 169:2304–2323
14. Liu X, Sun Y, Korner CJ et al (2015) Bacterial leaf infiltration assay for fine characterization of plant defense responses using the *Arabidopsis thaliana-Pseudomonas syringae* Pathosystem. J Vis Exp 104:53364
15. Mackinney G (1941) Absorption of light by chlorophyll solutions. J Biol Chem 140:315–322
16. Jacyn Baker C, Mock NM, Culture O (1994) An improved method for monitoring cell death in cell suspension and leaf disc assays using Evans blue. Plant Cell Tissue Organ Cult 39: 7–12
17. Roine E, Wei W, Yuan J et al (1997) Hrp pilus: an *hrp*-dependent bacterial surface appendage produced by *Pseudomonas syringae* pv. *Tomato* DC3000. Proc Natl Acad Sci U.S.A 94:3459–3464
18. Mackey D, Holt BF 3rd, Wiig A et al (2002) RIN4 interacts with *Pseudomonas syringae* type III effector molecules and is required for RPM1-mediated resistance in Arabidopsis. Cell 108:743–754
19. Liu PP, von Dahl CC, Klessig DF (2011) The extent to which methyl salicylate is required for signaling systemic acquired resistance is dependent on exposure to light after infection. Plant Physiol 157:2216–2226

Chapter 7

Assessing the High Temperature Effects on Stomatal Production

Rini Rahiman and On Sun Lau

Abstract

The production of stomata, the epidermal pores of plants, is influenced by diverse environmental signals including high temperature. To assess its impact on stomatal formation, researchers need to grow plants in a carefully designed regime under controlled conditions and capture clear, microscopic views of the epidermis. Here, we describe a procedure to study the effect of high temperature on stomatal formation. This method can generate high-quality epidermal images of cotyledons, leaves, and hypocotyl of young *Arabidopsis* seedlings, which allow the determination of the pattern, density, and index of stomata on these tissues. Besides temperature, the protocol can serve as a general approach to examine stomatal phenotype and the effect of other external signals on stomatal formation.

Key words Stomatal development, High temperature, Plant tissue clearing and fixation, Imaging of stomata, Phenotypic analysis

1 Introduction

The observable traits and characteristics of a plant offer valuable information about a plant's growth and development, as well as its adaptive responses to various environmental conditions. In this chapter, we outline a method for the phenotypic analysis of stomatal development in response to high temperature.

Stomata are microscopic pores, consisting of a pair of kidney-shaped guard cells, surrounding a central pore. They are formed on the epidermal layer of aerial organs such as leaves and hypocotyls, and they allow plants to exchange gases with the atmosphere for photosynthesis and transpiration. In *Arabidopsis*, stomata are the final product of the stomatal lineage, and recent studies have demonstrated that the activity of the lineage is influenced by various environmental factors, which leads to altered stomatal number [1–4]. This response allows plants to modify its physiology and adapt to the surroundings. For instance, as light intensity increases, plants

Meng Chen (ed.), *Thermomorphogenesis: Methods and Protocols*, Methods in Molecular Biology, vol. 2795,
https://doi.org/10.1007/978-1-0716-3814-9_7,

promote the formation of stomata to enhance its photosynthetic capacity [5].

High temperature is one of the major environmental factors that influences plant growth and development [6]. In response to high ambient temperature (e.g., 27–30 °C for *Arabidopsis*), plants undergo a suite of developmental and morphological changes, including altered stomatal production. At high temperature, *Arabidopsis* represses stomatal development, likely to prevent the excessive loss of water via transpiration [7]. Due to global warming, it is imperative to understand how increasing temperatures affect stomatal formation in plants, which could have a major impact on agriculture and global water and carbon cycle.

Here, we describe a simple method for the visualization of stomatal lineage cells and the quantification of stomatal developmental features, such as stomatal density and index, of seedlings exposed to high temperature. First, we detail how to establish a high temperature condition that yields consistent results and discuss the important variables involved. Then, we provide a protocol to clear and fix plant tissues for capturing high-quality images of the epidermis. This allows the counting of the number of stomatal lineage cells and pavement cells. The data are then used to calculate stomatal density and index. The protocol was developed for examining the seedlings of *Arabidopsis* grown at high temperature, but with minor modifications, it can be adapted to study the effect of other external cues and in other tissue types and plant species.

2 Materials

Preparation of the solutions mentioned in Subheadings 2.2 and 2.3 can be done when the seedlings are at the stage required for phenotypic analysis. For handling chemicals, refer to the respective Safety Data Sheet of the chemicals and follow all the necessary safety procedures.

2.1 Plant Growth

1. Standard half strength Murashige and Skoog (1/2MS) solid media: 0.22% MS salt, 2.56 mM MES, 1% Agar, pH 5.7. Weigh 2.17 g of MS salts and transfer it to a graduated flask. Add 0.5 g of MES to this flask and top it up with 900 mL of water. Stir the mixture well using a magnetic stir bar. Using 1 M KOH, adjust the pH of this solution to 5.7. Add 10 g of micropropagation agar to the solution and make up the volume in the flask to 1000 mL. Cover the flask top with aluminum foil and autoclave the media. After autoclave, pour the media into sterile petri plates inside a laminar flow hood. Alternatively, allow the media to solidify at room temperature within the flask. Whenever needed, melt the solidified media by microwaving it, then pour it into sterile petri plates as described.

2. Plant tissue culture chamber: Models with two independently controlled chambers on one footprint are particularly useful (e.g., Percival Model CU-22-L).

2.2 Seedling Fixation and Clearing

1. Fixing solution: Ethanol:acetic acid (7:1, v/v). In a 10 mL falcon tube, add 1 mL acetic acid into 7 mL absolute ethanol. Vortex to mix. Store the solution at room temperature.
2. Clearing solution: Chloral hydrate:glycerol:water (8:1:2, w/v/v). In a 10 mL falcon tube, add 8 g of chloral hydrate in 2 mL of distilled water and dissolve the chemical by shaking/vortexing the tube. Once the chloral hydrate granules are dissolved, add 1 mL of 100% glycerol to the mixture and vortex to obtain a homogenous solution. Store the solution at room temperature. It is best to prepare the required amounts fresh for each experiment.

2.3 Stomata Imaging Analysis

1. Compound light microscope with differential contrast interference (DIC) and imaging capabilities.
2. Glass slides and cover slip (18 mm × 18 mm).
3. Fine forceps.

3 Methods

The following procedure is focused on studying the effect of high temperature on stomatal production on the cotyledons of *Arabidopsis.* However, the method is also applicable to other tissues like hypocotyl and sepals and can be a starting point for older tissues and samples from other plant species.

Excluding seedling growth, the entire procedure of phenotyping takes 2–3 days.

3.1 Seeds Sterilization and Stratification (See Note 1)

1. Transfer the required amount of seeds in a 1.5 mL tube. Each sample should have at least 10 seedlings for fixation, but more is recommended (*see* **Note 2**).
2. Add 1 mL of 70% ethanol (EtOH) to the tube. Disperse the seeds by shaking and leave the tube on a rotator for 10–15 min.
3. Remove the 70% EtOH by pipetting and repeat Step 2 by adding fresh 70% EtOH (*see* **Note 3**). If needed, the washes can be repeated several times.
4. The final wash is done with 1 mL of absolute EtOH for 1 min (*see* **Note 4**).
5. After removing the EtOH, dry the seeds in the tube itself. Alternatively, add 1 mL of fresh absolute EtOH to the tube, and with a pipette, transfer the sterilized seeds along with the EtOH, onto a sterile filter paper inside the laminar flow hood.

This allows the seeds to dry more effectively (usually in 5–10 min).

6. After the seeds have dried, sprinkle the seeds from the tube or filter paper onto the plant growth media plates. Seal the plates with surgical tapes (e.g., Micropore™) and leave them in darkness at 4 °C for 2–3 days to stratify the seeds.

3.2 Germination and Growth of Seedlings in Standard and High Temperature Conditions

1. After stratification, transfer the plates to a plant tissue culture chamber. We generally allow all seeds to germinate and grow for 1–4 days at standard temperature (22 °C), before splitting half the samples to a high temperature chamber (28 °C) to start the treatment (*see* **Note 5**). Both the standard and high temperature chambers are set with the same light intensity (70 μmol m^{-2} s^{-1}; *see* **Note 6**) and the desired photoperiod (*see* **Note 7**). Under long-day conditions (16 h light/ 8 h dark), the 4-day-old seedlings are grown for an additional 6–10 days before sample fixation (*see* **Note 8**).
2. If continuous high temperature treatment is desired, after stratification, the plates can be directly moved to the standard and high temperature chambers for germination and growth (*see* **Note 9**).

3.3 Tissue Fixation of Seedlings

1. At the desired age of the seedlings, e.g., 10–14 days post germination (dpg), lift 8–10 seedlings gently from the plate using fine forceps and transfer them to 1.5 mL tubes containing 1 mL of the fixing solution (*see* **Note 10**).
2. Leave the seedlings in the fixing solution overnight (*see* **Note 11**).

3.4 Washing and Clearing of Seedlings

1. The next day, remove the fixing solution from the tubes with a pipette (*see* **Note 12**) and add 1 mL of absolute ethanol. Leave the tubes on the bench for 15 min to wash away the remaining fixing solution from the samples (*see* **Note 13**).
2. Discard the absolute ethanol and wash the seedlings for a second time by adding 1 mL of 70% ethanol and incubating them on the bench for 20 min (*see* **Note 14**).
3. After the last wash, remove the ethanol as much as possible and add 1 mL (*see* **Note 15**) of freshly prepared clearing solution to the tubes (*see* **Note 16**).
4. The samples can be kept in the clearing solution for 4–5 h (*see* **Notes 17–19**).

3.5 Imaging and Image Analysis

1. When the samples are ready, carefully pick the seedlings out from the tubes using fine forceps and transfer it to a microscopic slide for dissection.

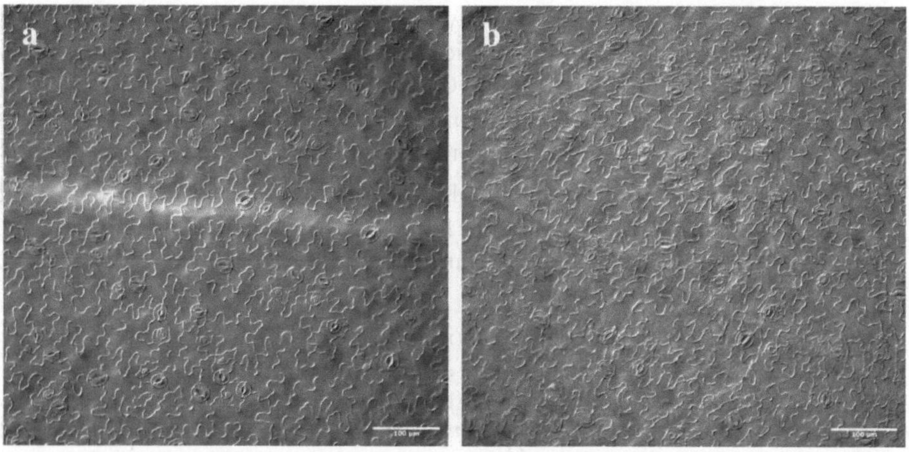

Fig. 1 DIC images of the abaxial cotyledons of wild-type *Arabidopsis* (8 dpg) grown at 22 °C that were cleared for 4 h (**a**) or overnight (**b**). Panel (**a**) shows a high-quality epidermal image with good contrast. Panel (**b**) shows the formation of wrinkles on the epidermis due to overnight clearing. Scale bar, 100 μm. Both images were taken at the center of the cotyledon's upper region (around the mid-vein) with the 20X objective

2. With the help of fine forceps, dissect and detach the cotyledon from the seedlings, which can be aided with a stereomicroscope.
3. Mount the flattened tissues on a microscope slide, using the clearing solution as the mounting media (*see* **Notes 20** and **21**), and proceed with differential interference contrast (DIC) imaging.
4. For hypocotyl samples, the air gaps formed between the cover-slip and slides will have to be filled with excess clearing solution to get clear images of the hypocotyl.
5. After placing the slide under the microscope, adjust the focus to the abaxial epidermis, which can be easily identified. Compared to the adaxial layer, it has more stomata and smaller pavement cells (*see* **Note 22**).
6. Adjust the DIC settings to obtain a good contrast for visualizing the outline of the epidermal cells (Fig. 1a). This is necessary to enable accurate identification and counting of different cell types.
7. For capturing images of the epidermis, select a defined region on the abaxial cotyledons across samples for consistency. We generally image the upper central part of the cotyledon that is in between the "outer" vein and the mid-vein (in 10–14 dpg cotyledons).
8. Capture the image using a camera-equipped compound microscope (e.g., 20X objective; field of view of image: 0.441 mm^{-2}; *see* **Note 23**) and save it as a jpg/jpeg/tiff file, in addition to the original image format, for subsequent analyses (Fig. 1).

9. For image analysis, we routinely use the ImageJ software (NIH). Quantification of different cell types can be performed using the Cell Counter plug-in (https://imagej.nih.gov/ij/plugins/cell-counter.html) on ImageJ, which is freely available online on the microscopy server under ImageJ plugins.
10. To install the plugin, copy the Cell Counter.jar file into the ImageJ plugins folder of your computer and restart ImageJ. The Cell Counter plugin will now appear under the Plugins menu.
11. For counting stomata, open the Cell Counter command (Plugins>Analyze> cell counter) and the image that you want to count in ImageJ. Click "Initialize" (on Cell Counter), then choose "Type 1" on the Cell Counter to represent "Stomata." Locate and click on a stoma in the image and a colored number will mark the stoma. Repeat until all the stomata are marked. Every time you click on a stoma, the number of the "Type 1" counter will be updated. In a similar fashion, select the "Type 2" counter to count the number of pavement cells in the image. Record the total number of stomata and pavement cells in each image.
12. In this way, process at least 15 images for each sample (22 °C and 28 °C grown seedlings).
13. For comparing the stomatal phenotype in the cotyledons of the 22 °C and 28 °C grown seedlings quantitatively, for example, we calculate and compare the Stomatal Density (SD) and Stomatal Index (SI) of each sample. SD is the number of mature stomata in the image divided by the area of the image, while SI is calculated as the number of stomata over the total number of epidermal cells (i.e., pavement cells + all stomata lineage cells) within the image view (*see* **Note 24**). The SD and SI of a sample can then be determined by averaging the respective values of all the individual images obtained from a sample.

4 Notes

1. Other commonly used protocols for seed sterilization can also be used.
2. As older seeds tend to germinate later, we advise using freshly harvested seeds for all experiments to ensure that all the seeds germinate at the same time.
3. Perform the sterilization steps inside a laminar flow hood to keep the seeds in sterile conditions.
4. Do not leave the seeds in absolute EtOH for too long as it may affect germination.

5. This minimizes the effect of high temperature on germination and early seedling development.
6. The high temperature-mediated effect on stomatal density and index is less evident when seedlings are grown at high light intensities (> 100 µmol m^{-2} s^{-1}). We attribute this observation to the suppression of PIFs under high light [8]. Therefore, for the assay, a relatively lower light intensity is used for seedling growth.
7. Among the two photoperiods, the stomatal developmental response under short-day conditions is more pronounced as reported in our previous study [7]. However, longer time is needed under this condition for the seedlings to grow and develop.
8. Although the high temperature response of hypocotyl elongation is obvious one day after high temperature treatment, significant differences in stomatal phenotypes are observable after at least 5 days of high temperature treatment. The exact number of days post-transfer can vary from 5 to 10 days (e.g., 10–14 dpg seedlings) since they yield similar results.
9. Older seed stocks might germinate later, especially at 28 ° C. Therefore, it is important to check their germination rate on plates at 22 and 28 °C when the plates are moved to the growth chambers.
10. Hold the seedlings carefully at the roots or the bottom region of the hypocotyls to prevent damaging the cotyledons or upper hypocotyls, which may be imaged later. If handling multiple samples, 12-well tissue culture plates can be used for this and the following steps.
11. After adding the fixing solution, the seedlings gradually lose all the chlorophyll and become white in color. To fully clear large leaves (8–14 dpg), you may have to change the fixing solution a couple times until the solution becomes light green. Tissues can stay in this solution for an extended time (~1 week or even longer) without issues.
12. While doing the washes, be careful not to damage the cotyledons with the pipette tips.
13. We do not have to rotate the tubes as this can tangle the seedlings inside and make it difficult to separate the seedlings.
14. After the ethanol washes, seedlings look fairly opaque and white.
15. We routinely clear 10 seedlings (10–14 dpg) in 1 mL of fixing or clearing solution, which has enough volume to cover the tissues.

16. After adding the clearing solution, the seedlings start losing their opacity. Clearing is completed when the seedlings become transparent and appear invisible.
17. Leaving the seedlings in the freshly prepared clearing solution for 4–5 h is sufficient to yield good quality samples for DIC imaging. To shorten the waiting time by half, vacuum infiltration of the solution can be done for 30 min immediately after adding the clearing solution to the fixed seedlings.
18. We recommend preparing the required amounts of clearing solution freshly each time, but the freshly prepared clearing solution can be stored and used within a week. It can also be reused one or two times. However, its clearing efficiency drops significantly afterward, which will affect image quality.
19. Although chloral hydrate is commonly used to optically clear plant tissues, leaving seedlings in the clearing solution for too long (e.g., overnight) can lead to formation of wrinkles on the epidermis (Fig. 1b). This should be avoided because it results in poor quality images that make identifying and distinguishing epidermal cells difficult. Storing the cleared seedlings at 4 °C may help slow down this process, however.
20. We do NOT recommend placing more than 1 cotyledons (10–14 dpg) under one coverslip (22 × 22 mm). This ensures that the sample is sufficiently flat for high-quality images.
21. Ensure that there are no large air gaps between the coverslip and slide, as it will make focusing of samples under the microscope difficult. Additional clearing solution can be applied to the edges of the slide to fill the gaps.
22. High temperature-induced changes in the stomatal density or index of the abaxial epidermis are also more readily observable compared to the adaxial side.
23. Capture all the images using the same objective and magnification settings to facilitate comparisons of the samples.
24. For stomatal density, it generally refers to the density of stomata only, not the earlier stomatal lineage cells. Thus, the other stomatal lineage cell types, if present, should be counted and quantified separately. Stomatal patterning defects can be identified as clusters of stomata or smaller-stomatal lineage cells violating the one-cell spacing rule.

Acknowledgments

We thank Dr. Zimin Zhou and Dr. Xin Yang for comments on the manuscript. This work was supported by Academic Research Fund (AcRF) Tier 2 (MOE-T2EP30120-0014) from the Ministry of Education, Singapore, to O.S.L. The authors declare no competing financial interests.

References

1. Pillitteri LJ, Torii KU (2012) Mechanisms of stomatal development. Annu Rev Plant Biol 63:591–614
2. Lau OS, Bergmann DC (2012) Stomatal development: a plant's perspective on cell polarity, cell fate transitions and intercellular communication. Development 139:3683–3692
3. Qi X, Torii KU (2018) Hormonal and environmental signals guiding stomatal development. BMC Biol 16:21
4. Yang X, Gavya SL, Zhou Z, Urano D, Lau OS (2022) Abscisic acid regulates stomatal production by imprinting a SnRK2 kinase-mediated phosphocode on the master regulator Speechless. Sci Adv 8:eadd2063
5. Wang S, Zhou Z, Rahiman R, Lee GSY, Yeo YK, Yang X, Lau OS (2021) Light regulates stomatal development by modulating paracrine signaling from inner tissues. Nat Commun 12:3403
6. Quint M, Delker C, Franklin KA, Wigge PA, Halliday KJ, van Zanten M (2016) Molecular and genetic control of plant thermomorphogenesis. Nat Plants 2:15190
7. Lau OS, Song Z, Zhou Z, Davies KA, Chang J, Yang X, Wang S, Lucyshyn D, Tay IHZ, Wigge PA, Bergmann DC (2018) Direct control of Speechless by PIF4 in the high-temperature response of stomatal development. Curr Biol 28:1273–1280
8. Park E, Kim Y, Choi G (2018) Phytochrome B requires PIF degradation and sequestration to induce light responses across a wide range of light conditions. Plant Cell 30:1277–1292

References

Chapter 8

Characterizing the Synergistic Response Induced by Warm Temperatures and Shade

Yogev Burko

Abstract

Plants exhibit an impressive capability to detect and respond to neighboring plants by closely monitoring changes in the light spectrum. They possess the ability to perceive adjustments in the ratio of red (R) to far-red (FR) light (R/FR) triggered by the presence of nearby plants, even before experiencing complete shading. When the R/FR ratio falls below 1, plants activate a shade avoidance response that manifests as hypocotyl elongation. Furthermore, elevated ambient temperatures can also stimulate hypocotyl elongation. As hypocotyl elongation is a visible characteristic, it is a valuable indicator for monitoring shade avoidance response, warm ambient temperature response, and the interplay between the two.

Key words Hypocotyl length, Cell length, Shade avoidance response, Thermomorpohgenesis

1 Introduction

The study of plant growth, particularly elongation, traces its roots back to the nineteenth century when plant physiologists began to dissect the growth and development of plants systematically. Since then, the hypocotyl, the embryonic stem of a plant, has emerged as the preferred organ for measuring plant growth. The hypocotyl's adaptability to various environmental and endogenous factors made it a valuable tool in physiology and genetics research.

The manipulation of hypocotyl elongation played a crucial role in numerous breakthroughs. For instance, mutant screens focused on short or long hypocotyls led to significant discoveries, including identifying key components involved in hormone and light signaling pathways [1–5]. This exploration of hypocotyl length was a foundation for understanding how plants respond to stimuli.

Even today, measuring hypocotyl length remains a fundamental trait employed in plant research. It offers a straightforward and reliable means of evaluating plant responses to their surroundings. From investigating plant development to studying stress responses,

Meng Chen (ed.), *Thermomorphogenesis: Methods and Protocols*, Methods in Molecular Biology, vol. 2795,
https://doi.org/10.1007/978-1-0716-3814-9_8,

hypocotyl length continues to be utilized in various research areas, contributing to our ongoing understanding of plant biology.

2 Materials

2.1 Plant Materials, Equipment, and Software

1. *Arabidopsis* seeds [Col-0 and mutant of interest; for example, *phytochrome-interacting factor* (*pif*) mutants [6]].
2. Square plates 10 cm × 10 cm.
3. Desiccator (or any sealed box).
4. 250 mL beaker.
5. 3 M Micropore Surgical Paper Tape.
6. Nylon mesh (ELKO filtering, 100 mm, 44% open area, 40-inch-width).
7. Plates without grids.
8. Microscope slides.
9. Coverslip.
10. Confocal microscope, Zeiss LSM 710. Excitation 561 and Emission 570–625 can be used for propidium iodide detection.
11. ImageJ software [7].
12. R platform [8] using ggplot2 [9], agricolae [10], Rmisc [11], and ggbeeswarm [12] packages.
13. LED growth chambers (Percival Scientific, E-30LED) emit red, blue, and far-red light with controllable temperatures.

2.2 Solutions

1. Seed sterilization solution: Commercial bleach (5–6% sodium hypochlorite) and 4 mL hydrochloric acid (HCl). For chlorine gas formation, pour 100 mL bleach into a 250 mL beaker. Carefully add 4 mL of HCl to the bleach in a fume hood.
2. Propidium iodide (PI) solution: 50 μg/mL PI and 0.4% Triton X-100 in distilled water. Dissolve 1 mg PI in 1 mL distilled water to make the stock PI solution. Aliquot and store at −20 ° C in darkness. To make 1 mL of 50 μg/mL PI working solution, mix 50 μL PI stock solution and 4 μL Triton X-100 with 946 μL of distilled water.
3. LS media (0.5 X): Dissolve one packet of Linsmaier and Skoog Basal Salts (Caisson Laboratories, LSP01) to 2 L of distilled water. Add 8 g of Micropropagation type I agar (Caisson Laboratories, A038) to 1 L of the 0.5 X LS media. Autoclave for 20 min at 121 °C and then allow the medium to cool to 45–50 °C. Pour 30–35 mL 0.5 X LS media for each plate in a laminar flow hood. Allow the plates to cool at room temperature for about 1 h. Store at 4 °C.

3 Methods

3.1 Seed Sterilization and Plating

1. Aliquot 40–100 seeds of each genotype into a 0.5 mL microcentrifuge labeled tube (*see* **Notes 1** and **2**).
2. Place the tubes with the seeds in a plastic rack with the lid open.
3. Place the plastic rack into the desiccator inside a fume hood.
4. Place inside the desiccator 250 mL beaker with 100 mL bleach.
5. Add 4 mL of HCl to the bleach beaker and close the desiccator immediately (*see* **Note 3**).
6. After 2–3 h, open the desiccator and immediately take the plastic rack with the seeds to a laminar flow hood. Allow the tubes to vent for 1–3 h.
7. Plate the sterile seeds on the 0.5 X LS. Seeds can be plated dry, using a sterile toothpick, or wet, by adding 1 mL sterile water and then plate one by one with 10 μL cut tips attached to a 20 μL pipette. Place two rows of seeds, following the arrangement depicted in Fig. 1a. Ensure a minimum of two replicates from each genotype are plated on separate plates.
8. Seal the plates with surgical paper tape and place them in the dark at 4 °C for 4–6 days.

3.2 Seedling Growth and Hormone Treatment

1. Place the plates vertically in LED chambers (we used 21 °C constant simulated white light), using red, blue, and far-red LED (70 μE, red/far-red = 17.8) for three days and then either keep them at 21WL or transfer them to 30 °C constant white light (30WL) or 21 °C constant white light supplemented with far-red light (21FR), using white light supplemented with far-red light (70 μE, red/far-red = 0.6), or 30 °C constant white light supplemented with far-red light (30FR), for an additional three days.
2. At the end of the experiment, place the plates on a scanner with the plate lid open, place a ruler next to the plate for scaling, scan with a dark cover on top, and measure the hypocotyl as described in Subheading 3.3.
3. To test the effect of hormones and chemicals on hypocotyl elongation:
   (a) Cut nylon mesh to the size of the plate surface.
   (b) Place the nylon mesh on plates with 0.5 X LS media using sterile tweezers.
   (c) Make sure all the nylon mesh is in contact with the LS. Take out air bubbles if needed.
   (d) On the nylon mesh, plate the seeds in two rows, as described in Subheading 3.1.

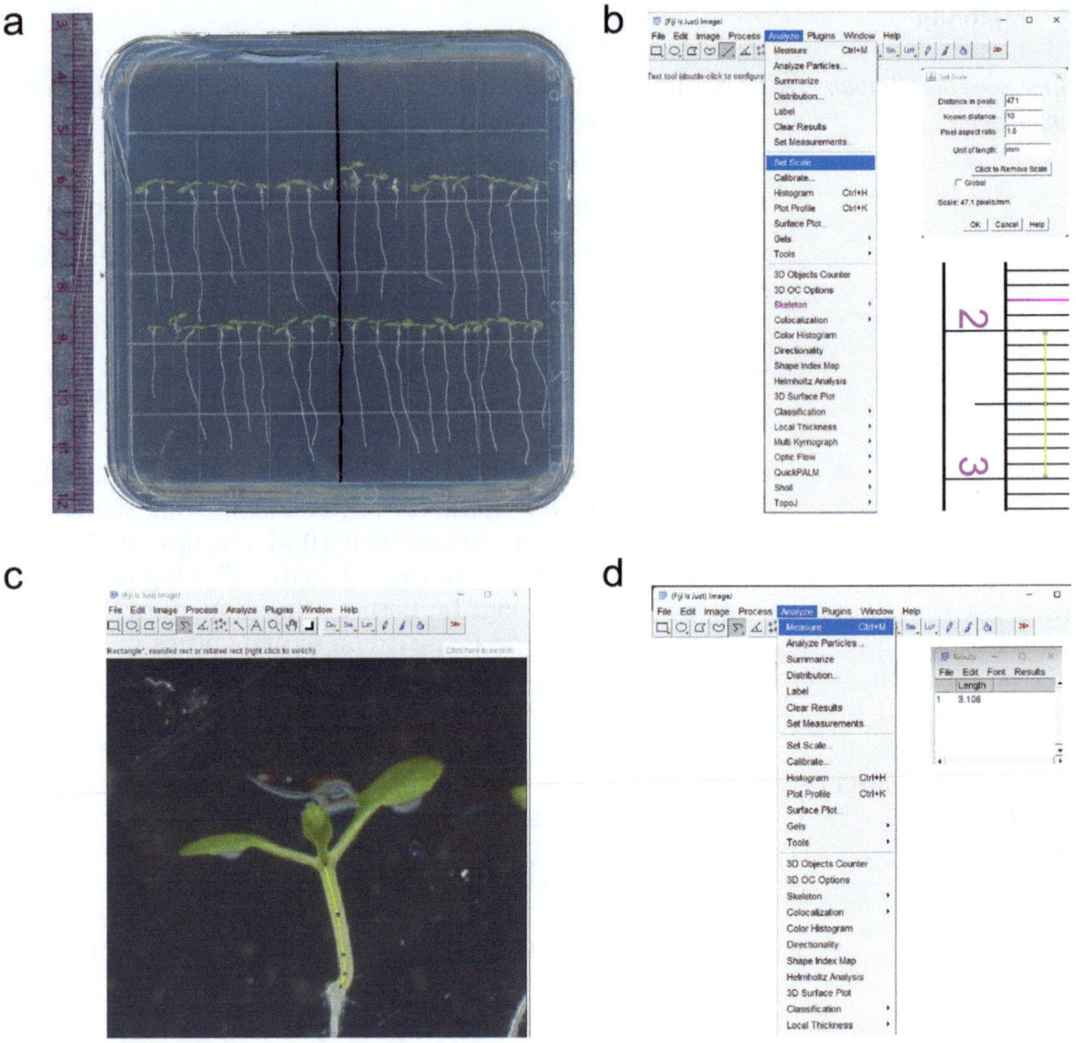

Fig. 1 Capture a screenshot of the ImageJ program displaying the hypocotyl length measurement process. (**a**) A scanned plate containing *Arabidopsis* seedlings that were used for quantifying hypocotyl length. (**b**) A snapshot demonstrating the process of setting the image scale. (**c**) A snapshot of an *Arabidopsis* seedling, where a yellow line was used to highlight the hypocotyl length for measurement. (**d**) A snapshot illustrating the process of recording the hypocotyl length

(e) Seal the plates and place them in the dark at 4 °C for 4–6 days.

(f) Place the plate vertically in the control growth conditions for three days.

(g) After three days, open the plates in a laminar flow hood and, using sterile tweezers, carefully transfer the nylon mesh with the seedlings to a new 0.5 X LS plate containing the hormone or the chemical in the media.

(h) Seal the plates and place them vertically in the tested growth condition (temperature, light, shade, etc.) for another 2–4 days.

(i) At the end of the experiment, carefully move all the seedlings to a new plate (without grids) with a thin layer of LS media.

(j) Scan and measure the hypocotyl as described in Subheading 3.3.

3.3 Measurements of Hypocotyl Cell Length Using ImageJ

(a) Use the "straight line" tool to mark 10 mm on the ruler (Fig. 1b).

(b) Scale the image by setting the known distance to 10 (Fig. 1b, analysis tab- set scale -known distance).

(c) Use the "segmented line" tool to mark the hypocotyl (Fig. 1c, the segment that starts where the two cotyledons connect and ends at the shoot-root junction).

(d) Take the length measurement using the measure tool in the analysis tab (Fig. 1d, a shortcut can be used).

(e) Export the data to Excel.

3.4 Epidermal Cell Measurement

Arabidopsis seedlings can be treated and grown as described in the previous sections. At the end of the experiment, propidium iodide (PI) is used to stain the seedlings, and a confocal microscope is used for imaging.

1. Collect 9–11 seedlings from each condition.
2. Trim the cotyledons and the roots (*see* **Note 4**).
3. Incubate the seedlings in 50 μg/mL PI solution for 10–20 min.
4. Wash the seedlings 1–2 times with sterile water.
5. Place 2–3 hypocotyls on a microscope slide.
6. Place a drop of water on top of the hypocotyl.
7. Cover the slide with a cover slip to flatten the hypocotyls.
8. Dry exceeded water.
9. Image hypocotyls apex, middle, and base segments using a confocal microscope with 20X objective.
10. Use ImageJ to measure 2–3 cells in each hypocotyl segment (apex, middle, and base; Fig. 2a, b) as described in Subheading 3.3.
11. Export the data to Excel.

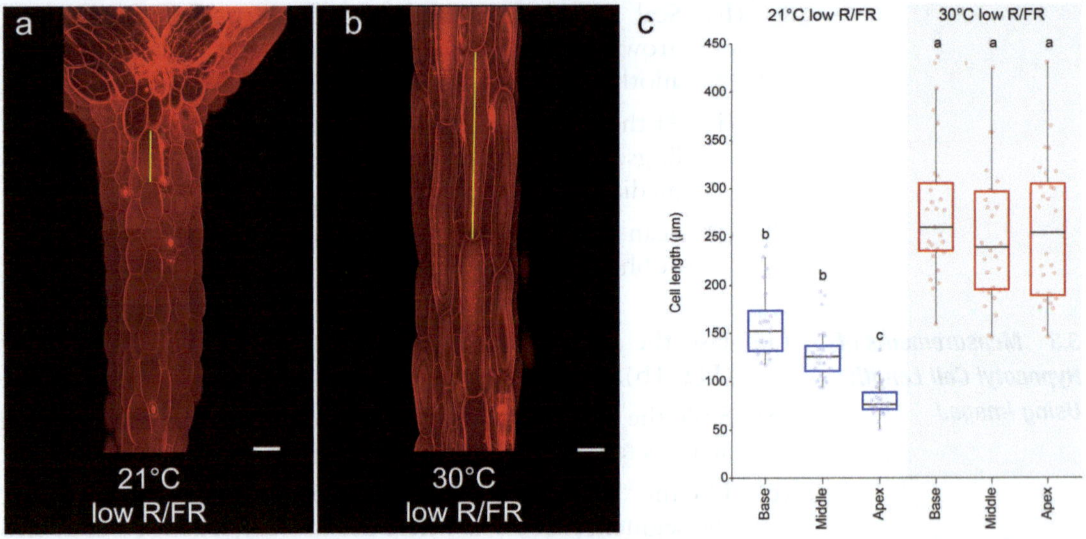

Fig. 2 Measurement of hypocotyl cell length. (**a**, **b**) Confocal images of the *Arabidopsis* hypocotyl apex of seedlings grown under white light supplemented with far-red light at (**a**) 21 °C and (**b**) 30 °C. Images obtained using a confocal microscope, Zeiss LSM 710; red-propidium iodide; the yellow line mark the length of the epidermal cells which were measured; scale bar = 50 μm. (**c**) An example of a box plot represents the epidermal cell length measurements. The raw data was sourced from Burko et al. [6] and the box plot was generated using R

3.5 Data Analysis

Use your favorite statistical analysis platform to generate boxplots (Fig. 2c) and to perform statistical analyses such as *t*-test, one-way analysis of variance (ANOVA), and Tukey HSD test. For example, the R platform [8] can be used with ggplot2 [9], agricolae [10], Rmisc [11], and ggbeeswarm [12] packages or web tools such as BoxPlotR [http://shiny.chemgrid.org/boxplotr/ [13]].

4 Notes

1. Testing the labeling pen before seed sterilization is strongly advised to ensure it is not cleared from the tube by chlorine.
2. While the chlorine gas sterilization method outlined below can be utilized, other effective methods can also be used [14].
3. *Caution:* When performing seed sterilization, work with bleach and HCl separately. Use appropriate personal protective equipment. Use a beaker size at least twice the total liquid volume of bleach with HCl to prevent splashes, which can damage seeds, bleach clothes, or cause burns.
4. When cotyledons and roots are trimmed for imaging (*see* Subheading 3.4), keep the shoot apical meristem and the root-shoot junction for orientation.

References

1. Bleecker AB, Estelle MA, Somerville C, Kende H (1988) Insensitivity to ethylene conferred by a dominant mutation in Arabidopsis thaliana. Science 241(4869):1086–1089. https://doi.org/10.1126/science.241.4869.1086
2. Chory J, Peto C, Feinbaum R, Pratt L, Ausubel F (1989) Arabidopsis thaliana mutant that develops as a light-grown plant in the absence of light. Cell 58(5):991–999. https://doi.org/10.1016/0092-8674(89)90950-1
3. Deng XW, Caspar T, Quail PH (1991) cop1: a regulatory locus involved in light-controlled development and gene expression in Arabidopsis. Genes Dev 5(7):1172–1182. https://doi.org/10.1101/gad.5.7.1172
4. Estelle MA, Somerville C (1987) Auxin-resistant mutants of Arabidopsis thaliana with an altered morphology. Mol Gen Genet MGG 206(2):200–206. https://doi.org/10.1007/bf00333575
5. Koornneef M, Rolff E, Spruit CJP (1980) Genetic control of light-inhibited hypocotyl elongation in Arabidopsis thaliana (L.). Heynh Zeitschrift für Pflanzenphysiologie 100(2):147–160. https://doi.org/10.1016/s0044-328x(80)80208-x
6. Burko Y, Willige BC, Seluzicki A, Novák O, Ljung K, Chory J (2022) PIF7 is a master regulator of thermomorphogenesis in shade. Nature. Communications 13(1). https://doi.org/10.1038/s41467-022-32585-6
7. Schindelin J, Arganda-Carreras I, Frise E, Kaynig V, Longair M, Pietzsch T, Preibisch S, Rueden C, Saalfeld S, Schmid B, Tinevez JY, White DJ, Hartenstein V, Eliceiri K, Tomancak P, Cardona A (2012) Fiji: an open-source platform for biological-image analysis. Nat Methods 9(7):676–682. https://doi.org/10.1038/nmeth.2019
8. R Core Team (2022) R: a language and environment for statistical computing. R Foundation for Statistical Computing, Vienna. https://www.R-project.org/
9. Wickham H (2016) ggplot2: elegant graphics for data analysis. Springer-Verlag, New York. ISBN 978-3-319-24277-4. https://ggplot2.tidyverse.org
10. Mendiburu Fd (2020) Agricolae: Statistical Procedures for Agricultural Research. https://cran.r-project.org/package=agricolae
11. Hope RM (2022) Rmisc. https://CRAN.R-project.org/package=Rmisc
12. Clarke E, Sherrill-Mix S, Dawson S (2023) ggbeeswarm. https://CRAN.R-project.org/package=ggbeeswarm
13. Spitzer M, Wildenhain J, Rappsilber J, Tyers M (2014) BoxPlotR: a web tool for generation of box plots. Nat Methods 11(2):121–122. https://doi.org/10.1038/nmeth.2811
14. Espinosa-Ruiz A, Martínez C, Prat S (2015) Protocol to treat seedlings with Brassinazole and measure hypocotyl length in Arabidopsis thaliana. Bio-Protocol 5(16). https://doi.org/10.21769/BioProtoc.1568

Part II

Temperature Sensing in Thermomorphogenesis

Chapter 9

Measurement of Phytochrome B Thermal Reversion Rates In Vivo

Cornelia Klose and Andreas Hiltbrunner

Abstract

Thermal reversion of phytochromes is the light-independent but strongly temperature-dependent relaxation of the light-activated Pfr form of phytochromes back into the inactive Pr ground state. The thermal reversion rates of different phytochromes vary considerably. For phytochrome B (phyB), thermal reversion represents a critical parameter affecting phyB activity as it reduces the active phyB Pfr pool, accelerated by increasing temperatures. Phytochromes are dimers existing in three different states: Pfr-Pfr homodimer, Pfr-Pr heterodimer, and Pr-Pr homodimer. Consequently, thermal reversion occurs in two steps, with Pfr-Pfr to Pfr-Pr reversion being much slower than reversion from Pfr-Pr to Pr-Pr. To measure thermal reversion in vivo, the relative proportion of Pfr in relation to the total amount of phytochrome (Ptot) must be determined in living samples. This is accomplished by in vivo spectroscopy utilizing dual wavelength ratiospectrophotometers, optimized for assaying phytochromes in highly scattering plant material. The method is depending on the photoreversibility of phytochromes displaying light-induced absorbance changes in response to actinic irradiation. In this chapter, we describe the experimental design and explain step-by-step the calculations necessary to determine the thermal reversion rates of phyB in vivo, taking into account phytochrome dimerization.

Key words Phytochrome, Thermal reversion, Ratiospectrophotometer, In vivo spectroscopy

1 Introduction

Phytochromes are dimeric chromoproteins, with each subunit harboring a covalently attached bilin chromophore. Upon light absorption, they interconvert between the red light (R)-absorbing, physiologically inactive Pr form and the far-red light (FR)-absorbing, physiologically active Pfr form, which is repeatedly reversible. The change in absorbance between Pr ($\lambda_{max} = 665$ nm) and Pfr ($\lambda_{max} = 730$ nm) can be directly correlated to the phytochrome concentration in a sample. Consequently, phytochrome photo-interconversion can be utilized to measure the amount of photoreversible phytochromes in spectrophotometric assays.

Meng Chen (ed.), *Thermomorphogenesis: Methods and Protocols*, Methods in Molecular Biology, vol. 2795,
https://doi.org/10.1007/978-1-0716-3814-9_9,

In addition to the Pfr to Pr photo-reversion, Pfr can spontaneously relax back to the Pr ground state in a strongly temperature-dependent process termed thermal reversion (Fig. 1a) [1]. Thermal reversion is an intrinsic molecular property of phytochromes but is also modified in vivo by a number of molecular mechanisms including intra- and intermolecular interactions or posttranslational modifications [2]. Thermal reversion can vary considerably among different phytochromes but is most influential and best characterized for phytochrome B (phyB) from *Arabidopsis*. The combination of a fast thermal reversion rate and the strong temperature dependence of this rate enables phyB to act as temperature sensitive light sensor in physiologically relevant ambient light and temperature conditions (Fig. 1b) [2 –5]. As predicted from the phytochrome dimer model, thermal reversion of phyB occurs in two steps: a slow reversion from the Pfr-Pfr homodimer to the Pfr-Pr heterodimer (at rate k_{r2}) and a much faster reversion from the Pfr-Pr heterodimer to the Pr-Pr homodimer ground state (at rate k_{r1}) (Fig. 1c) [6]. Whereas the slow Pfr-Pfr reversion determines the persistence of physiologically active phyB Pfr after transfer to darkness, the 100x faster Pfr-Pr reversion competes efficiently with the Pr-to-Pfr photoactivation, affecting Pfr levels under light conditions in a temperature-dependent manner [2 , 3 , 6 , 7]. Both thermal reversion processes can be quantitatively determined using in vivo spectroscopic measurements as described in this chapter.

In vivo spectroscopy of phytochromes has to overcome some major challenges: (i) it requires extremely high sensitivity due to the low abundance of the photoreversible phytochrome pigments and (ii) it has to deal with highly inhomogeneous samples (e.g., light scattering tissues, heterogeneous pigment distribution, additional light absorbing pigments) [8 , 9]. Dual wavelength ratiospectrophotometers are specifically designed instruments for in vivo spectroscopy, capable of accurately measuring small absorbance differences in highly scattering samples with strong background absorption, like plant tissues, but also other tissue types or cell cultures [9 –12]. However, since pigments absorbing in the red light spectral range such as chlorophyll can interfere with the measurement [13], only non-green plant tissue (e.g., etiolated seedlings or light grown tissue that has been bleached) can be used. Another limitation is the relatively low abundance of native phyB in plant tissue. In *Arabidopsis thaliana* and other plants, phyA is the most abundant phytochrome in etiolated seedlings and can be measured by in vivo spectroscopy. The other phytochromes, phyB, C, D or E, are below detection limit at endogenous levels. To spectroscopically measure only phyB in vivo, transgenic overexpression lines (PHYBox) are required, preferentially in a *phyA-phyB* mutant background.

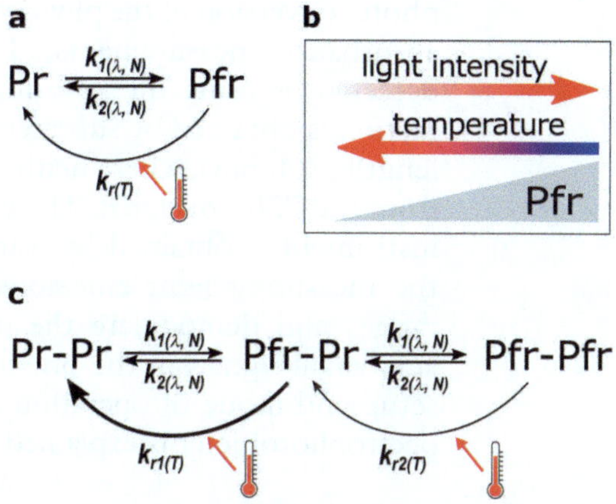

Fig. 1 Phytochrome reaction schemes. (**a**) Monomer model of the phytochrome photoconversion and thermal reversion reactions. The photoconversion rates k_1 and k_2 for Pr-to-Pfr and Pfr-to-Pr photoconversions, respectively, are dependent on the wavelength λ and intensity N of light, whereas k_r is the temperature (T)-dependent thermal reversion rate. (**b**) The amount of the active Pfr form of phyB is determined by an interaction of light and temperature. Higher light intensities of R favor Pfr accumulation, whereas increasing temperatures accelerate thermal reversion, thereby counteracting Pr-to-Pfr photoconversion and reducing Pfr levels. (**c**) Dimer model of the phytochrome photoconversion and thermal reversion reactions. PhyB exists in three different states: Pr-Pr homodimer, Pfr-Pr heterodimer, and Pfr-Pfr homodimer. The photoconversion rates k_1 and k_2 are similar to the monomer model depicted in (a), and k_{r1} and k_{r2} are the temperature-dependent thermal reversion rates of the Pfr-Pr heterodimer and the Pfr-Pfr homodimer, respectively

In vivo spectroscopy depends on the photoreversibility of phytochromes displaying light-induced absorbance changes in response to actinic irradiation that induces photochemical reactions of phytochromes. The general setup, operation mode, and limitations of a custom built, fully automated dual wavelength ratiospectrophotometer is described in [9]. The method not only allows to detect total amounts of photoreversible phytochromes (Ptot) but also to determine the proportion of phytochrome that initially was in the active Pfr or inactive Pr conformation. This chapter focuses on measurements of relative phyB Pfr/Ptot ratios in living *Arabidopsis thaliana* PHYBox seedlings to determine the thermal reversion rates of phyB.

2 Materials

2.1 Dual Wavelength Ratiospectrophotometer

A dual wavelength ratiospectrophotometer uses two different types of light sources: (i) low intensity measuring light (730 nm and approx. 800 nm) and (ii) strong actinic light (R and FR) to execute photoconversion of the phytochromes in the sample in between the absorbance measurements. The Pfr form of phytochromes is detected by using the 730 nm measuring light. Background correction is obtained by subtracting the signal at the reference wavelength (800 nm), where neither Pfr nor Pr can absorb significantly, from the 730 nm signal. The extraordinarily high sensitivity of the instrument is obtained by using lock-in-amplifiers that modulate the measuring light emissions at a high frequency (in the kHz range) and demodulate the detected signal in order to extract very small signals in the presence of high background noise. The setup and mode of operation of a typical dual wavelength ratiospectrophotometer is explained in more detail in [9].

2.2 Plant Growth

1. *Arabidopsis thaliana* seeds of PHYBox in *phyAphyB* mutant background (*see* **Note 1**).
2. Petri dishes.
3. Filter paper fitting to the petri dishes.
4. Autoclaved distilled water.
5. Ethanol 70% and 100%.

2.3 Sample Preparation

1. Temperature-controlled incubator equipped with light sources for light and temperature treatments.
2. Sample cuvettes wrapped in film and precooled on ice.
3. Scalpel, forceps, paper towel, metal, or glass rod fitting into the cuvette.

3 Methods

3.1 Experimental Design for Light and Temperature Treatments to Determine phyB Thermal Reversion Rates k_{r1} and k_{r2}

Using in vivo spectroscopy, it is possible to detect phyB Pfr in living plant material, but not to directly discriminate between the two different Pfr-containing dimer species, Pfr-Pfr and Pfr-Pr. Consequently, thermal reversion rates k_{r2} of the Pfr-Pfr homodimer and k_{r1} of the Pfr-Pr heterodimer cannot be measured directly; only measurement of apparent thermal reversion rates is possible. Determination of k_{r2} and k_{r1} requires an experimental design taking account of the different dimer properties of the phyB dimer. The Pfr-Pfr thermal reversion is slow and determines the persistence of physiologically active phyB Pfr after transfer to darkness [6]. Hence, k_{r2} can be approximated from typical thermal reversion kinetics after saturating R treatment and transfer of the sample to darkness. The Pfr-Pr thermal reversion is 100× faster and therefore able to compete with the fast Pr-to-Pfr photoactivation, affecting Pfr levels especially in weak light (Fig. 2) [6]. Consequently, k_{r1} can be calculated from Pfr/Ptot measurements under light conditions (e.g., low intensity R) that establish a steady-state Pfr/Ptot level at

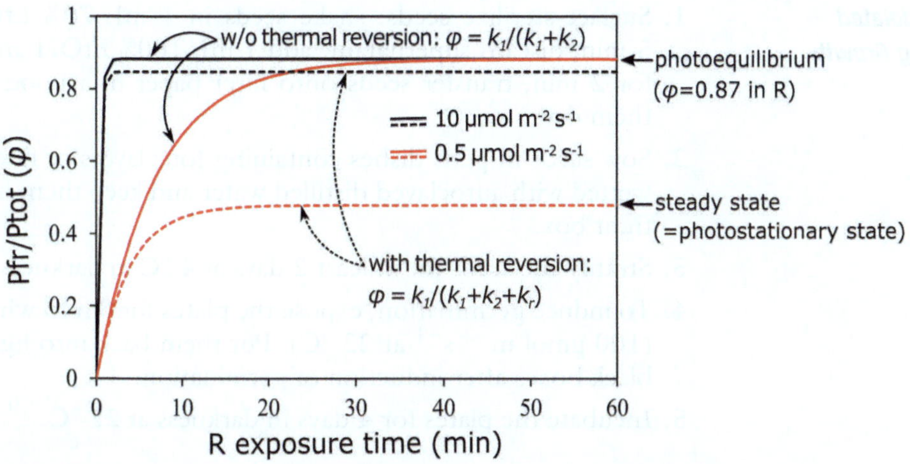

Fig. 2 Phytochrome photoconversion kinetics for two different R intensities with or without thermal reversion. Increase in Pfr/Ptot ratio φ as a function of red light (R) exposure time in high R (10 μmol m^{-2} s^{-1}) and low R (0.5 μmol m^{-2} s^{-1}) calculated w/o thermal reversion (solid lines) or with thermal reversion (dashed lines). For simplicity, the calculations are based on the phytochrome monomer model as depicted in Fig. 1a. The Pfr/Ptot photoequilibrium φ is determined solely by the light-dependent photoconversion rates k_1 and k_2. In the absence of thermal reversion, the photoequilibrium φ = 0.87 in R is established irrespective of the R intensity used, only dependent on the irradiation time. The steady state (also photostationary state according to Mancinelli (1994) [1]) represents the Pfr/Ptot equilibrium resulting from the interaction of light-dependent photoconversion (rates k_1 and k_2) and light-independent but temperature-dependent thermal reversion (k_r) and has to be distinguished from the photoequilibrium. At a constant k_r, the Pfr/Ptot steady state is dependent on the light intensity only and is generally lower compared with the respective photoequilibrium, and this difference is more pronounced at lower light intensities

an equilibrium that is lower than the photoequilibrium for R (Fig. 2). To be able to determine both rates, k_{r1} and k_{r2}, two different types of thermal reversion experiments are required:

1. For measuring thermal reversion kinetics in darkness, irradiate plates containing 4-day-old etiolated seedlings for 20 min with saturating R (50 μmol m^{-2} s^{-1}) to establish a high Pfr/Ptot ratio (Pfr/Ptot at the R photoequilibrium is approx. 0.87). Then transfer seedlings to darkness at the respective temperature. Measure Pfr/Ptot at selected timepoints during dark incubation (for phyB typically between 0 and 8 h).
2. For determining steady-state Pfr/Ptot ratios established under subsaturating light conditions (light intensities not sufficient to reach the photoequilibrium) or in different light and temperature combinations, treat plates containing 4-day-old etiolated seedlings for at least 1 h with the respective light and temperature conditions to establish steady-state conditions (*see* **Note 2**). Remove the plates from light treatment and immediately transfer the filter paper with seedlings to ice/water in darkness to inhibit unintended thermal reversion and proceed with sample preparation.

3.2 Etiolated Seedling Growth

1. Surface-sterilize seeds: shake seeds in 1 mL 70% EtOH for 5 min, discard supernatant, add 1 mL 100% EtOH and shake for 2 min, transfer seeds onto filter paper in a hood and let them dry.
2. Sow seeds in petri dishes containing four layers of filter paper wetted with autoclaved distilled water and keep them in a light tight box.
3. Stratify the seeds for at least 2 days at 4 °C in darkness.
4. To induce germination, expose the plates for 8 h to white light (100 μmol m^{-2} s^{-1} at 22 °C). Put them back into light tight black boxes after induction of germination.
5. Incubate the plates for 4 days in darkness at 22 °C.

3.3 Sample Preparation

1. Under dim green-safelight: carefully scrape the seedlings from the filter paper, collect them on a tissue, and thoroughly pat the seedlings dry (*see* **Note 3**).
2. Under dim green-safelight: transfer the seedlings into pre-cooled sample cuvettes and determine the fresh weight.
3. Carefully compress the seedlings in the cuvette using a pre-cooled metal rod in order to obtain an even sample geometry.
4. Transfer sample to the ratiospectrophotometer to measure Pfr/Ptot.

3.4 Measuring Procedure and Calculation of Pfr/Ptot

The ratiospectrophotometer measures the optical density (OD) of the sample at 730 nm (OD730), representing the maximum of Pfr absorption, and at 800 nm (OD800) as reference wavelength, were no light absorption by phytochromes is expected. The difference in light absorbance ΔA between OD730 and OD800 is determined by subtracting the reference OD800 from OD730 ($\Delta A = OD730 - OD800$). After an initial ΔA determination of the sample ($\Delta A_{initial}$), a strong FR pulse is used to revert all Pfr in the sample to Pr and ΔA is measured again (ΔA_{FR0}). The resulting difference $\Delta\Delta A_0$ between before and after the first reverting FR treatment represents the initial Pfr amount that was present in the sample ($\Delta\Delta A_0 = \Delta A_{initial} - \Delta A_{FR0}$). To determine the total amount of photoreversible phytochrome (Ptot), the sample is then treated alternately with strong R and FR pulses to photoconvert all phytochrome from Pr to the Pfr form back and forth multiple times. Between the light treatments, ΔA_R (ΔA after R treatments) and ΔA_{FR} (ΔA after FR treatments) is measured for multiple cycles. Then, the average $\Delta\Delta A_{mean}$ is calculated over all R-FR cycles, reflecting the total photoreversible phytochrome amount ($\Delta\Delta A_{mean} = |\Delta A_R - \Delta A_{FR}|$). Ptot in the sample is calculated from $\Delta\Delta A_{mean}$ ($Ptot = 1.15\ \Delta\Delta A_{mean}$), taking into account the photoequilibrium of 0.87 in R according to Mancinelli (1994) [1]. The initial Pfr amount ($\Delta\Delta A_0$) is then calculated relative to the total amount of photoreversible phytochrome Ptot and given as relative Pfr/Ptot ratio φ ($\varphi = \Delta\Delta A_0/\text{Ptot}$); for more details, see [9].

3.5 Calculation of Thermal Reversion Rates k_{r1} and k_{r2}

A typical Pfr thermal reversion kinetics is measured after R treatment of the sample and transfer to darkness for increasing time intervals. For each temperature tested, the apparent thermal reversion rate k_r can be calculated using approximate single exponential decay functions of the following form:

$$\varphi_{(t)} = \varphi_0 e^{-k_r t}$$

where $\varphi_{(t)}$ is the Pfr/Ptot ratio measured at time t and φ_0 is the Pfr/Ptot ratio at time $t = 0$.

The rate constant of the exponential decay function represents the apparent thermal reversion rate k_r and is calculated by:

$$-k_r = \ln\left(\frac{\varphi(t)}{\varphi_0}\right)\frac{1}{t}.$$

According to the three-state dimer model of phytochrome B dimers it is predicted that, after exposure of a phytochrome sample to R followed by darkness, the majority of Pfr is present in the Pfr-Pfr homodimer pool, whereas the Pfr-Pr heterodimer pool is rapidly depleted due to the fast thermal reversion rate k_{r1} of the Pfr-Pr heterodimer (Fig. 1c) [3, 6]. Therefore, the rate-limiting step for the apparent thermal reversion kinetics in darkness is the slow thermal reversion rate k_{r2} of the Pfr-Pfr homodimer. Consequently, the apparent thermal reversion rate k_r can be related to the reversion rate k_{r2} of the Pfr-Pfr homodimer in the following way:

$$k_r = 2k_{r2}.$$

Estimation of k_{r2} is prerequisite to subsequently calculate the thermal reversion rate k_{r1} of the Pfr-Pr heterodimer. Because k_{r1} is about 100× faster compared with k_{r2}, temperature effects on Pfr steady state levels in light predominantly depend on the fast thermal reversion rate k_{r1}. Using temperature-dependent Pfr/Ptot steady-state measurements in subsaturating light conditions (where Pfr/Ptot is below the respective photoequilibrium), the thermal reversion rate k_{r1} can be calculated for different temperatures using the following analytical equation based on the three-state dimer model [3 , 6]:

$$k_{r1} = \frac{\frac{k_1(2k_2+2k_{r2})+2k_1{}^2}{\varphi} - 2k_1{}^2 - 2k_1(2k_2+2k_{r2})}{2k_2+2k_{r2}} - k_2$$

where k_1 and k_2 are the photoconversion rates for Pr-to-Pfr and Pfr-to-Pr photoconversions, respectively, and φ is the Pfr/Ptot ratio measured in steady-state conditions. The wavelength-dependent photoconversion rates to be inserted in the equation above can be determined according to Mancinelli (1994) [1] by:

$$k_{1,\lambda} = N_\lambda \sigma_{R,\lambda} \text{ and } k_{2,\lambda} = N_\lambda \sigma_{FR,\lambda}$$

where N_λ is the photon fluence rate (in mol of photons m^{-2} s^{-1}) at wavelength λ, and $\sigma_{R,\ \lambda}$ or $\sigma_{FR,\ \lambda}$ are the photoconversion cross-sections (in m^2 mol^{-1}) of Pr or Pfr at wavelength λ, respectively (can be derived from Ref. [1]).

4 Notes

1. To spectroscopically detect phyB in vivo, overexpression of phyB is necessary (PHYBox) because endogenous phyB levels in *Arabidopsis* are below the detection limit for in vivo spectroscopy. It is recommendable to use a *phyA* mutant background for PHYBox in order to avoid overlaying signals of PHYBox and the highly abundant endogenous phyA. Alternatively, the light labile phyA in the background could be depleted by 3 h pretreatment with R, not affecting PHYBox abundance.
2. Do not irradiate seedlings longer than 4 h with continuous red, blue, or white light to prevent chlorophyll accumulation. Chlorophyll can interfere with light absorption measurement.
3. Work as fast and consistent as possible when preparing the samples. Cool down harvested seedlings immediately to avoid any unintended changes in Pfr/Ptot ratios.

Acknowledgments

This study was supported by the German Research Foundation (DFG) under Germany's Excellence Strategy (CIBSS-EXC 2189—Project ID 390939984). Construction of a new custom-made dual wavelength ratiospectrophotometer was supported by CIBSS and realized in collaboration with the Fraunhofer Institute for Physical Measurement Techniques in Freiburg and the workshop at the Institute of Biology II, University of Freiburg.

References

1. Mancinelli AL (1994) The physiology of phytochrome action. In: Kendrick RE, Kronenberg GHM (eds) Photomorphogenesis in plants. Springer, Dordrecht, pp 211–269
2. Klose C, Nagy F, Schäfer E (2020) Thermal reversion of plant phytochromes. Mol Plant 13:386–397. https://doi.org/10.1016/j.molp.2019.12.004
3. Legris M, Klose C, Burgie ES et al (2016) Phytochrome B integrates light and temperature signals in Arabidopsis. Science 354:897–900. https://doi.org/10.1126/science.aaf5656
4. Viczián A, Klose C, Ádám É et al (2017) New insights of red light-induced development. Plant Cell Environ 40:2457–2468. https://doi.org/10.1111/pce.12880
5. Burgie ES, Gannam ZTK, McLoughlin KE et al (2021) Differing biophysical properties underpin the unique signaling potentials

within the plant phytochrome photoreceptor families. PNAS 118. https://doi.org/10.1073/pnas.2105649118

6. Klose C, Venezia F, Hussong A et al (2015) Systematic analysis of how phytochrome B dimerization determines its specificity. Nature Plants 1:15090. https://doi.org/10.1038/nplants.2015.90
7. Jung J-H, Domijan M, Klose C et al (2016) Phytochromes function as thermosensors in Arabidopsis. Science 354:886–889. https://doi.org/10.1126/science.aaf6005
8. Gross J, Seyfried M, Fukshansky L et al (1984) In: Smith H, Holmes MG (eds) In vivo spectrophotometry. Methods in photomorphogenesis, biological technique series. Academic Press, London, pp 131–157
9. Klose C (2019) In vivo spectroscopy: in: Hiltbrunner, A. (eds) Phytochromes. Methods Mol Biol 2026:113–120. https://doi.org/10.1007/978-1-4939-9612-4_8
10. Butler WL, Norris KH, Siegelman HW et al (1959) Detection, assay, and preliminary purification of the pigment controlling photoresponsive development of plants. PNAS 45:1703–1708
11. Pratt LH, Wampler JE, Rich ES (1984) An automated dual-wavelength spectrophotometer optimized for phytochrome assay. Instrum Sci Technol 13:269–287. https://doi.org/10.1080/10739148408544183
12. Lamparter T, Hughes J, Hartmann E (1994) A fully automated dual-wavelength photometer for phytochrome measurements and its application to phytochrome from chlorophyll-containing extracts. Photochem Photobiol 60:179–183. https://doi.org/10.1111/j.1751-1097.1994.tb05088.x
13. Grill R (1977) Influence of chlorophyll content on phytochrome measurements in turnip cotyledons. Planta 134:11–16. https://doi.org/10.1007/BF00390087

Chapter 10

Characterization of Thermoresponsive Photobody Dynamics

Juan Du and Meng Chen

Abstract

Photobodies (PBs) are subnuclear membraneless organelles that self-assemble via the condensation of the plant photoreceptor and thermosensor phytochrome B (phyB). Changes in the light and temperature environment directly modulate PB formation and maintenance by altering the number and size of PBs. In thermomorphogenesis, increases in the ambient temperature incrementally reduce the number of PBs, suggesting that individual PBs possess distinct thermostabilities. Here, we describe a detailed protocol for characterizing cell type-specific PB dynamics induced by warm temperatures in *Arabidopsis*.

Key words Photobody, Phytochrome B, PhyB, Thermomorphogenesis

1 Introduction

Temperature is an important environmental cue that influences almost all aspects of plant development, growth, metabolism, and immunity [1, 2]. Plants sense changes in ambient temperature primarily via the red (R) and far-red (FR) photoreceptor phytochrome B (phyB) [3, 4]. PhyB monitors light quality, quantity, and periodicity through the photoconversion between two relatively stable conformers, a R-light-absorbing inactive Pr and a FR-light-absorbing active Pfr conformer [5, 6]. In addition to photoconversion, the active Pfr can spontaneously revert to the inactive Pr in a light-independent process called dark or thermal reversion [7]. The rate of phyB's thermal reversion, which is rapid enough to counteract photoactivation, accelerates with temperature increases between 10 °C and 30 °C [3]. These intrinsic properties make phyB a thermosensor in addition to a photoreceptor [3, 4]. At the cellular level, photoactivation of phyB triggers its rapid nuclear accumulation and subsequent condensation into discrete subnuclear membraneless organelles or biomolecular condensates named photobodies (PBs) [8–10]. Because the assembly and maintenance of PBs are driven by the active form of phyB, perturbing the Pfr/Pr

Meng Chen (ed.), *Thermomorphogenesis: Methods and Protocols*, Methods in Molecular Biology, vol. 2795,
https://doi.org/10.1007/978-1-0716-3814-9_10,

equilibrium by either light or temperature alters the steady-state pattern of PBs [3, 11–14]. The light-dependent assembly of phyB to PBs has been extensively studied using *Arabidopsis* transgenic lines expressing fluorescent protein-tagged phyB (phyB-FP). Under strong light conditions, which keep the majority of phyB in the active form, phyB-FP forms several large PBs of 0.7–2 μm in diameter [11, 14, 15]. Proteomic studies demonstrate that PBs also contain many critical phyB signaling components [16, 17]. The formation of large PBs correlates closely with phyB signaling functions [9, 12, 15, 18–22].

Characterization of the effects of warm temperatures on PBs revealed two features of temperature-dependent PB dynamics [14]. First, although both warm temperatures and shade destabilize the active phyB, warm temperatures elicit a largely different PB response from the one induced by shade. Shade triggers a transition from large to small PBs. In contrast, increasing temperature incrementally reduces the number of PBs. Second, individual PBs from distinct subnuclear locations may possess different thermostabilities. PBs can be divided into two classes based on their positions relative to the nucleolus: (1) nucleolar-associated PBs (Nuo-PBs) that localize to the periphery of the nucleolus and (2) non-nucleolar-associated PBs (nonNuo-PBs) that localize away from the nucleolus. The warm temperature-dependent reduction of PB number is conserved in various organs and across different ecotypes. The thermoresponsive PB dynamics are best exemplified in the cotyledon pavement epidermal cells of the *PBG* line, which expresses phyB-GFP in the *phyB-5* background [23]. In 4-day-old *PBG* seedlings at 21 °C, most pavement cell nuclei in the cotyledon contain two PBs, a Nuo-PB, and a nonNuo-PB; in contrast, at 27 °C, the majority of pavement cell nuclei lose the nonNuo-PB, leaving only one Nuo-PB (Fig. 1). The disappearance of the nonNuo-PB takes place within 6 h upon the transition from 21 °C to 27 °C (Fig. 2). These results suggest that while the Nuo-PB was thermally stable, the nonNuo-PB was thermally unstable and disassembled in the warmer temperature. The thermosensitive PB response represents one of the early readouts of warm temperature signaling. Future investigations of the mechanism of temperature-induced PB dynamics hold great promise for elucidating the mechanism and function of PB formation in temperature signaling. Here we describe a detailed protocol to characterize the temperature-induced PB dynamics based on a previously published method for characterizing steady-state PB patterns [24].

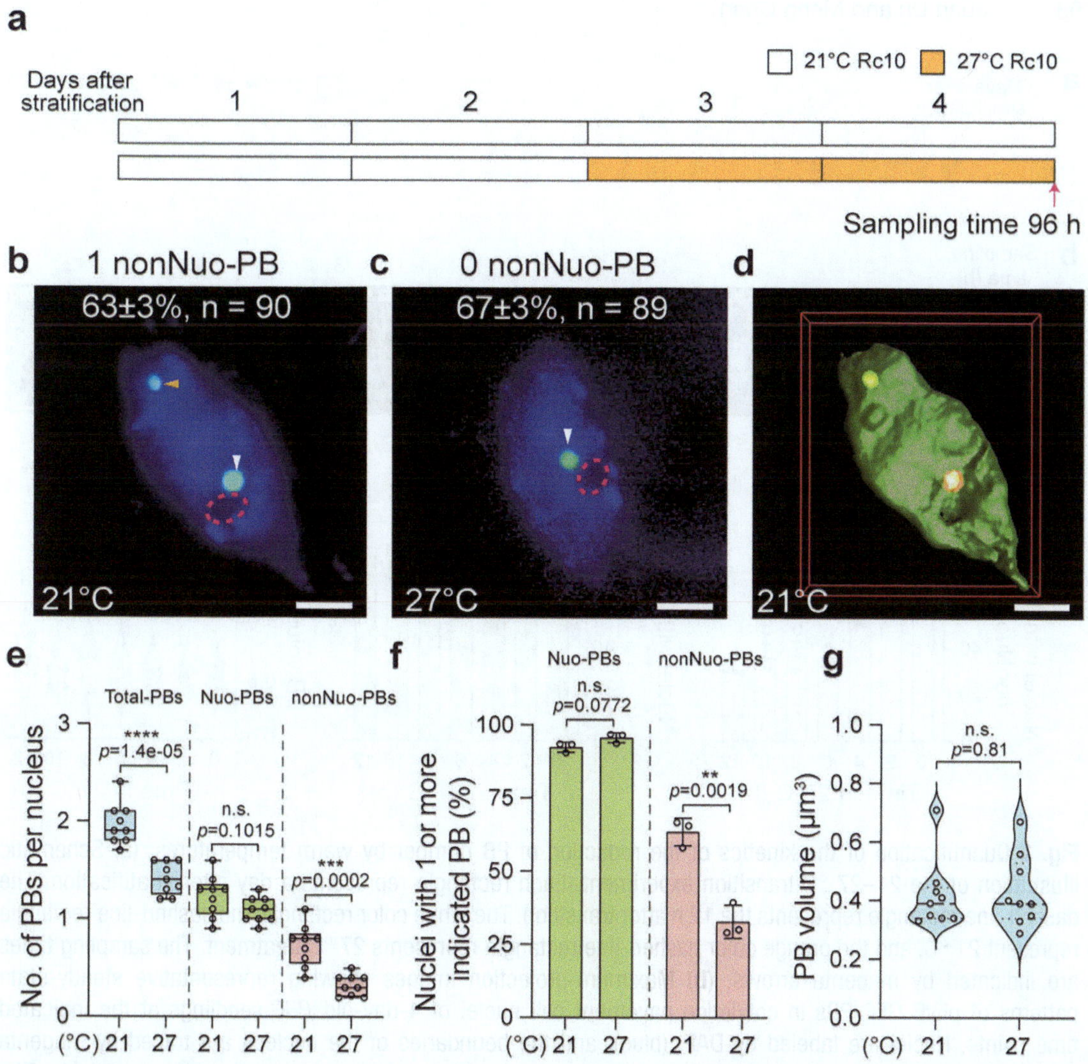

Fig. 1 Quantification of warm temperature-dependent reduction in PB number. (**a**) Schematic illustration of the warm temperature treatment experiment. Each rectangle represents a day after stratification. The white and orange color rectangles represent 21 °C and 27 °C temperature treatment, respectively. The sampling time is labeled by a magenta arrow. (**b**, **c**) Maximum-projection, deconvolved fluorescence microscopy images showing representative steady-state patterns of phyB-GFP (green) PBs in cotyledon pavement cell nuclei from 4-day-old *PBG* seedlings grown under 10 μmol m^{-2} s^{-1} red light at 21 °C and 27 °C. Nuclei are labeled by DAPI (blue), and the boundaries of nucleoli are traced by magenta dashed lines. Nuo-PBs and nonNuo-PBs are indicated by white and orange arrowheads, respectively. The percentage (with s.e.) of nuclei with the indicated number of nonNuo-PB is shown in each image; n indicates the total number of nuclei analyzed. Scale bars represent 2 μm. (**d**) Three-dimensional image of the nucleus and PBs reconstructed by Huygens Essential Object Analyzer. The scale bar represents 2 μm. (**e**) Quantification of the numbers of total, Nuo, and nonNuo PBs per nucleus in cotyledon pavement cell nuclei of the *PBG* seedlings ($n = 9$) described in (**b**, **c**). The boxes represent the 25% to 75% quantiles, and the bars are equal to the mean. (**f**) Quantification of the percentage of nuclei in cotyledon pavement cell nuclei containing one or more either Nuo- or nonNuo-PB in the *PBG* seedlings described in (**b**, **c**). The error bars represent s.e. calculated from groups of seedlings ($n = 3$). (**g**) Quantification of the volume of PBs in cotyledon epidermal-cell nuclei of the *PBG* seedlings ($n = 9$) described in (**b**, **c**) For e. g., the *p* values between the data at 21 °C and 27 °C are calculated using Student's two-tailed *t*-test; asterisks indicate statistical significance (* $p \le 0.05$, ** $p \le 0.01$, *** $p \le 0.001$, **** $p \le 0.0001$), and n.s. indicates not significant

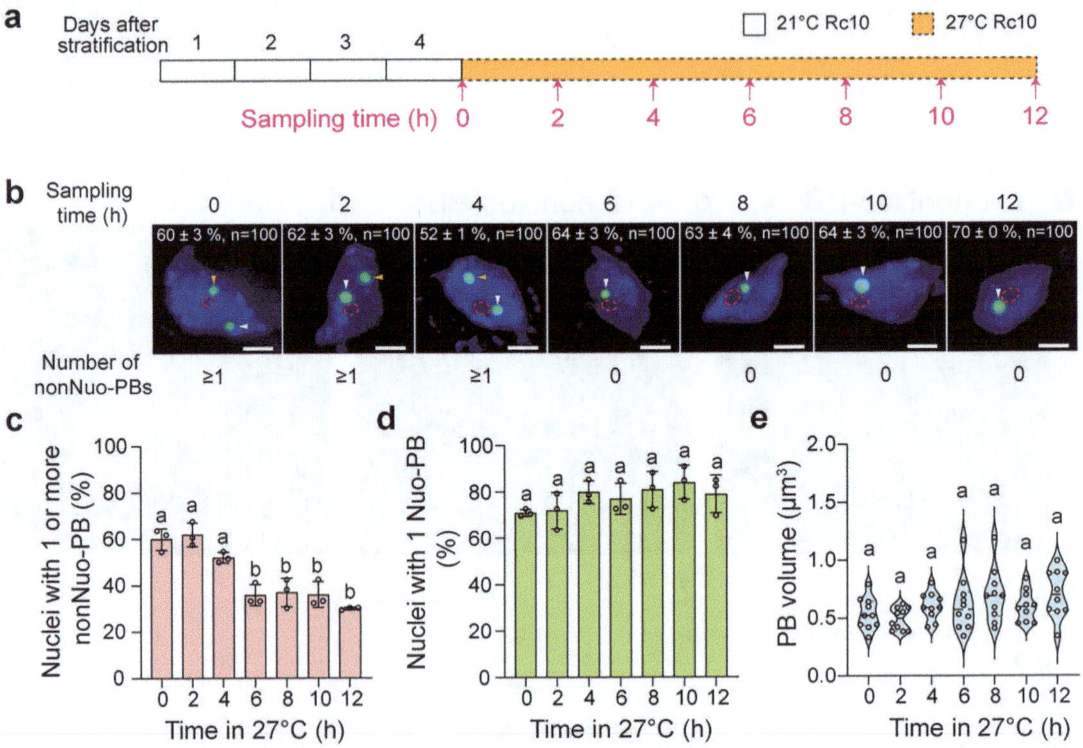

Fig. 2 Quantification of the kinetics of the reduction of PB number by warm temperatures. (**a**) Schematic illustration of the 21–27 °C transition experiment. Each rectangle represents a day after stratification. The dashed-line rectangle represents the 12 h after transition. The white color rectangle and dashed-line rectangle represent 21 °C, and the orange color dashed-line rectangle represents 27 °C treatment. The sampling times are indicated by magenta arrows. (**b**) Maximum-projection images showing representative steady-state patterns of phyB-GFP PBs in cotyledon pavement cell nuclei of 4-day-old *PBG* seedlings at the indicated time points. Nuclei are labeled by DAPI (blue), and the boundaries of the nucleoli are traced by magenta dashed lines. Nuo-PBs and nonNuo-PBs are indicated by white and orange arrowheads, respectively. The percentage of nuclei with s.e. showing the indicated PB pattern is included in each image; n indicates the total number of nuclei analyzed. Scale bars represent 2 μm. (**c**) Quantification of the percentage of cotyledon pavement cell nuclei with one or more nonNuo-PB in the *PBG* seedlings described in (**a**). (**d**) Quantification of the percentage of cotyledon pavement cell nuclei with one Nuo-PB in the *PBG* seedlings described in (**a**). For **b** and **c**, error bars represent s.e. calculated from groups of seedlings. Different letters denote statistically significant differences (ANOVA, Tukey's HSD, $P \leq 0.05$, $n = 3$). (**e**) Quantification of the volume of PBs in cotyledon epidermal-cell nuclei of the *PBG* seedlings during the 21 °C–27 °C transition described in (**a**). Samples with the same letter exhibit no statistically significant difference in PB volume (ANOVA, Tukey's HSD, $P \leq 0.05$, $n = 10$)

2 Materials

2.1 Plant Materials and Seedling Growth

1. *PBG* (*phyB-GFP*) line: a transgenic line expressing phyB-GFP under the Cauliflower Mosaic Virus 35S promoter to complement the *phyB-5* mutant in the *Arabidopsis thaliana* accession Landsberg *erecta* [23].

2. Half-strength Murashige and Skoog (1/2 MS) medium: Mix 2.2 g Murashige and Skoog with Gamborg's vitamins (Caisson Labs, Smithfield, UT), 0.5 g MES, 0.8% (w/v) agar. Add water to a total volume of 1 L, adjust pH to 5.7, and autoclave.
3. Percival LED chamber (Percival Scientific, Perry, IA) for seedling growth under monochromatic R light.

2.2 Reagents

1. Seed sterilization solution: 0.01% (v/v) Triton X-100 in 50% (v/v) bleach.
2. 10× PBS (phosphate-buffered saline): 100 mM Na_2HPO_4, 18 mM KH_2PO_4, 1.37 M NaCl, 27 mM KCl. Dissolve 14.4 g Na_2HPO_4, 2.4 g KH_2PO_4, 80 g NaCl, 2 g KCl in 800 mL ddH_2O. Adjust pH to 7.4 with HCl and then add ddH_2O to a final volume of 1 L.
3. Fixation solution: 1% (v/v) paraformaldehyde in 1× PBS. Add 6.25 mL 16% paraformaldehyde (Electron Microscopy Sciences, Hatfield, PA) and 10 mL 10× PBS, then add water to a total volume of 100 mL (*see* **Note 1**).
4. Quenching solution: 50 mM NH_4Cl in 1× PBS.
5. Permeabilization solution: 0.5% (v/v) Triton X-100 in 1× PBS.
6. DAPI (4′,6-diamidino-2-phenylindole) staining solution: 5 μg/mL DAPI in 1× PBS.
7. ProLong Diamond Antifade Mountant (Thermo Fisher Scientific, Waltham, MA).

2.3 Imaging and Image Analysis

1. Fluorescence microscope or confocal microscope equipped with a 63× or 100× oil immersion objective (*see* **Note 2**).
2. Superfrost® Plus micro slides.
3. Micro cover glasses, 22 × 40 mm, No. 1.5.
4. Zeiss ZEN 2.3 software (Carl Zeiss AG, Jena, Germany).
5. Huygens Essential software (Scientific Volume Imaging, Hilversum, The Netherlands).
6. Adobe Photoshop CC (Adobe, San Jose, CA).
7. Adobe Illustrator (Adobe, San Jose, CA).
8. Prism 9.0 (GraphPad Software, La Jolla, CA).

3 Methods

3.1 Warm Temperature Treatments

1. Sterilize *PBG* seeds in seed sterilization solution for 10 min and wash four times in sterile ddH_2O before plating on two 1/2 MS plates.
2. Stratify seeds in the dark at 4 °C for 5 days.

3. Place the two plates containing stratified *PBG* seeds under 10 μmol m^{-2} s^{-1} monochromatic R light at 21 °C in a Percival LED chamber for 2 days (48 h) to induce seed germination.
4. For continuous warm temperature treatments, keep one plate at 21 °C and transfer the other plate to 27 °C under 10 μmol m^{-2} s^{-1} R light for an additional 2 days (48 h). Collect samples at 96 h after stratification (Fig. 1a).
5. For the 21–27 °C transition experiment, grow *PBG* seedlings under 10 μmol m^{-2} s^{-1} R light at 21 °C for 2 more days (48 h), and then transfer seedlings to 27 °C under 10 μmol m^{-2} s^{-1} R light. Collect samples at 2 h intervals for a duration of 12 h post-temperature elevation (Fig. 2a) (*see* **Note 3**).

3.2 Seedling Fixation and Slide Preparation

1. Collect seedlings into a bottle containing the fixation solution.
2. Apply vacuum infiltration for 10 min under the respective growth conditions and then slowly release the vacuum (*see* **Note 4**).
3. Wash seedlings three times with the quenching solution for 5 min each time to quench the residual amount of fixative.
4. Permeabilize seedlings with the permeabilization solution for 20 min.
5. Stain nuclei with DAPI staining solution for 20 min.
6. Wash seedlings three times with 1× PBS, 5 min each time.
7. Apply a drop of ProLong Diamond Antifade to the seedling on a slide and then slowly lower a cover glass onto the seedling to avoid trapping any air bubbles.
8. Seal the edges of the cover glass with nail polish and cure the sample overnight in the dark.
9. Store slides at 4 °C (*see* **Note 5**).

3.3 Imaging and Photobody Quantification

1. Load the glass slide onto the microscope stage, select an appropriate objective lens, and set the filter or laser parameters (*see* **Note 6**).
2. Identify elongated spindle-shaped pavement epidermal nuclei from the cotyledon using DAPI signals (*see* **Note 7**).
3. To acquire Z-stack images, use the First/Last mode to set the first and last plane of the Z-stack to encompass the top and bottom boundaries of a nucleus. Next, select the optimal intervals to achieve good Z resolution (*see* **Note 8**). Then, initiate the acquisition to take three-dimensional image stacks of individual pavement cell nuclei using either widefield microscopy or confocal microscopy. Typically, we image at least 10 nuclei from each seedling and analyze a minimum of 5 seedlings. For

statistical analysis, each seedling can be used as a biological replicate.

4. For image processing, load images into ZEN 2.3 or equivalent software. Process the image stacks with orthogonal projection using the maximum method. Incorporate the scale bar into the images and export the images as TIFF files in individual channels. Import TIFF files to Adobe Photoshop CC, adjust the brightness and contrast, and merge the layers of the GFP and the DAPI channel using the screen mode. Use the "place" function to import the images into Adobe Illustrator, and then annotate the image in Adobe Illustrator. Delineate the nucleolus, which appears as a black cavity in the DAPI channel, with a magenta dashed line. Label the PB at the nucleolus periphery as the Nuo-PB and the PB away from the nucleolus as the nonNuo-PB by distinct colors of arrows (Figs. 1b, c and 2b).
5. To quantify the number and the volume of PBs within each nucleus, first load the image stacks into Huygens Essential, verify the microscopic parameters, and then start the Deconvolution Wizard (*see* **Note 9**); then use Object Analyzer to analyze the deconvoluted image stack. Initiate the object analyzer and set the threshold and seed as default for object segmentation (*see* **Note 10**). Select the "experiment preset" button and then select the small particle geometry in basic modes to retrieve the morphological parameters of PBs. The parameter preset in this mode includes the parameter IsoVol which represents the volume of PB. Click on the "analyze all" button, and individual PBs in the GFP channel are segmented (Fig. 1d) and a table of each designated PBs with its corresponding volumes is produced. Export the table to Excel for further analysis.
6. To generate a combined box plot and scatter plot showing the number of PBs, Nuo-PBs, and nonNuo-PBs per nucleus under 21 and 27 °C, first calculate the total number of PBs in each nucleus using the data generated from Step 5. Distinguish the Nuo-PBs and nonNuo-PBs in the maximum projected, deconvolved images manually and count the number of Nuo-PBs and nonNuo-PBs in each nucleus. Calculate the average number of PBs, Nuo-PBs, and nonNuo-PBs per nucleus in each replicate under 21 °C and 27 °C and then import data into Prism 9.0. Use the same set of data to create a "Box-and-Whiskers" graph with "Tukey" whisker style and a "Scatter dot plot" with "No line or error bar" setting. Merge the box plot and scatter dot plot in a new layout (Fig. 1e) and then copy the merged graph into Adobe Illustrator for further annotation.
7. To create bar graphs presenting the percentage of nuclei with an indicated number of Nuo-PBs or nonNuo-PBs, first use

every three replicates as a group, and then calculate the percentage of nuclei with an indicated number of Nuo-PBs or nonNuo-PBs in each group. Each group can be used as a biological replicate. Create "Bar" graphs with "Mean with SD" style with imported data calculated from each group under 21 °C and 27 °C (Fig. 1f) or across different sampling time points during the 21–27 °C transition (Fig. 2c, d).

8. To generate the violin plot showing the PB volume, first calculate the average PB volume in each replicate. Generate the violin plot with the setting of the "Violin" plot. Show all points with the imported PB volume data at 21 °C and 27 °C (Fig. 1g) or across different sampling time points during the 21–27 °C transition (Fig. 2e).
9. For statistical analysis, compare the data from 21 °C and 27 °C with a Student's two-tailed t-test in Excel and then annotate the p value and significance in Adobe Illustrator (Fig. 1e–g). Use ANOVA analysis in Prism 9.0 to compare the differences across sampling time points during the 21–27 °C transition and label significance with letters in Adobe Illustrator (Fig. 2c–e).

4 Notes

1. Methanol is known to immobilize and denature proteins [25]. Therefore, it is critical to use methanol-free paraformaldehyde to preserve the fluorescent proteins during fixation. Prepare the fixation solution freshly within a chemical hood to ensure its effectiveness, as paraformaldehyde can polymerize over time, reducing its ability to fix the tissue effectively.
2. We used a Zeiss Axio Observer.Z1 inverted microscope equipped with a Plan-Apochromat 100×/1.4 oil-immersion objective and an Axiocam mono camera (Carl Zeiss AG, Jena, Germany) and a Zeiss LSM800 inverted confocal microscope equipped with a Plan-Apochromat 100×/1.4 oil-immersion 100×/1.4 Oil DIC objective.
3. The temperature-dependent PB dynamics is characterized under continuous R light for two reasons. First, the thermomorphogenic response at the 4-d-old seedling stage is attenuated by the blue light and therefore is best seen under monochromatic R light [14, 26, 27]. Second, the temperature response is monitored in the light to highlight the daytime temperature response, which is different from the warm temperature response in the nighttime under short-day conditions [14, 26, 28].
4. To avoid any alterations in PB morphology during fixation, we recommend fixing seedlings under the growth condition.

5. The FP signal gradually diminishes when stored at 4 °C. Therefore, we image PBs within a week after fixation.
6. For widefield fluorescence microscopy, FPs and DAPI were excited using a broad spectrum X-Cite 120LED Boost high-power LED illumination system (Excelitas Technologies, Waltham, MA, USA) and monitored using the following Zeiss filter sets: GFP, exciter 470/40 nm/nm, emitter 525/50 nm/nm (Zeiss Filter Set 38); DAPI, exciter 365 nm, emitter 445/50 nm/nm (Zeiss Filter Set 49). For confocal microscopy, DAPI was excited by a 405 nm laser and monitored using a 410–470 nm bandpass emission; GFP was excited by a 488 nm laser and detected using a 495–550 nm bandpass emission.
7. In *Arabidopsis*, the shape of nuclei varies among cell types. Differentiated pavement epidermal cells have distinct large, elongated spindle-shaped nuclei [29].
8. When taking three-dimensional image stacks, we take sections at 0.70 μm intervals for confocal microscopy and use a finer interval of 0.25 μm for widefield fluorescence microscopy.
9. We recommend performing deconvolution before analysis using Object Analyzer. The deconvolution process will reduce noise and blurring significantly and will thereby improve the analysis quality.
10. To ensure consistency in measuring PBs from different samples, it is important to use the default threshold and seed.

Acknowledgments

This work was supported by National Institute of Health grant R01GM087388 to M.C.

References

1. Quint M, Delker C, Franklin KA et al (2016) Molecular and genetic control of plant thermomorphogenesis. Nat Plants 2:15190
2. Casal JJ, Balasubramanian S (2019) Thermomorphogenesis. Annu Rev Plant Biol 70:321–346
3. Legris M, Klose C, Burgie ES et al (2016) Phytochrome B integrates light and temperature signals in *Arabidopsis*. Science 354:897–900
4. Jung J-H, Domijan M, Klose C et al (2016) Phytochromes function as thermosensors in *Arabidopsis*. Science 354:886–889
5. Burgie ES, Vierstra RD (2014) Phytochromes: an atomic perspective on photoactivation and signaling. Plant Cell 26:4568–4583
6. Rockwell NC, Lagarias JC (2020) Phytochrome evolution in 3D: deletion, duplication, and diversification. New Phytol 225:2283–2300
7. Klose C, Nagy F, Schäfer E (2019) Thermal reversion of plant phytochromes. Mol Plant 13:386–397
8. Chen M, Chory J (2011) Phytochrome signaling mechanisms and the control of plant development. Trends Cell Biol 21:664–671

9. Van Buskirk EK, Decker PV, Chen M (2012) Photobodies in light signaling. Plant Physiol 158:52–60
10. Chen D, Lyu M, Kou X et al (2022) Integration of light and temperature sensing by liquid-liquid phase separation of phytochrome B. Mol Cell 82:3015–3029.e6
11. Chen M, Schwab R, Chory J (2003) Characterization of the requirements for localization of phytochrome B to nuclear bodies. Proc Natl Acad Sci USA 100:14493–14498
12. Van Buskirk EK, Reddy AK, Nagatani A et al (2014) Photobody localization of phytochrome B is tightly correlated with prolonged and light-dependent inhibition of hypocotyl elongation in the dark. Plant Physiol 165: 595–607
13. Trupkin SA, Legris M, Buchovsky AS et al (2014) Phytochrome B nuclear bodies respond to the low red to far-red ratio and to the reduced irradiance of canopy shade in *Arabidopsis.* Plant Physiol 165:1698–1708
14. Hahm J, Kim K, Qiu Y et al (2020) Increasing ambient temperature progressively disassemble *Arabidopsis* phytochrome B from individual photobodies with distinct thermostabilities. Nat Commun 11:1660
15. Chen M, Galvao RM, Li M et al (2010) *Arabidopsis* HEMERA/pTAC12 initiates photomorphogenesis by phytochromes. Cell 141:1230–1240
16. Huang H, Yoo CY, Bindbeutel R et al (2016) PCH1 integrates circadian and light-signaling pathways to control photoperiod-responsive growth in *Arabidopsis.* elife 5:e13292
17. Kim C, Kwon Y, Jeong J et al (2023) Phytochrome B photobodies are comprised of phytochrome B and its primary and secondary interacting proteins. Nat Commun 14:1708
18. Pardi SA, Nusinow DA (2021) Out of the dark and into the light: a new view of phytochrome photobodies. Front Plant Sci 12:732947
19. Shi H, Zhong S (2023) Light and temperature perceptions go through a phase separation. Curr Opin Plant Biol 74:102397
20. Willige BC, Zander M, Yoo CY et al (2021) Phytochrome-interacting factors trigger environmentally responsive chromatin dynamics in plants. Nat Genet 53:955–961
21. Yoo CY, Pasoreck EK, Wang H et al (2019) Phytochrome activates the plastid-encoded RNA polymerase for chloroplast biogenesis via nucleus-to-plastid signaling. Nat Commun 10:2629
22. Yang EJ, Yoo CY, Liu J et al (2019) NCP activates chloroplast transcription by controlling phytochrome-dependent dual nuclear and plastidial switches. Nat Commun 10:2630
23. Yamaguchi R, Nakamura M, Mochizuki N et al (1999) Light-dependent translocation of a phytochrome B-GFP fusion protein to the nucleus in transgenic *Arabidopsis.* J Cell Biol 145:437–445
24. Yoo CY, Williams D, Chen M (2019) Quantitative analysis of photobodies. Methods Mol Biol 2026:135–141
25. Nybo K (2012) GFP imaging in fixed cells. BioTechniques 52:359–360
26. Qiu Y, Li M, Kim RJ-A et al (2019) Daytime temperature is sensed by phytochrome B in *Arabidopsis* through a transcriptional activator HEMERA. Nat Commun 10:140
27. Ma D, Li X, Guo Y et al (2016) Cryptochrome 1 interacts with PIF4 to regulate high temperature-mediated hypocotyl elongation in response to blue light. Proc Natl Acad Sci USA 113:224–229
28. Qiu Y, Pasoreck EK, Yoo CY et al (2021) RCB initiates *Arabidopsis* thermomorphogenesis by stabilizing the thermoregulator PIF4 in the daytime. Nat Commun 12:2042
29. Goto C, Hara-Nishimura I, Tamura K (2021) Regulation and physiological significance of the nuclear shape in plants. Front Plant Sci 12:673905

Chapter 11

Phase Separation of Phytochrome B in HEK293T Cells

Di Chen and Shangwei Zhong

Abstract

In this method, we employed HEK293T cells to express the plant photoreceptor phytochrome B (phyB). Through the application of various treatments such as phycocyanobilin (PCB) supplementation, red light exposure, and temperature adjustments, the phyB proteins exhibited liquid-liquid phase separation, leading to the formation of biomolecular condensates. Here, we present a comprehensive description of the protein expression, cell treatment, and imaging capture procedures. This detailed guide provides step-by-step instructions on how to induce phase separation of phyB proteins in HEK293T cells. By utilizing this approach, researchers can investigate the physicochemical characteristics and dynamic formation process of phyB photobodies with precision.

Key words Phytochrome B, Phase separation, Biomolecular condensates, Photobody, Temperature sensation, Light signal

1 Introduction

In *Arabidopsis*, phytochrome B (phyB) serves as the primary red light photoreceptor. The photosensory capacity of phyB is determined by its reversible conformational changes between two distinct forms: the inactive red light-absorbing Pr form and the active far-red light-absorbing Pfr form [1]. In etiolated seedlings, phyB is synthesized as the Pr form and accumulates in the cytoplasm. Upon exposure to red light, the Pr form phyB proteins are photoactivated to the Pfr form and phosphorylated by calcium-dependent protein kinases CPKs, followed by a rapid translocation into the nucleus [2–4]. Within the nucleus, phyB proteins undergo compartmentalization into discrete speckle-like structures known as photobodies. These photobodies exhibit high sensitivity to changes in environmental light conditions and temperature variations [5–7].

Recently, our research has unveiled the mechanism behind the formation of phyB photobodies through a process called liquid-liquid phase separation [7]. To further explore the characteristics of

Meng Chen (ed.), *Thermomorphogenesis: Methods and Protocols*, Methods in Molecular Biology, vol. 2795,
https://doi.org/10.1007/978-1-0716-3814-9_11,

phyB biomolecular condensates, we conducted experiments involving the expression of a phyB variant called YHB, which contains a point mutation (Tyr276 to His) that locks it in the Pfr conformation, in human embryonic kidney (HEK)293 T cells [7]. In this method, we provide a detailed description of the procedures employed to induce phyB phase separation in HEK293T cells. Additionally, we utilized a confocal laser scanning microscope for dynamic tracking and recording of the behaviors exhibited by phyB biomolecular condensates.

2 Materials

2.1 Cell Culture and Transfection

1. HEK293T cells (ATCC, CRL-3216™).
2. Expression plasmids of pcDNA3.1 (ABclonal, RK21020).
3. Polyethylenimine (PEI) transfection reagent (*see* **Note 1**):1 mg/mL PEI 4000 in water, pH 6.8–6.9. Weigh 1 g PEI 4000 powder and transfer to 900 mL Milli-Q ultrapure water, pH 6.8–6.9. Add ultrapure water to a final volume of 1 L. Filter the solution through a 0.2 μm filter membrane. Store at 4 °C (*see* **Note 2**).
4. Cell culture incubator for cell growth, CO_2 levels set at 5% (vol/vol) and temperature set at 37 °C.
5. Cell culture solution: Supplement Dulbecco's modified Eagle's medium (DMEM) with 10% (vol/vol) fetal bovine serum, 100 IU penicillin, and 100 mg/L streptomycin. Store at 4 °C.
6. Phosphate buffer saline (PBS) buffer: 137 mM sodium chloride (NaCl), 2.7 mM potassium chloride (KCl), 10 mM disodium hydrogen phosphate (Na_2HPO_4), 1.76 mM potassium dihydrogen phosphate (KH_2PO_4). Weigh 8 g NaCl, 0.2 g KCl, 1.44 g Na_2HPO_4, and 0.24 g KH_2PO_4 and transfer to 800 mL Milli-Q ultrapure water, pH 7.4. Add ultrapure water to a final volume of 1 L. Filter with 0.2 μm filter membrane. Store at 4 °C (*see* **Note 3**).
7. Glass bottom cell culture dishes (NEST, 801001).

2.2 Cell Pretreatment and Imaging

1. PCB (phycocyanobilin) solution: isolated from blue-green algae using a previously described method [8].
2. Light box with 5% (vol/vol) CO_2 and adjustable temperature (Panasonic).
3. A1R-Si + confocal microscope (Nikon).
4. NIS-Elements AR imaging software (Nikon).
5. High Speed Spinning Disk Confocal Microscope (Andor Dragonfly).

6. Imaris software (Bitplane).
7. Ultra VIEW VoX spinning disc confocal microscope (PerkinElmer).
8. Volocity software (PerkinElmer).

3 Methods

3.1 Material Preparation

1. Culture HEK293T cells in glass-bottom cell culture dishes (*see* **Note 4**).
2. Place the HEK293T cells in a cell culture solution and incubate them in a 5% (vol/vol) CO_2 incubator at 37 °C for 24 h (*see* **Note 5**).
3. Use seamless cloning/In-Fusin cloning technology to construct pcDNA3.1-phyB-YFP and pcDNA3.1-YHB-YFP plasmids (*see* **Note 6**).
4. Isolate plasmid using a plasmid extraction kit (such as CWBIO, GoldHi EndoFree Plasmid Maxi kit) to obtain a plasmid solution with a minimum concentration of 2000 ng/μL (*see* **Note 7**).

3.2 Cell Transfection and Protein Expression

1. Mix 1.5 μL DNA, 1 μL PEI, and 160 μL DMEM at 37 °C for 15 min.
2. Add the transfection mixture to the HEK293T cells and gently shake by hand to ensure proper distribution.
3. Incubate the cells for 6–8 h at 37 °C, and then remove as much of the solution as possible (*see* **Note 8**).
4. Add 1 mL cell culture solution to the HEK293T cells.
5. Culture the cells in a 5% (vol/vol) CO_2 incubator at 37 °C for 24 h (*see* **Note 9**).

3.3 Cell Treatment

1. Add 10 μL PCB solution to 1 mL PBS buffer to make the imaging solution (*see* **Note 10**).
2. Remove the culture solution and add the imaging solution.
3. Put the cells in red light (30 μmol m^{-2} s^{-1}) box with 5% (vol/vol) CO_2 at the specified temperature for 1.5 h (for YHB) and 3–5 h (for phyB) (*see* **Note 11**).

3.4 Cell Imaging

1. The fluidity capacity of the phyB biomolecular condensates

   Fluorescence recovery after photobleaching (FRAP) assays were conducted using a Nikon A1R-Si + confocal microscope equipped with a 60× oil-immersion objective. The phyB-YFP biomolecular condensates were subjected to bleaching at a laser intensity of 15% and a wavelength of 514 nm for 927.336 ms

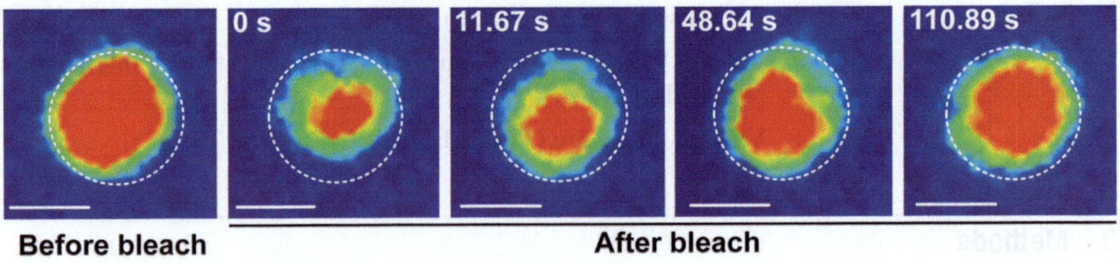

Fig. 1 FRAP images of a phyB biomolecular condensate in HEK293T cells. Scale bars, 1 μm

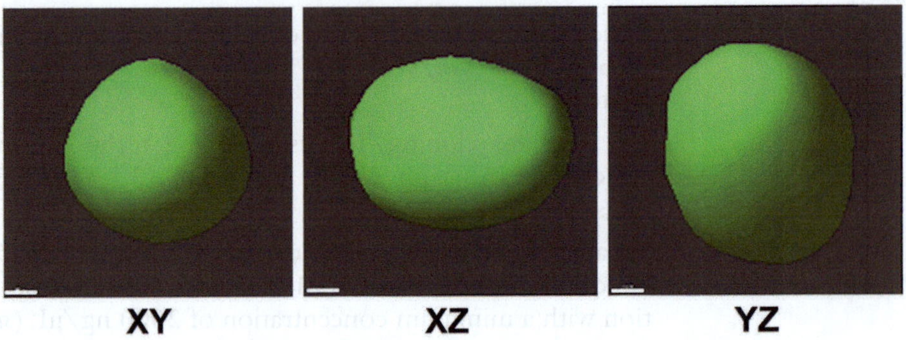

Fig. 2 Rendered 3D images of a phyB biomolecular condensate in HEK293T cells. The XY, XZ, and YZ planes are labeled in the bottom of each image. Scale bars, 0.2 μm

(Fig. 1). Fluorescence recovery was detected at 505–575 nm and recorded every 2 s for 90 time points, and the resulting images were color-coded using NIS-Elements AR imaging software (Nikon) (*see* **Note 12**).

2. The geometric capacity of phyB biomolecular condensates

   The phyB biomolecular condensates in HEK293T cells were analyzed with a high-speed spinning disk confocal microscope (Andor Dragonfly) (*see* **Note 13**). The 3D rendering analysis of the data was performed using Imaris software (Bitplane) (Fig. 2) (*see* **Note 14**).

3. The fusion capacity of phyB biomolecular condensates

   For fusion events, HEK293T cells were observed using a Nikon A1R-Si + microscope with a 60× oil-immersion objective. Yellow fluorescent protein (YFP) was stimulated at a wavelength of 514 nm and fluorescence images were acquired every 1 min for 30 min (Fig. 3) (*see* **Note 15**).

4. Modulation of phyB biomolecular condensates in response to increasing temperature

   To investigate the impact of elevated temperature on phyB biomolecular condensates, HEK293T cells are subjected to a red light exposure for 90 min. During this period, the temperature is incrementally raised from 15 °C to 30 °C, with an

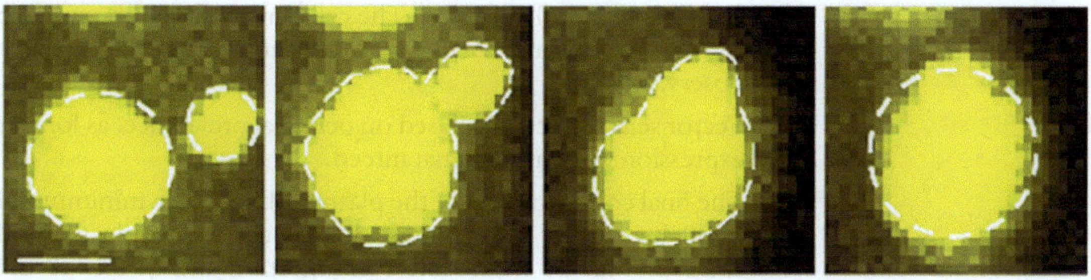

Fig. 3 Fusion events of phyB biomolecular condensates in HEK293T cells. Scale bars, 1 μm

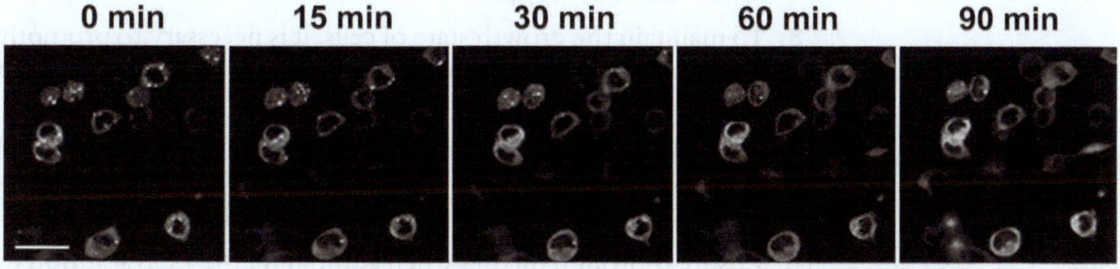

Fig. 4 YHB biomolecular condensates dissociate with increased temperature. Scale bar, 30 μm

increase of 5 °C per 10 min (Fig. 4). The cells were imaged using an Ultra VIEW VoX spinning disc confocal microscope (PerkinElmer) with a 60× oil-immersion objective (*see* **Note 16**).

4 Notes

1. Based on personal preference, alternative transfection reagents to the PEI transfection reagent can be employed to guarantee efficient transfection.
2. Prior to reaching the final volume, the pH level can be adjusted using HCl or NaOH. The solution can be divided into smaller portions and stored at 4 °C for a maximum of 3 months.
3. Commercially available pre-made PBS buffer can also be procured to ensure sterility and accuracy of the solution. After use, any remaining solution should be stored at 4 °C.
4. When utilizing cells for long-term imaging and manipulation, it is advisable to use culture dishes specifically designed for confocal microscopy, which will ensure optimal growth conditions and convenience for subsequent observations.
5. The HEK293T cells employed for transfection should be maintained in a favorable growth state and at an appropriate density. The growth state can be assessed by examining the cell shape

and degree of cell adhesion. The cell density should be suitable to ensure optimal expression efficiency and fluorescence observation.

6. Vector selection can be based on personal preference, as long as expression efficiency is guaranteed.
7. The final concentration of the plasmid should be a minimum of 2000 ng/μL. It is recommended to use a suitable plasmid extraction kit to ensure the concentration and quality of the plasmid. Concentration can be increased using concentration methods, but the initial extraction concentration should not be lower than 500 ng/μL.
8. To maintain the growth state of cells, it is necessary to promptly replace the transfection solution with fresh cell culture medium within 6–8 h after transfection.
9. The cell culture duration can be adjusted as per the desired protein expression effect. Typically, a 24-hour culture is sufficient to achieve optimal observation conditions.
10. Ensure thorough mixing when combining the PCB solution to the PBS buffer.
11. It is advisable to select a compact and stable incubator that provides suitable CO_2 concentration and humidity for cell growth.
12. For laser quenching of small areas, it is preferable to use high magnification objectives. The intensity for laser quenching may need to be adjusted based on the FRAP program parameters of different microscope models. Preliminary experiments can be conducted to achieve optimal ablation and image capture results.
13. In confocal microscopy, the resolution along the Z-axis for capturing 3D images is often lower than that along the XY-axis. To improve image capture results, it is recommended to use a microscope with higher resolution.
14. The microscope's built-in software or Imaris software can be utilized for 3D reconstruction when processing images.
15. To account for randomness in the fusion process, it is advisable to select shooting parameters with larger observation areas and longer shooting times.
16. Precise temperature adjustments can be made using a live cell workstation.

Acknowledgments

This work was supported by a grant from the National Natural Science Foundation of China (81891014).

References

1. Quail PH, Boylan MT, Parks BM, Short TW, Xu Y, Wagner D (1995) Phytochromes: photosensory perception and signal transduction. Science 268(5211):675–680. https://doi.org/10.1126/science.7732376
2. Sakamoto K, Nagatani A (1996) Nuclear localization activity of phytochrome B. Plant J 10(5): 859–868. https://doi.org/10.1046/j.1365-313x.1996.10050859.x
3. Kircher L, Kozma-Bognar L, Kim L, Adam E, Harter K, Schafer E, Nagy F (1999) Light quality-dependent nuclear import of the plant photoreceptors phytochrome A and B. Plant Cell 11(8):1445–1456. https://doi.org/10.1105/tpc.11.8.1445
4. Zhao Y, Shi H, Pan Y, Lyu M, Yang Z, Kou X, Deng XW, Zhong S (2023) Sensory circuitry controls cytosolic calcium-mediated phytochrome B phototransduction. Cell 186(6): 1230–1243. https://doi.org/10.1016/j.cell.2023.02.011
5. Van Buskirk EK, Reddy AK, Nagatani A, Chen M (2014) Photobody localization of Phytochrome B is tightly correlated with prolonged and light-dependent inhibition of hypocotyl elongation in the dark. Plant Physiol 165(2): 595–560. https://doi.org/10.1104/pp.114.236661
6. Casal JJ, Balasubramanian S (2019) Thermomorphogenesis. Annu Rev Plant Biol 70:321–346. https://doi.org/10.1146/annurev-arplant-050718-095919
7. Chen D, Lyu M, Kou X, Li J, Yang Z, Gao L, Li Y, Fan L-M, Shi H, Zhong S (2022) Integration of light and temperature sensing by liquid-liquid phase separation of phytochrome B. Mol Cell 82(16):3015–3029. https://doi.org/10.1016/j.molcel.2022.05.026
8. Shimizu-Sato S, Huq E, Tepperman JM, Quail PH (2002) A light-switchable gene promoter system. Nat Biotechnol 20(10):1041–1044. https://doi.org/10.1038/nbt734

Chapter 12

Isolation of Phytochrome B Photobodies

Yongmin Kwon, Chanhee Kim, and Giltsu Choi

Abstract

Phytochrome B (phyB), a plant photoreceptor, forms a membraneless organelle known as a photobody. Here, we present a protocol for the isolation of phyB photobodies through fluorescence-activated particle sorting from mature transgenic *Arabidopsis* leaves expressing *phyB-GFP*. This protocol involves the isolation of nuclei from frozen ground leaves using sucrose gradient centrifugation, the disruption of nuclear envelopes by sonication, and the subsequent isolation of phyB photobodies through fluorescence-activated particle sorting. We include experimental tips and notes for each step.

Key words Phytochrome, Photobody, *Arabidopsis thaliana*, Fluorescence-activated particle sorting

1 Introduction

Phytochromes (phy, phyA to phyE in *Arabidopsis*) are red and far-red light photoreceptors that regulate various light responses in plants [1]. Phys reversibly take one of two spectral forms called Pr and Pfr driven by the reversible photoisomerization of covalently linked chromophores by absorbing red (670 nm) and far-red (730 nm) light [2]. The Pr localizes in the cytosol with little biological activities, whereas the Pfr enters the nucleus and promotes the light responses by interacting with and inhibiting various proteins, including PIFs and COP1/SPA complexes [1, 3–10]. One interesting phenomenon associated with the Pfr is the formation of nuclear condensate, referred to as a photobody, through liquid-liquid phase separation (LLPS) [3, 5, 11–13]. The photobody has been suggested to sequester negative light signaling components, regulate transcription, and control post-translational modifications including ubiquitination and degradation [11, 12, 14, 15]. However, further investigation is needed to elucidate the composition and function of the photobody.

Yongmin Kwon and Chanhee Kim contributed equally with all other contributors.

Meng Chen (ed.), *Thermomorphogenesis: Methods and Protocols*, Methods in Molecular Biology, vol. 2795,
https://doi.org/10.1007/978-1-0716-3814-9_12,

Fluorescence-activated particle sorting (FAPS) is a modified method of fluorescence-activated cell sorting (FACS) tailored to sort small particles (sub-micron to a few micrometers) which sizes are smaller than typical eukaryotic cells [16]. Though the number and size of photobody within a nucleus vary depending on the amount of Pfr which is influenced by phy protein level, light intensity, and temperature [17–21], the typical size of a photobody falls within a range suitable for the FAPS. A recently developed method for isolating phyB photobody by the FAPS provides a good opportunity to dissect this enigmatic membraneless organelle [15]. We describe a detailed step-by-step protocol for the isolation of phyB photobodies from transgenic *Arabidopsis* leaves expressing phyB-GFP by the FAPS.

2 Materials

Prepare the following solutions using Milli-Q quality ultrapure water (18 MΩ/cm at 25 °C).

2.1 Plant Materials

1. Transgenic *Arabidopsis* leaves expressing phyB-GFP.

2.2 Reagents

1. Photobody fractionation buffer 1: 10 mM Tris, 10 mM $MgCl_2$, 0.4 M sucrose, 160 μM MG132, 1 mM phenylmethylsulfonyl fluoride (PMSF), with protease inhibitor cocktail, pH 8.00, titrated using HCl (*see* **Note 1**).
2. Photobody fractionation buffer 2: 10 mM Tris, 10 mM $MgCl_2$, 1% (v/v) Triton X-100, 0.25 M sucrose, 160 μM MG132, 1 mM PMSF, with protease inhibitor cocktail, pH 8.00, titrated using HCl.
3. Photobody fractionation buffer 3: 10 mM Tris, 2 mM $MgCl_2$, 0.15% (v/v) Triton X-100, 1.7 M sucrose, 160 μM MG132, 1 mM PMSF, with protease inhibitor cocktail, pH 8.00, titrated using HCl.
4. 10 X PBS: 1.37 M NaCl, 27 mM KCl, 100 mM disodium monohydrogen phosphate, 18 mM monopotassium dihydrogen phosphate (*see* **Note 2**).
5. Blocking buffer: PBS with 0.2% (v/v) Triton X-100, and 1% (w/v) BSA, 160 μM MG132, 1 mM PMSF, with protease inhibitor cocktail.
6. Wash buffer: PBS with 0.2% (v/v) Triton X-100, 160 μM MG132, 1 mM PMSF, with protease inhibitor cocktail.
7. Fluorescent-dye conjugated antibody for immunostaining such as Alexa Fluor 488-conjugated anti-GFP antibody.
8. Flow cytometry reference bead such as Flow Cytometry Sub-micron Particle Size Reference Kit (Invitrogen, F13839).

2.3 Equipment

1. Swing bucket centrifuge.
2. Refrigerated microcentrifuge.
3. Rotary shaker.
4. Sonicator such as Bioruptor (Diagenode, B01020001).
5. Flow cytometry with cell sorter such as MoFlo Astrios EQ (Beckman Coulter) (*see* **Note 3**).
6. Fluorescence microscope.
7. Confocal microscope.

2.4 Plastic Ware and Others

1. Mortar and pestle.
2. Liquid nitrogen.
3. Miracloth (Merck, 475855).
4. Conical tubes, 15 mL and 50 mL.
5. Microfuge tubes, 1.7 mL.
6. Sonication tubes, 14 mL.
7. Polystyrene round-bottom tube with cell-strainer cap, 5 mL.

3 Methods

3.1 Preparation of Nuclear Extract

1. Grow transgenic *Arabidopsis* expressing *35S::phyB-GFP* in a growth room under a 16-hour/8-hour light/dark cycle at 22 °C (Fig. 1a).
2. Collect 4-week-old mature *Arabidopsis* rosette leaves. Gather 7 g of rosette leaves and flash-freeze using liquid nitrogen (*see* **Note 4**).

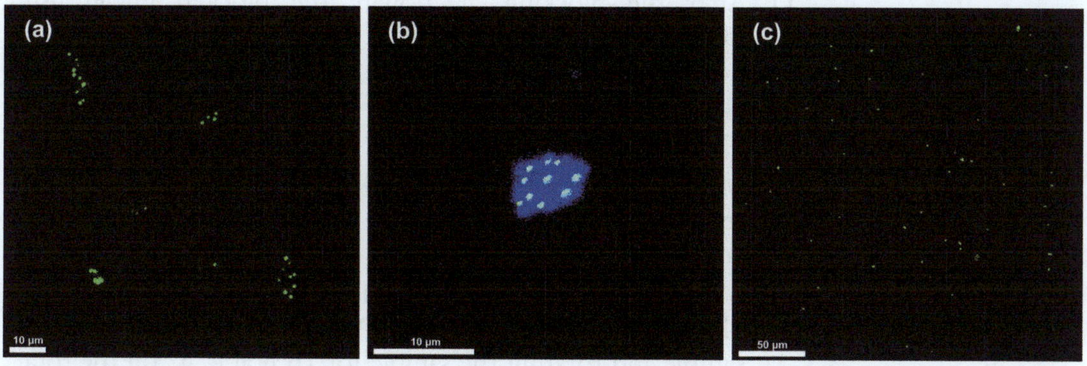

Fig. 1 Isolation of phyB photobodies by FAPS. Mature rosette leaves of *Arabidopsis PHYB-GFP* (**a**) are sampled and ground. Nuclei are isolated by sucrose gradient centrifugation (**b**). The nuclei are ruptured by sonication and photobodies are isolated via FAPS (**c**). Images are captured using confocal microscopy with DAPI staining. Scale bars = 10 μm

3. Grind the sample into a fine powder with a mortar and pestle while chilling with liquid nitrogen (*see* **Note 5**). After grinding, transfer the powder from the mortar into a 50-mL conical tube using a liquid nitrogen-chilled spatula. Add 20 mL of ice-cold photobody fractionation buffer 1 to the sample-containing conical tube (*see* **Note 6**).
4. Filtrate the resuspended sample through two layers of Miracloth by the low-speed centrifugation in a swing bucket centrifuge (4 °C, 3500 × g for 20 min). After the centrifugation, immediately remove the Miracloth layers and the supernatant (*see* **Note 7**).
5. Wash the pellet by resuspending in 15 mL of the ice-cold photobody fractionation buffer 2 and then precipitate it again by a low-speed centrifugation in a swing bucket centrifuge (4 °C, 3500 × g for 20 min). Repeat this washing step four more times.
6. Gently resuspend the pellet with a micropipette in 400 μL of the photobody fractionation buffer 2.
7. Make a sucrose step-gradient using the photobody fractionation buffer 3 and the sample in the photobody fractionation buffer 2 in a 1.7 mL microfuge tube (*see* **Note 8**).
8. Fractionate nuclei through high-speed centrifugation using a table-top microcentrifuge (4 °C, 15,000 × g for 30 min).
9. Remove the supernatant and gently resuspend the pellet with a micropipette in 400 μL of the photobody fractionation buffer 2. Repeat the sucrose step-gradient fractionation one more time.
10. Remove the supernatant and resuspend the pellet in 1 mL of the blocking buffer. Incubate the resuspended sample for 1 h at 4 °C on a rotary shaker at 12 rpm.
11. After the blocking, fractionate the blocked nuclei using low-speed fractionation on a table-top microfuge (4 °C, 1500 × g for 3 min).
12. Gently resuspend the blocked nuclei pellet by 1 mL of the blocking buffer.
13. Monitor the resuspended nuclei under a fluorescence microscope using 5 μL of the sample to assess the quality of nuclei and photobodies (Fig. 1b).
14. Immunostain the sample by adding 10 μL (2 μg/mL) of Alexa Fluor 488-conjugated anti-GFP antibody to the resuspended sample and rotating the sample for 16 h at 4 °C on the rotary shaker rotating at 12 rpm (*see* **Note 9**).

15. Fractionate the stained nuclei by low-speed centrifugation (4 °C, 1500 × g for 3 min). Carefully, remove the supernatant. The pellet contains immunostained nuclei.
16. Wash the stained nuclei faction by resuspending the pellet in 1 mL of the wash buffer and precipitate the nuclei fraction by low-speed centrifugation (4 °C, 1500 × g for 3 min). Repeat this washing step two more times, for a total of three washes.
17. Precipitate the nuclei fraction by low-speed centrifugation (4 °C, 1500 × g for 3 min).
18. Resuspend the pellet in 1 mL of the Wash buffer. Monitor the staining quality of the resuspended sample using 5 μL in comparison with non-stained samples imaged before the immunostaining process. After immunostaining, the signal intensity of the sample will be amplified.
19. Set the sonicator to high-power output for a 30-second on/30-second off cycle. Fill the sonicator chamber with ice-cold water.
20. Transfer the immunostained nuclei samples into a 14-mL sonication tube.
21. Rupture the resuspended nuclei with 15 cycles of the sonication in a chamber filled with ice-cold water.
22. Fractionate the nuclei debris using low-speed centrifugation in table-top microfuge (4 °C, 1500 × g for 3 min). Carefully transfer the supernatant to a round-bottom tube by passing through a cell-strainer cap. Dilute the supernatant to fourfold in PBS buffer.

3.2 Estimation of Photobody Sizes with Reference Beads

1. Estimate the sizes of photobodies by comparing them in leaf cells with green-fluorescent reference beads (0.2, 0.5, 1 μm) under a fluorescence microscope. Observe reference beads and photobodies under the same magnification and use the same intensity gains to take images (*see* **Note 10**).

3.3 Fluorescence-Activated Particle Sorting (FAPS)

1. Set up the sorter according to the sorter instruction manual.
2. Load the reference beads and click the acquisition tab to generate flow cytometry plots for reference beads. During acquisition, adjust the voltage for forward scatter (FSC), side scatter (SSC), and 488-513/26 nm in the acquisition parameters session to center the bead signals in the plot (Fig. 2). The reference beads mixture is made by mixing one drop of each type of reference beads (0.2, 0.5, 1 μm beads) in 2 mL of the wash buffer. The sampler pressure is set to 10%.
3. Rinse the sorter with Milli-Q quality water for 5 min. Set the sampler pressure to 50% during the washing process. Ensure an adequate supply of water throughout the procedure.

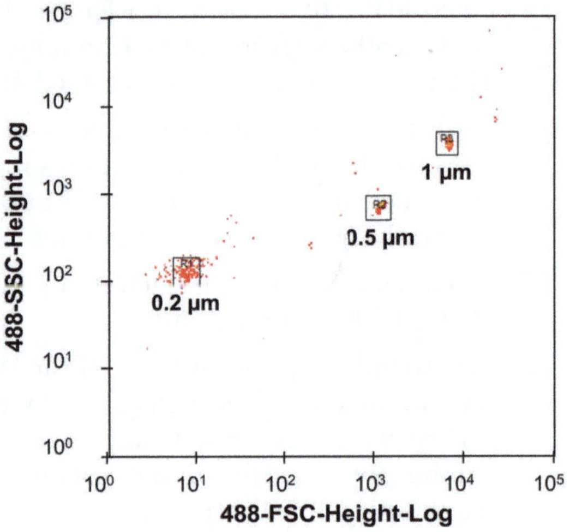

Fig. 2 The flow cytometry plot of green-fluorescent reference beads. The mix of reference beads contain 0.2, 0.5, and 1 μm sized reference beads. Position of each reference beads is indicated by a square

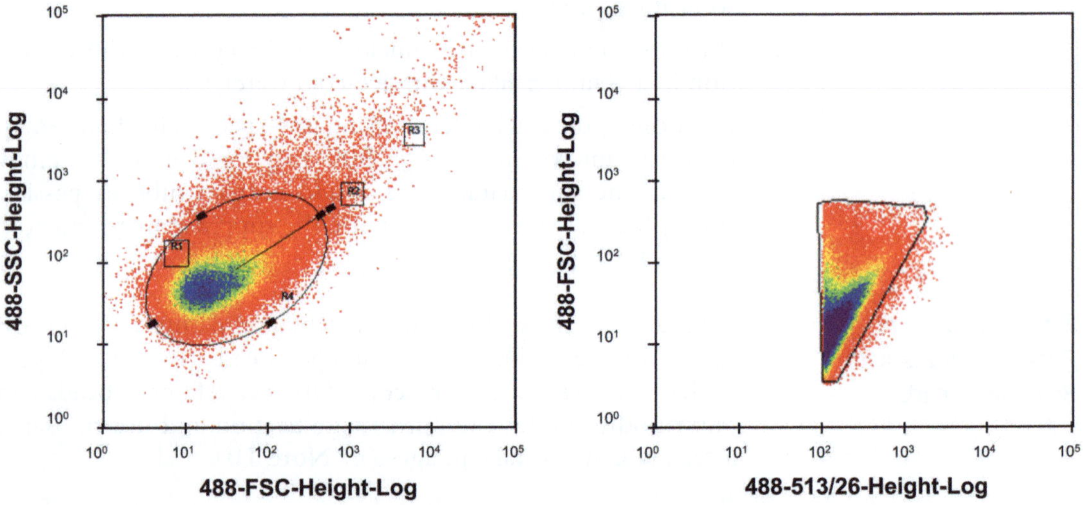

Fig. 3 Flow cytometry plots showing the forward and side scatter profile (left) and the fluorescence intensity profile (right) of phyB photobodies in FAPS. The sorting position for phyB photobodies is indicated by an ellipse (left) and triangle (right). Positions of reference beads are marked by squares (left)

3. Load the sample. Click the acquisition tab to generate flow cytometry plots of photobodies (Fig. 3). Readjust the voltage values for 488-513/26 nm, FSC, and SSC to center the sample signals on the plot (*see* **Note 11**). Set the threshold value of 488-513/26 to 0.3% to minimize the sorting of auto-fluorescent particles (*see* **Note 12**).
4. Check the event per second (EPS) values on the control panel. If the EPS value exceeds 10,000, dilute the sample to reduce the EPS.

5. Define a region for the photobodies and establish a gating method that specifically sorts the photobodies in both 488-FSC-H-Log × 488-SSC-H-Log plot and 488-513/26-H-Log × 488-FSC-H-Log plot (*see* **Note 13**).
6. Place a 50 mL conical tube into the collection chamber.
7. Execute the sorting of photobodies (*see* **Note 14**).
8. After the sorting, assess the quality of sorted photobodies (Fig. 1c).
9. Collect the sorted photobodies by transferring the sorted sample from the conical tube into several 1.7-mL microfuge tubes and precipitating them using high-speed centrifugation (4 °C, 15,000 × g 10 min).
10. Discard the supernatant of all microfuge tubes. Resuspend the sorted photobodies in a total 1 mL of the wash buffer by vortexing for at least 30 sec for each tube.
11. Precipitate the photobodies using high-speed centrifugation (4 °C, 15,000 × g 10 min) and discard the supernatant.
12. Wash the photobodies three times by resuspending them in 1 mL of wash buffer and precipitating using high-speed centrifugation (4 °C, 15,000 × g 10 min).
13. After the washing step, discard the supernatant and store the sample in −80 °C. The sample is now ready for further analysis, such as immunoblot and LC-MS/MS analysis.

4 Notes

1. PMSF and protease inhibitor cocktail are light-sensitive. Tubes or bottles containing PMSF or protease inhibitor cocktail should be covered with aluminum foil.
2. Do not autoclave to sterilize 10X PBS solutions. We recommend sterilizing 10X PBS by filtering it through a 0.2 μm syringe filter. Do not adjust pH of either 10X or 1X PBS using any other acids or alkalis. The pH of the 1X PBS solution should fall within the range of pH 7.30 and pH 7.50 (typically around pH 7.42). Note that the pH of the 10X PBS will be slightly acidic. The phosphate buffers are sensitive to dilution.
3. Check the sorting size limitation of the device. Some cell sorters do not support the sorting of sub-micron particles. The size limitation is usually specified in the equipment's specification sheet.
4. Seven grams of leaves from the same batch of plants were designated and statistically treated as one of biological replicate. Please ensure that samples are not contaminated with soil particles during collection.

5. Grind the sample at least 5 min to obtain a fine powder. This step is critical for obtaining a large number of sorted photobodies.
6. Ensure the sample is well resuspended by continuous vortexing for at least 1 min. If the sample is not adequately resuspended, rotate the sample at 4 °C for 10 min on a rotary shaker at 12 rpm and vortex it again.
7. Place two layers of Miracloth (average pore size: about 20 micrometers) in a cup shape (>20 mL) at the top of an open conical tube. Slowly load the sample onto the cup-shaped Miracloth layers and close the cap to secure the Miracloth layers. The low-speed centrifugation will filter the nuclei through the layers and precipitate them at the bottom of a tube.
8. To make the sucrose step-gradient, slowly add 1 mL of photobody fractionation buffer 3 at the bottom of a 1.7 mL microfuge tube. Make sure not to introduce any bubbles. Lay the sample solution onto the top of the bottom layer (buffer 3) by slowly rotating a volume adjusting knob in a micropipette. Be very careful not to disturb the bottom layer.
9. Photobodies can be sorted without immunostaining. However, it has been observed that immunostaining with Alexa Fluor 488-conjugated antibodies enhances sorting efficiency by increasing the brightness of photobodies. The optimal incubation time for immunostaining may vary depending on the antibodies used.
10. Estimating fluorescent particle sizes under a fluorescence microscope is influenced by the optical properties of objects, so the estimated photobody sizes using this method may not be accurate. Nevertheless, it provides a rough size range for sorting.
11. The following parameters were used for sorting phyB photobodies in our experiment [15]. The threshold for 488-513/26 nm was set to 0.3%. Acquisition parameters: 488 nm-FSC was set to 360 V; 488 nm-SSC was set to 400 V; 488-513/26 nm was set to 620 V. The differential pressure ranged from 0.4 to 0.5 psi. The drop envelope setting was set to 1–2.
12. Determine the flow cytometry plots of negative control samples that do not express phyB-GFP. Negative controls should not exhibit a significant number of sorting signals in the plots or should have distinct profiles during acquisition.
13. Define a region outlining the photobodies in 488-FSC-H-Log × 488-SSC-H-Log plot and label this region as R1. Similarly, define a region for the photobodies in 488-513/26-H-Log × 488-FSC-H-Log plot and designate this region as R2.

Configure the gating method to include both R1 and R2, signifying that the device sorts the particles that exclusively appear in both regions and reject others.

14. During FAPS process, monitor the EPS, the drop delay, the differential pressure, the remaining sample volume, the remaining Sheath fluid volume, the waste volume, and the number of aborted particles from each sample. EPS should gradually increase until it reaches a plateau. The drop delay and the differential pressure should remain stable. If the drop delay significantly decreases, it is necessary to restart the sorting from the beginning. The differential pressure should not exceed 0.7 psi. Careful attention should be paid to the sheath fluid tank volume and the waste tank volume.

References

1. Legris M, Ince Y, Fankhauser C (2019) Molecular mechanisms underlying phytochrome-controlled morphogenesis in plants. Nat Commun 10(1):5219. https://doi.org/10.1038/s41467-019-13045-0
2. Rockwell NC, Su YS, Lagarias JC (2006) Phytochrome structure and signaling mechanisms. Annu Rev Plant Biol 57:837–858. https://doi.org/10.1146/annurev.arplant.56.032604.144208
3. Kircher S, Kozma-Bognar L, Kim L, Adam E, Harter K, Schäfer E, Nagy F (1999) Light quality–dependent nuclear import of the plant photoreceptors phytochrome A and B. Plant Cell 11(8):1445–1456. https://doi.org/10.1105/tpc.11.8.1445
4. Sakamoto K, Nagatani A (1996) Nuclear localization activity of phytochrome B. Plant J 10(5):859–868. https://doi.org/10.1046/j.1365-313x.1996.10050859.x
5. Yamaguchi R, Nakamura M, Mochizuki N, Kay SA, Nagatani A (1999) Light-dependent translocation of a phytochrome B-GFP fusion protein to the nucleus in transgenic Arabidopsis. J Cell Biol 145(3):437–445. https://doi.org/10.1083/jcb.145.3.437
6. Park E, Kim Y, Choi G (2018) Phytochrome B requires PIF degradation and sequestration to induce light responses across a wide range of light conditions. Plant Cell 30(6):1277–1292. https://doi.org/10.1105/tpc.17.00913
7. Bauer D, Viczian A, Kircher S, Nobis T, Nitschke R, Kunkel T, Panigrahi KC, Adam E, Fejes E, Schafer E, Nagy F (2004) Constitutive photomorphogenesis 1 and multiple photoreceptors control degradation of phytochrome interacting factor 3, a transcription factor required for light signaling in Arabidopsis. Plant Cell 16(6):1433–1445. https://doi.org/10.1105/tpc.021568
8. Podolec R, Ulm R (2018) Photoreceptor-mediated regulation of the COP1/SPA E3 ubiquitin ligase. Curr Opin Plant Biol 45 (Pt A):18–25. https://doi.org/10.1016/j.pbi.2018.04.018
9. Ponnu J, Hoecker U (2021) Illuminating the COP1/SPA ubiquitin ligase: fresh insights into its structure and functions during plant Photomorphogenesis. Front Plant Sci 12:662793. https://doi.org/10.3389/fpls.2021.662793
10. Pham VN, Kathare PK, Huq E (2018) Phytochromes and phytochrome interacting factors. Plant Physiol 176(2):1025–1038. https://doi.org/10.1104/pp.17.01384
11. Pardi SA, Nusinow DA (2021) Out of the dark and into the light: a new view of phytochrome photobodies. Front Plant Sci 12:732947. https://doi.org/10.3389/fpls.2021.732947
12. Van Buskirk EK, Decker PV, Chen M (2012) Photobodies in light signaling. Plant Physiol 158(1):52–60. https://doi.org/10.1104/pp.111.186411
13. Chen D, Lyu M, Kou X, Li J, Yang Z, Gao L, Li Y, Fan LM, Shi H, Zhong S (2022) Integration of light and temperature sensing by liquid-liquid phase separation of phytochrome B. Mol Cell 82(16):3015–3029.e3016. https://doi.org/10.1016/j.molcel.2022.05.026
14. Kaiserli E, Paldi K, O'Donnell L, Batalov O, Pedmale UV, Nusinow DA, Kay SA, Chory J (2015) Integration of light and photoperiodic signaling in transcriptional nuclear foci. Dev Cell 35(3):311–321. https://doi.org/10.1016/j.devcel.2015.10.008

15. Kim C, Kwon Y, Jeong J, Kang M, Lee GS, Moon JH, Lee H-J, Park Y-I, Choi G (2023) Phytochrome B photobodies are comprised of phytochrome B and its primary and secondary interacting proteins. Nat Commun 14(1): 1708. https://doi.org/10.1038/s41467-023-37421-z
16. Hubstenberger A, Courel M, Benard M, Souquere S, Ernoult-Lange M, Chouaib R, Yi Z, Morlot JB, Munier A, Fradet M, Daunesse M, Bertrand E, Pierron G, Mozziconacci J, Kress M, Weil D (2017) P-body purification reveals the condensation of repressed mRNA regulons. Mol Cell 68(1): 144–157 e145. https://doi.org/10.1016/j.molcel.2017.09.003
17. Hahm J, Kim K, Qiu Y, Chen M (2020) Increasing ambient temperature progressively disassembles Arabidopsis phytochrome B from individual photobodies with distinct thermostabilities. Nat Commun 11(1):1660. https://doi.org/10.1038/s41467-020-15526-z
18. Trupkin SA, Legris M, Buchovsky AS, Tolava Rivero MB, Casal JJ (2014) Phytochrome B nuclear bodies respond to the low red to far-red ratio and to the reduced irradiance of canopy shade in Arabidopsis. Plant Physiol 165(4):1698–1708. https://doi.org/10.1104/pp.114.242438
19. Kim C, Kim SJ, Jeong J, Park E, Oh E, Park YI, Lim PO, Choi G (2020) High ambient temperature accelerates leaf senescence via Phytochrome-Interacting Factor 4 and 5 in Arabidopsis. Mol Cells 43(7):645–661. https://doi.org/10.14348/molcells.2020.0117
20. Legris M, Klose C, Burgie ES, Rojas CC, Neme M, Hiltbrunner A, Wigge PA, Schafer E, Vierstra RD, Casal JJ (2016) Phytochrome B integrates light and temperature signals in Arabidopsis. Science 354(6314): 897–900. https://doi.org/10.1126/science.aaf5656
21. Chen M, Schwab R, Chory J (2003) Characterization of the requirements for localization of phytochrome B to nuclear bodies. Proc Natl Acad Sci USA 100(24):14493–14498. https://doi.org/10.1073/pnas.1935989100

Chapter 13

Analysis of Phase Separation of EARLY FLOWERING 3

Maolin Peng, Stephanie Hutin, Aleksandra Mironova, Chloe Zubieta, and Philip A. Wigge

Abstract

Phase separation is an important mechanism for regulating various cellular functions. The EARLY FLOWERING 3 (ELF3) protein, an essential element of the EVENING COMPLEX (EC) involved in circadian clock regulation, has been shown to undergo phase separation. ELF3 is known to significantly influence elongation growth and flowering time regulation, and this is postulated to be due to whether the protein is in the dilute or phase-separated state. Here, we present a brief overview of methods for analyzing ELF3 phase separation in vitro, including the generation of phase diagrams as a function of pH and salt versus protein concentrations, optical microscopy, fluorescence recovery after photobleaching (FRAP), and turbidity assays.

Key words Phase separation, EARLY FLOWERING 3 (ELF3), Prion-like domain, FRAP, Turbidity assays, Phase diagrams

1 Introduction

Prion proteins are a subset of amyloid proteins, which can function as heritable components when aggregated. They exhibit self-replication and possess the ability to endure and be transmitted across generations [1]. Prion proteins typically feature prion-forming domains abundant in glutamine and asparagine (Q/N). Multi-domain proteins may also possess prion-like domains (PrD), which are defined as intrinsically disordered regions of low amino acid complexity with similarity to infectious yeast prions. While some PrD-containing proteins form amyloid-like structures and highly stable aggregates or fibrils, many PrDs sample reversibly form condensed states via liquid-liquid phase separation (LLPS). LLPS is metastable, dynamic, and often highly reversible, creating membrane-less compartments within cells, thereby facilitating the

Meng Chen (ed.), *Thermomorphogenesis: Methods and Protocols*, Methods in Molecular Biology, vol. 2795,
https://doi.org/10.1007/978-1-0716-3814-9_13,

organization and control of diverse cellular functions [2]. PrDs are highly enriched in proteins that undergo phase separation, and phase separation is hypothesized to play a key role in a variety of biological processes, including stress responses, RNA metabolism, and protein degradation [3–6]. Prion-like domains have been identified in many different organisms, including plants. For example, the *Arabidopsis thaliana* proteome contains 474 prion-like domain proteins and the rice proteome contains 201 prion-like domain proteins [7, 8]. While the physiological role of many of these PrD-containing proteins is not yet clear, some are implicated in critical transcriptional regulation and environmental response associated with the circadian clock.

The Evening Complex (EC), composed of three proteins, LUX ARRHYTHMO (LUX), EARLY FLOWERING 3 (ELF3), and ELF4, is a fundamental and highly conserved component of the plant circadian clock [9]. *LUX*, a single MYB domain transcription factor, was identified in a genetic screen for long hypocotyl mutants and found to regulate circadian-regulated genes [10]. *ELF3* and *ELF4* were initially identified in genetic screens for photoperiodism mutants and found to regulate circadian rhythms [11–13]. Interestingly, ELF3 contains a PrD and this domain is critical for EC function.

ELF3 regulates plant growth and development, particularly in response to temperature [14]. The ELF3 PrD is required for thermoresponsiveness and functions as a direct thermosensor in vivo and in vitro [15–17]. At temperatures below 22 °C, ELF3, as a part of the EC, forms a transcriptional repressor complex, targeting genes, such as the gene encoding the elongation growth-promoting transcription factor, PIF4 (Box et al., 2015; Franklin et al., 2011). When the temperature increases to 27 °C, the EC becomes inactive and loses its DNA binding activity [14, 16, 18]. The temperature-dependent activity of the EC is controlled by the PrD in ELF3, which causes the protein to undergo a phase change in response to temperature and form liquid droplets at elevated temperature. This change is reversible, and lowering the temperature returns ELF3 to the dilute phase and restores EC activity [15, 17]. This chapter describes techniques used to investigate the in vitro phase separation behavior of ELF3, such as light microscopy, phase diagrams, in vitro fluorescence recovery after photobleaching (FRAP), and turbidity assays. These basic techniques are useful to characterize phase separation and to determine the dynamics of the phase-separated state, particularly of PrD proteins that may undergo multiple transitions from liquid to hydrogel to fibrils. Examples of liquid and hydrogels of ELF3-PrD are shown in example figures.

2 Materials

2.1 Constructs, Buffers, and Reagents

1. pESPRIT2 bacterial expression vector or similar encoding ELF3 PrD with and without GFP-tag.
2. Bacterial strains suitable for protein expression such as Rosetta™ 2(DE3)pLysS Competent Cells, *E.coli* BL21-Codon-Plus-RIL, or similar.
3. Luria–Bertani (LB) medium: To prepare 1 L of LB media (Roth), weigh 20 g of LB power, add 1 L distilled H_2O, and autoclave at 121 °C for 20 min. Prepare 10 mL of 1000 x antibiotic stock solutions: 50 mg/mL kanamycin stock solution in ultrapure water and 34 mg/mL chloramphenicol in 100% ethanol. Prepare 10 mL of a 1 M solution of IPTG in ultrapure water and filter sterilize.
4. Resuspension/lysis buffer: 100 mM BTP pH 9.4, 300 mM NaCl, 20 mM imidazole, 1 mM TCEP, and complete protease-inhibitor cocktail. Mix 100 mL of 1 M BTP pH 9.4 with 60 mL NaCl and 1 mL TCEP in a 1 L graduated cylinder. Add distilled H_2O to a final volume of 1 L. Protease inhibitors may be added just before use.
5. Equilibration buffer: 100 mM BTP pH 9.4, 300 mM NaCl, 20 mM imidazole, 1 mM TCEP.
6. High salt buffer: 100 mM BTP pH 9.4, 1 M NaCl, 20 mM imidazole, and 1 mM TCEP.
7. Elution buffer: 100 mM BTP pH 9.4, 300 mM NaCl, 300 mM imidazole, and 1 mM TCEP.
8. Dialysis buffer: 50 mM BTP pH 9.4, 500 mM NaCl, and 1 mM TCEP. Store at 4 °C.
9. Dialysis buffers for phase diagrams (pH): 200 mL each, 50 mM BTP, 500 mM NaCl, 1 mM TCEP. Adjust the buffer pH to 9.4, 9.2, 9.0, 8.8, 8.6, 8.4, 8.2, 8.0, 7.8, 7.6, 7.4, 7.2, 6.8 with concentrated HCl. Store at 4 °C.
10. Dialysis buffers for phase diagrams (salt, 200 mL each): 50 mM BTP, 1 mM TCEP, × mM NaCl with x = 500, 400, 300, 200, and 100. Select the pH of all buffers to 0.2 units higher than the tested onset of liquid droplets for the highest protein concentration. Store at 4 °C.
11. 2× SDS loading dye: 4% SDS, 20% glycerol, 120 mM Tris–HCl pH 6.8, and 0.02% weight/volume bromophenol blue.
12. 1× Laemmli running buffer: 25 mM Tris–HCl, pH 8.3, 0.192 M glycine, and 0.1% SDS.
13. 12% SDS denaturing gels.
14. Ready-to-use Coomassie blue solution.

15. Size exclusion chromatography (SEC) buffer: 500 mL: 50 mM Bis-Tris-propane, pH 9.4, 1.5 M NaCl, 1 mM TCEP, filtered and degassed.
16. cOmplete™, EDTA-free Protease Inhibitor Cocktail (Roche).
17. Ni Sepharose 6 Fast Flow (Sigma-Aldrich) or similar.
18. Spectrum Spectra/Por 1 RC Dialysis Membrane Tubing 12–14 kDa MWCO (Fisher Scientific) or similar.

2.2 Equipment

1. Shaker-incubator for bacterial expression, Infors Multitron or similar.
2. Tabletop centrifuge for 1.5-mL reaction tubes.
3. Beckman Coulter Avanti J-26 XP centrifuge with rotors JLA 8.1000 and JA-25.25 or similar.
4. Sonicator (Branson Sonifer 250) or similar.
5. Fast Protein Liquid Chromatography (FPLC) system with fraction collector.
6. Superdex™ 200 Increase 10/300 GL column (Cytiva) or similar.
7. Magnetic stirrer and mixing platform.
8. Polyacrylamide gel rig.
9. Inverted light microscope (Olympus CKX41 model) equipped with a pE-300 Cool-Snap cf^2 camera or similar.
10. Spinning disk confocal microscope equipped with a high-intensity laser for photobleaching. EclipseTi-E Nikon inverted microscope coupled to a CSUX1-A1 Yokogawa confocal head, an Evolve EMCCD camera (Roper Scientific, Princeton Instruments). Use a Nikon CFI Plan-APO VC 60×, 1.4 NA, oil-immersion objective. Control of the microscope is via MetaMorph (Universal Imaging) software with autofocus function enabled or similar.
11. Cary 100 UV-vis spectrometer (Agilent Technologies) or similar.
12. Image analysis software: ImageJ, Mathlab, GraphPad Prism, or online image analysis tools.
13. Data analysis and graphing software: Sigmaplot 11, Systat Software.

3 Methods

3.1 Protein Expression and Purification of ELF3PrD and ELF3PrD-GFP

1. Transform competent bacteria optimized for protein expression with the plasmid of interest (e.g., *pESPRIT002 ELF3PrD-GFP*) following the manufacturer's protocol. Briefly, add 1 μL of plasmid to a ~ 50 μL aliquot of chemically competent

bacteria and incubate the reaction tube on ice for 15 min. Perform a heatshock by placing the tube at 42 °C for 30-45 s and then quickly on ice for 2 min. Add 1 mL of sterile Luria Bertani (LB) media without antibiotics and let the cells recover at 37 °C for 1 h. Spin down the cells at 10,000 × g for 3 s in a tabletop microtube centrifuge and remove most of the media by decanting. Resuspend the cells in the remaining media and spread the cells onto an LB agar plate with appropriate antibiotics (e.g., 50 μg/mL kanamycin and 34μg/mL chloramphenicol) and incubate at 37 °C overnight.

2. Grow a starting culture for large-scale protein expression. Inoculate 100 mL of LB with appropriate antibiotics in a 500-mL flask with one colony of the transformed bacteria and grow the culture at 37 °C, 150 rpm overnight.
3. Express the protein. Inoculate 1 L of LB with appropriate antibiotics with 20 mL of the starter culture and incubate at 37 °C and 150 rpm until an OD600 = 0.8 is reached. Reduce the temperature to 18 °C and induce expression with 1 mM IPTG. Incubate the cultures at 18 °C and 150 rpm overnight. Harvest the bacteria by centrifugation at 4500 × g and at 4 °C for 20 min (*see* **Note 1**).
4. Resuspend the bacterial pellets in a 100-mL beaker in approximately 50 mL of resuspension buffer (100 mM BTP pH 9.4, 300 mM NaCl, 20 mM imidazole, 1 mM TCEP, and EDTA-free protease inhibitors) and place the beaker in an ice-water slurry. Lyse the bacteria via sonication. For a 1-inch probe sonicator, 70% power, 1-s pulse, 1-s pause, 2 × 5 min pulse time may be used (*see* **Note 2**).
5. Separate the soluble from the insoluble components via centrifugation of the lysate at 45000 × g at 4 °C for 30 min.
6. Load the supernatant onto a nickel affinity column (Ni-column, 1.5-mL bed volume, washed with 10 CV of water and equilibrated with 10 CV of equilibration buffer), wash with 50 mL of equilibration buffer and 50 mL of high salt buffer, and elute the protein in 500 μL steps in elution buffer (*see* **Note 3).**
7. Inspect the different purification steps via sodium dodecyl sulfate (SDS)-polyacrylamide gel electrophoresis (PAGE): Load 1.5 μL of each elution fraction mixed with 1× loading dye on a 12% SDS gel and run at 200 V in 1× Laemmli running buffer. Stain the gel with a ready-to-use Coomassie blue solution.
8. Pool the fractions of interest and dialyze for 2 h at 4 °C in 250–500 mL dialysis buffer at pH 9.4, with gentle stirring using a magnetic stir bar and stirplate.

9. (Optional) perform SEC: Connect the SEC column to the system and wash with 1.5 CV of water and equilibrate with 1.5 CV of filtered and degassed SEC buffer. Choose the flow rate depending on the column in use (0.5 mL/min for the S200 increase 10/300 GL) and define the maximum pressure according to the column specifications. Spin down the protein sample at 13000 x g at 4 °C for at least 5 min prior to injecting onto the column. Inject the sample into the loop. Program the FPLC system to run 1.2 CV isocratic elution and collect 1 mL fractions (*see* **Note 4**).
10. Inspect the different fractions on a SDS PAGE and pool fractions of interest (*see* **Note 5**).

3.2 Determination of Phase Diagrams

1. Dilute the protein to minimal volume of 300 μL at different concentrations (i.e., 4 mg/mL, 2 mg/mL, 1 mg/mL, and 0.5 mg/mL) and place in mini dialysis cassettes with a 10 kDa MWCO or use dialysis tubing and clamps, equilibrated in dialysis buffer pH 9.4 and dialyze the protein against dialysis buffer at 4 °C for 2 h slightly agitating the buffer using a magnetic stir bar.
2. Carefully open the dialysis tubing and take a 10-μL aliquot or remove using a syringe from the dialysis cassette a sample and place 10 μL onto a pre-cooled glass slide. Visualize potential LLPS under the microscope. LLPS is defined based on the following criteria: at least 5 liquid droplets in the field of view, no gel formation, and fusion of droplets indicating liquid droplet phase (Fig. 1a).
3. Seal the sample back in the dialysis tubing or cassette and place the samples in the next dialysis buffer with a pH reduced by 0.2 pH units and incubate at 4 °C for 1 h.
4. Repeat steps 2 and 3 over the pH range from 9.4 to 6.8 or until all concentrations of protein exhibit precipitation.
5. Set up a new dialysis with fresh samples starting at 50 mM Bis-Tris-propane pH 7, 500 mM NaCl, and 1 mM TCEP and proceed as before but reducing the NaCl concentration by 100 mM at each dialysis step.
6. Plot the pH and salt concentration versus protein concentration to generate a phase diagram (Fig. 1b, c).

3.3 In Vitro Fluorescence Recovery After Photobleaching (FRAP)

1. Dialyze 4 mg/mL AtELF3-PrD-GFP protein against dialysis buffer at pH 9.4 and 500 mM NaCl for 1 h at 4 °C. Then reduce the pH in 0.2-unit steps until 0.2 units above the indicated LLPS undergoing value determined in the predetermined phase diagrams.

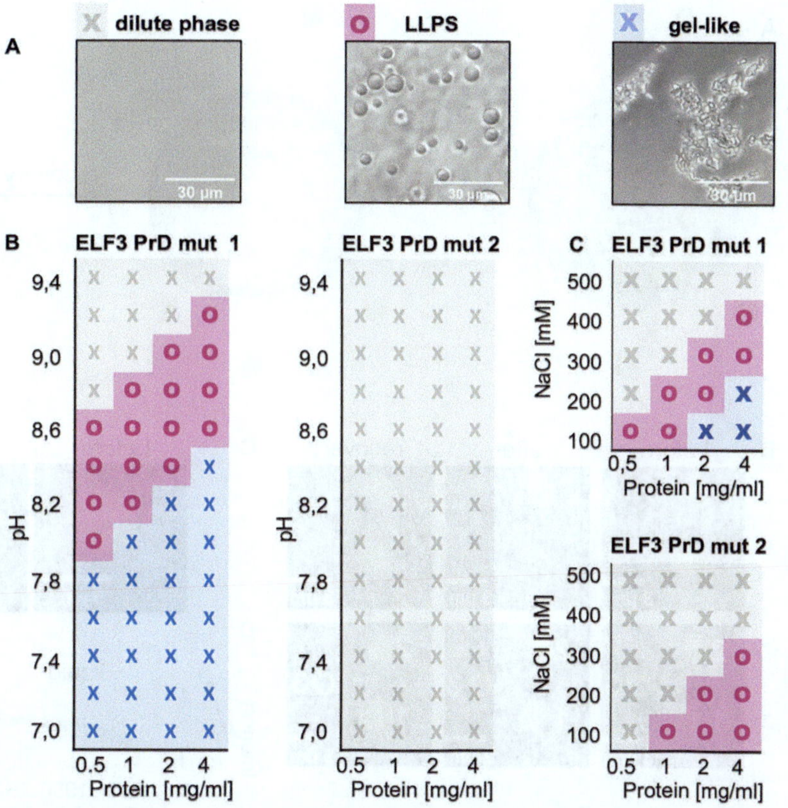

Fig. 1 Phase diagrams illustrate under which conditions a protein undergoes LLPS and gelation or precipitation. (**a**) Light-microscopy images of ELF-3 PrD in the dilute (left panel), liquid droplet (middle), and gel-like phase (right). (**b**) Examples of phase diagrams examining pH versus protein concentration (gray x = dilute phase, magenta o = LLPS, blue x = gel-like phase) and (**c**) Salt concentration versus protein concentration

2. Place 10 μL of sample onto a glass slide, cover it with a coverslip and quickly mount onto an inverted microscope as part of the confocal spinning disk system.
3. Allow droplets to adhere to the slide to minimize droplet movement during the experiment and start the scan.
4. Calibrate the laser (wavelength 488 nm) to reach 60–100% bleaching in a part of a 5–10 μm adhered droplet (e.g.,1 μm bleach diameter, 40% laser power, pulse of 2822 ms, and 166 repetitions).
5. Change position to an area not used for calibration.
6. Collect images at 530 nm: for example, every 1 s for 5 s before the photobleaching (before FRAP reference), then every 1 s for 3 min, or until a recovery plateau is reached (*see* **Note 6**) (Fig. 2a).

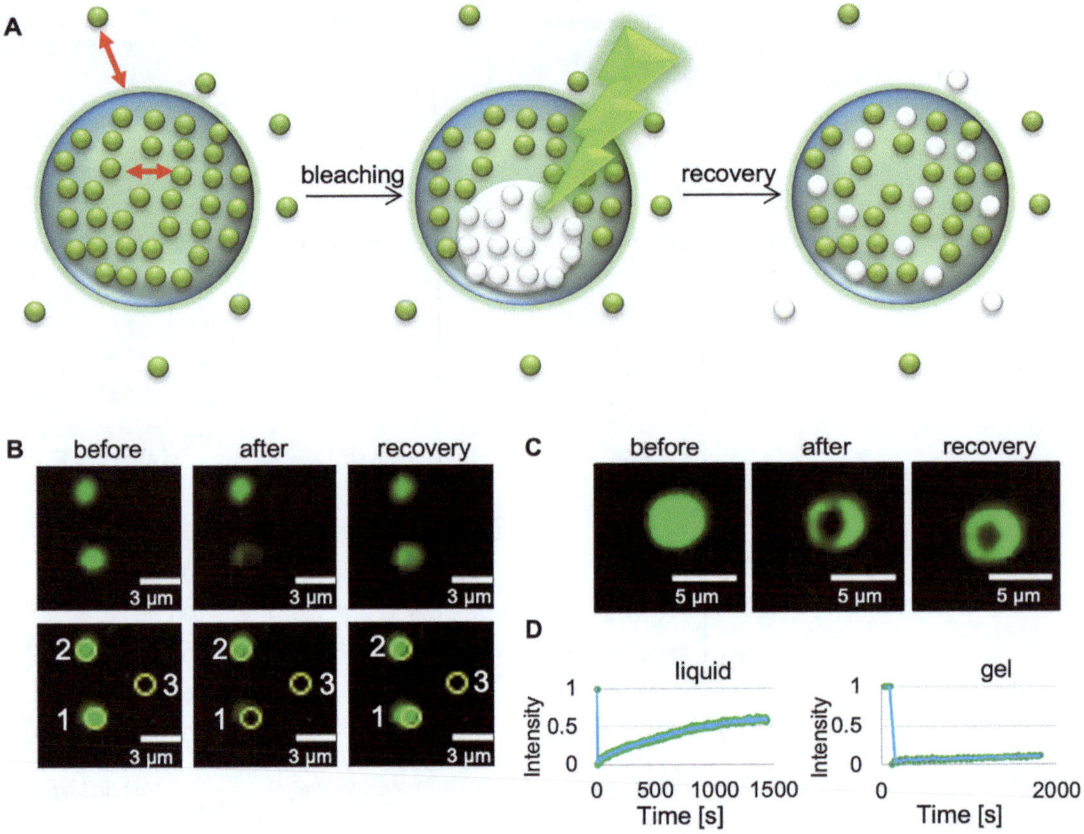

Fig. 2 Fluorescent recovery after photobleaching (FRAP) (**a**) Schematic overview of FRAP experiments. (**b**) fluorescent microscopy images of ELF3 PrD-GFP liquid droplet (upper panels) and with selected regions for analysis (lower panel). (**c**) Droplets in the gel-like phase. (**d**) FRAP recovery curves (left panel = liquid, right panel = gel-like phase)

7. Change to fresh sample after 2 FRAP experiments as samples tend to exhibit reduced mobility over time when adhered to the glass surface and repeat steps 2, 3, and 5. Ideally, 10–15 individual droplets should be measured for each construct of interest and a droplet of the same size used as an unbleached reference droplet in the same focal view of each FRAP experiment.
8. Analyze the FRAP data. Load the images from the experiment into ImageJ (Fiji) [19]. Drag the folder containing all images onto the ImageJ taskbar and open as a virtual stack. Go to Image → Properties and set the pixel width according to the objective and camera used. Zoom into an area with photobleached droplet and non-bleached droplets. Go to Image → Crop to select the area of interest. Optional: images can be false colored if desired in the Image → Lookup Tables by selecting the desired color. In the Analyze → Tools, select Scale Bar and choose the scale bar width and legend parameters. Save desired images as jpg or tif (Fig. 2b).

9. Select a region of interest (ROI_b) corresponding to the bleached area using the selection tool and appropriate shape to match the photobleached area as closely as possible. Use the ROI manager to add a selection for the background area (BG) using the same size and shape selection as per the ROI_b. From the same image, select an unbleached droplet of the same size as the bleached droplet as an unbleached reference droplet control (ROI_u).
10. Verify that the droplets and ROIs remain aligned throughout the stack of images. If necessary, use the align tool to correct for drift.
11. Measure the fluorescence intensity of the ROIs and the BG areas using multimeasure. Data may be exported for further processing in excel or other software for normalization (*see* **Note 7**). 12–14 provide an overview of data analysis normalization steps to be performed.
12. Subtract the BG fluorescence intensity from the ROI_B and ROI_U fluorescence intensity values to correct for background.
13. Normalize the fluorescence intensity data to account for photobleaching during the time course measurement of the experiment. Calculate the photobleaching rate, *r*, by dividing the fluorescence of the ROI_U at $t = 0$ from step 12 with the fluorescence intensity at each timepoint. Use this value for normalization of the fluorescence intensity of the photobleached region (ROI_B), with normalized fluorescence intensity of $ROI_{Bnorm} = ROI_B/r$.
14. Divide the ROI_{Bnorm} by ROI_{Unorm} to calculate fractional normalized recovery.
15. Plot the fractional normalized recovery as a function of time (t).
16. Perform curve fit to a single exponential curve in any available software.
17. Average FRAP curves from ~10 droplets and calculate the standard errors.

3.4 Turbidity Assay

1. Dialyze the protein at 4 °C as for the FRAP experiments at 15 μM AtELF3-PrD until before LLPS occurs (*see* **Note 8**).
2. Place the sample or only buffer, as a negative control in quartz cuvettes with a path length 10 mm.
3. Monitor the absorbance at 440 nm and increase the temperature from 4 °C to −40 °C at 1 °C min^{-1}) using a Cary 100 UV-vis spectrometer (Agilent Technologies) (*see* **Note 9**).
4. To assess reversibility, measure the turbidity while decreasing the temperature (from 40 °C to 10 °C; 1 °C min^{-1}).

5. Finally, increase the temperature again (from 10 °C to 40 °C; 1 °C min^{-1}) and measure the turbidity.
6. Normalize the AtELF3-PrD spectra with respect to the buffer.
7. Determine the transition temperature (*Tm*) by fitting the spectrum with a four-parameter sigmoidal equation with a data analysis and graphing software such as Sigmaplot.

4 Notes

1. For each 1 L of bacterial expression, the yield of final purified protein is ~1–2 mg. Generally, 2–4 L of bacterial culture is prepared. Cell pellets may be frozen at −20 °C until use.
2. Test the temperature of the lysate to avoid overheating of the sample. Lower sonicator power, regular mixing, and longer "off" times may be required.
3. All proteins are eluted after approximately 3–5 CV.
4. Due to the sensitivity of ELF3-PrD and ELF3-PrD-GFP proteins to aggregation, freshly purified samples are preferred for downstream experiments.
5. Concentration of the ELF3-PrD constructs using membrane concentrators is difficult due to adherence to the membrane. It is recommended to avoid a concentration step. If required, to obtain ELF3-PrD proteins at high concentration, Ni-affinity resin may be used as a concentration step.
6. The acquisition time will vary according to the time the FRAP requires. For a fast-recovering protein, 200 μs in the first 30 s might be needed.
7. Normalization may be performed in Fiji or ImageJ using the appropriate FRAP plug ins. For a detailed overview of how to use Fiji or Image J to process FRAP data, see http://rsbweb.nih.gov/ij/, https://imagej.nih.gov/ij/docs/index.html. Alternatively, online tools such as EasyFRAP [20] may be used to normalize and analyze the data.
8. For the turbidity measurements, the protein may be purified as described above but the buffer BTP pH 9.4 was replaced by CAPS pH 9.7 due to better protein stability and reversibility between the dilute and phase-separated states.
9. The low temperature within the spectrometer can result in condensation on the cuvettes, which can affect the accuracy of the result. To mitigate this issue, it is advised to store cuvettes at 4 °C prior to use.

Acknowledgments

We thank Ina Talke and the IMPRS program for providing support to MP during this project. PAW is supported by the Leibniz Society and State of Brandenburg. Work in PAW's lab is supported by a European Research Council Advanced Grant TIP-TOP (ERC-101021246). Work in CZ's lab is supported by the ANR (ANR-19-CE20-0021 and ANR-21-CE11-0037) and GRAL, a program from the Chemistry and Biology Health Graduate School of the University Grenoble Alpes (ANR-17- EURE- 0003). We thank Dr. Laëtitia Kurzawa and the microscopy facility MuLife of IRIG/DBSCI, funded by CEA Nanobio and GRAL LabEX (ANR-10-LABX-49-01) financed within the University Grenoble Alpes graduate school CBH-EUR-GS (ANR-17-EURE-0003).

References

1. Prusiner SB (1998) Prions. Proc Natl Acad Sci 95:13363–13383. https://doi.org/10.1073/pnas.95.23.13363
2. Lau Y, Oamen HP, Caudron F (2020) Protein phase separation during stress adaptation and cellular memory. Cell 9:1302. https://doi.org/10.3390/cells9051302
3. Bolognesi B, Faure AJ, Seuma M, Schmiedel JM, Tartaglia GG, Lehner B (2019) The mutational landscape of a prion-like domain. Nat Commun 10:4162. https://doi.org/10.1038/s41467-019-12101-z
4. Khan T, Kandola TS, Wu J, Venkatesan S, Ketter E, Lange JJ, Rodríguez Gama A, Box A, Unruh JR, Cook M, Halfmann R (2018) Quantifying nucleation in vivo reveals the physical basis of prion-like phase behavior. Mol Cell 71:155–168.e7. https://doi.org/10.1016/j.molcel.2018.06.016
5. Huai Y, Mao W, Wang X, Lin X, Li Y, Chen Z, Qian A (2022) How do RNA binding proteins trigger liquid-liquid phase separation in human health and diseases? BST 16:389–404. https://doi.org/10.5582/bst.2022.01449
6. Kim J, Lee H, Lee HG, Seo PJ (2021) Get closer and make hotspots: liquid–liquid phase separation in plants. EMBO Rep 22:e51656. https://doi.org/10.15252/embr.202051656
7. Chakrabortee S, Kayatekin C, Newby GA, Mendillo ML, Lancaster A, Lindquist S (2016) Luminidependens (LD) is an Arabidopsis protein with prion behavior. Proc Natl Acad Sci 113:6065–6070. https://doi.org/10.1073/pnas.1604478113
8. Garai S, Citu S-PSL, Sopory SK, Kaur C, Yadav G (2021) Complex networks of prion-like proteins reveal cross talk between stress and memory pathways in plants. Frontiers in Plant Sci 12
9. Nusinow DA, Helfer A, Hamilton EE, King JJ, Imaizumi T, Schultz TF, Farré EM, Kay SA (2011) The ELF4-ELF3-LUX complex links the circadian clock to diurnal control of hypocotyl growth. Nature 475:398–402. https://doi.org/10.1038/nature10182
10. Hazen SP, Schultz TF, Pruneda-Paz JL, Borevitz JO, Ecker JR, Kay SA (2005) LUX ARRHYTHMO encodes a Myb domain protein essential for circadian rhythms. Proc Natl Acad Sci 102:10387–10392. https://doi.org/10.1073/pnas.0503029102
11. Hicks KA, Millar AJ, Carré IA, Somers DE, Straume M, Meeks-Wagner DR, Kay SA (1996) Conditional circadian dysfunction of the Arabidopsis early-flowering 3 mutant. Science 274:790–792. https://doi.org/10.1126/science.274.5288.790
12. Zagotta MT, Hicks KA, Jacobs CI, Young JC, Hangarter RP, Meeks-Wagner DR (1996) The Arabidopsis ELF3 gene regulates vegetative photomorphogenesis and the photoperiodic induction of flowering. Plant J 10:691–702. https://doi.org/10.1046/j.1365-313x.1996.10040691.x
13. Doyle MR, Davis SJ, Bastow RM, McWatters HG, Kozma-Bognár L, Nagy F, Millar AJ, Amasino RM (2002) The ELF4 gene controls circadian rhythms and flowering time in Arabidopsis thaliana. Nature 419:74–77. https://doi.org/10.1038/nature00954
14. Box MS, Huang BE, Domijan M, Jaeger KE, Khattak AK, Yoo SJ, Sedivy EL, Jones DM, Hearn TJ, Webb AAR, Grant A, Locke JCW,

Wigge PA (2015) ELF3 controls thermoresponsive growth in Arabidopsis. Curr Biol 25: 194–199. https://doi.org/10.1016/j.cub.2014.10.076

15. Jung J-H, Barbosa AD, Hutin S, Kumita JR, Gao M, Derwort D, Silva CS, Lai X, Pierre E, Geng F, Kim S-B, Baek S, Zubieta C, Jaeger KE, Wigge PA (2020) A prion-like domain in ELF3 functions as a thermosensor in Arabidopsis. Nature 585:256–260. https://doi.org/10.1038/s41586-020-2644-7
16. Silva CS, Nayak A, Lai X, Hutin S, Hugouvieux V, Jung J-H, Lopez-Vidriero I, Franco-Zorrilla JM, Panigrahi K, Wigge PA, Zubieta C (2020) Molecular mechanisms of evening complex activity in Arabidopsis, vol 117. Proceedings of the National Academy of Sciences, p 6901
17. Hutin S, Kumita JR, Strotmann VI, Dolata A, Ling WL, Louafi N, Popov A, Milhiet P-E, Blackledge M, Nanao MH, Wigge PA, Stahl Y, Costa L, Tully MD, Zubieta C (2023) Phase separation and molecular ordering of the prion-like domain of the Arabidopsis thermosensory protein Early Flowering 3. Proc Natl Acad Sci 120:e2304714120. https://doi.org/10.1073/pnas.2304714120
18. Ezer D, Jung J-H, Lan H, Biswas S, Gregoire L, Box MS, Charoensawan V, Cortijo S, Lai X, Stöckle D, Zubieta C, Jaeger KE, Wigge PA (2017) The evening complex coordinates environmental and endogenous signals in Arabidopsis. Nature Plants 3:17087. https://doi.org/10.1038/nplants.2017.87
19. Fiji: an open-source platform for biological-image analysis | Nature Methods. https://www.nature.com/articles/nmeth.2019. Accessed 21 May 2023
20. Koulouras G, Panagopoulos A, Rapsomaniki MA, Giakoumakis NN, Taraviras S, Lygerou Z (2018) EasyFRAP-web: a web-based tool for the analysis of fluorescence recovery after photobleaching data. Nucleic Acids Res 46: W467–W472. https://doi.org/10.1093/nar/gky508

Chapter 14

In Vitro Determination of Temperature-Dependent DNA Binding of the Evening Complex Using Electrophoretic Mobility Shift Assays

Stephanie Hutin, Philip A. Wigge, and Chloe Zubieta

Abstract

Electrophoretic mobility shift assays (EMSAs) of DNA-binding proteins and labeled DNA allow the qualitative and quantitative characterization of protein-DNA complex formation using native (nondenaturing) polyacrylamide or agarose gel electrophoresis. By varying the incubation temperature of the protein-DNA binding reaction and maintaining this temperature during electrophoresis, temperature-dependent protein-DNA interactions can be investigated. Here, we provide examples of the binding of a transcriptional repressor complex called the Evening Complex, comprising the DNA-binding protein LUX ARRYTHMO (LUX), the scaffold protein EARLY FLOWERING 3 (ELF3), and the adapter protein ELF4, to its cognate DNA and demonstrate direct detection and visualization of thermoresponsive binding in vitro. As negative controls we use the LUX DNA-binding domain and LUX full length protein, which do not exhibit temperature-dependent DNA binding.

Key words Electrophoretic mobility shift assay, Protein-DNA binding, Temperature dependence, Native polyacrylamide gel electrophoresis, Native agarose gel electrophoresis, Evening Complex, LUX ARRYTHMO, EARLY FLOWERING 3, EARLY FLOWERING 4

1 Introduction

Proper gene regulation relies on the specific binding of different transcription factors (TF) and transcriptional complexes to their cognate DNA. In multicellular organisms, 6–8% of protein coding genes encode TFs. Understanding the interplay between TFs, DNA binding, and gene regulation as a function of environmental variables, including temperature, is a fundamental challenge in biology [1]. In vivo techniques such as chromatin immunoprecipitation followed by polymerase chain reaction (ChIP-PCR) or deep sequencing (ChIP-seq) have been commonly used to identify TF binding sites (TFBS) in vivo [2, 3]. In situ techniques including Cleavage Under Targets and Release Using Nuclease

Meng Chen (ed.), *Thermomorphogenesis: Methods and Protocols*, Methods in Molecular Biology, vol. 2795,
https://doi.org/10.1007/978-1-0716-3814-9_14,

(CUT&RUN) [4–6] and Cleavage Under Targets and Tagmentation (CUT&Tag) [7] have further refined the identification of in vivo DNA binding sites by investigating TF binding in the native chromatin environment and can be performed under different growth conditions including different temperatures. However, these techniques have distinct challenges including the need for specific antibodies or the generation of transgenic lines expressing a tagged version of the target TF, dependence on the chromatin environment and accessibility of putative binding sites, competition with other DNA-binding proteins, the presence or lack of cofactors, poor signal to noise ratios due to low endogenous TF expression levels or artifacts due to TF overexpression if a 35S transgene is used, for example. In order to simplify the complexity of in vivo systems and determine the molecular mechanisms of protein-DNA interactions, in vitro techniques are often used. In vitro genome-wide techniques include DNA-affinity purification and sequencing (DAP-seq) in which purified immobilized TF or TF complexes and fragmented genomic DNA (gDNA) are mixed and bound DNA fragments are sequenced and mapped to the genome [8–13]. Generation of a binding site model using systematic evolution of ligands by exponential enrichment and sequencing (SELEX-seq) is a similar approach in which a synthetic randomized DNA library is used in the pull-down step [14]. Often these in vivo and in vitro techniques are used to identify important putative target genes and specific DNA-binding sites for a given TF. In order to explore the specificity of DNA binding and perform systematic variation of binding conditions, purified TFs and a single DNA-binding site contained in a short DNA oligomer are commonly used.

Examination of binding to a single region by a given TF or TF complex and quantification of the interaction are often carried out to validate direct binding and identify the exact nucleotide sequence responsible for high-affinity site-specific interactions. Various methods are available to examine, quantitatively and qualitatively, these protein-DNA interactions including isothermal titration calorimetry [15], thermal shift assays [16], microscale thermophoresis [17], size exclusion chromatography [18], fluorescence anisotropy [19] and, most commonly, electrophoretic mobility shift assays (EMSAs) [20]. EMSAs, or band shift assays, provide a simple, rapid, and sensitive method to determine protein-DNA interactions both qualitatively and quantitatively. These experiments require nanomolar concentrations of labeled DNA and low micromolar concentrations of protein and rely on equipment present in most biochemistry laboratories. The TF or TF complex is generally recombinantly overexpressed, purified, and incubated with a short fluorescently labeled dsDNA fragment containing a known binding sequence or motif. The TF-DNA complex is then run on a polyacrylamide or agarose gel under nondenaturing conditions, allowing the separation of unbound and bound DNA, with the TF-DNA complex exhibiting a slower migration than the unbound DNA oligomer. The fluorescently tagged DNA allows for visualization of the different DNA-containing species (unbound DNA, DNA-TF complex(es)). As different variables such as ionic strength, pH, and temperature will affect protein-DNA interactions, these may be systematically changed to examine their effect on protein-DNA interactions. In this protocol, we focus on the effects of temperature on TF-DNA binding.

Temperature is a critical variable affecting plant growth and development. Changes in temperature alter the transcriptome by tuning the activity of thermosensitive transcription factors and complexes. One example of a transcriptional complex that exhibits thermosensitive binding in vivo and in vitro is the Evening Complex, a component of the circadian clock and a regulator of thermosensitive growth and flowering [21, 22]. The EC is comprised of three proteins—LUX ARRYTHMO (LUX), a Myb domain DNA-binding protein; EARLY FLOWERING 3 (ELF3), a plant-specific scaffold protein; and EARLY FLOWERING 4 (ELF4), a small alpha-helical adaptor protein [23]. At lower temperatures, the EC binds its cognate DNA and represses its target genes, while at higher temperatures, DNA binding is reduced, resulting in derepression of EC target genes such as *PHYTOCHROME INTERACTING FACTOR 4 (PIF4)*, a hub of stem/hypocotyl elongation and seedling thermomorphogenic growth [24, 25]. In order to determine whether the EC acts as a direct thermosensor and exhibits changes in DNA-binding affinity as a function of temperature [26], in vitro studies of its temperature-responsive DNA binding are required. Here, we provide a detailed protocol using EMSAs to characterize the temperature-dependent DNA binding of the EC over a physiologically relevant temperature range in which the EC exhibits differential DNA binding. This protocol is generally applicable to any TF or TF complex of interest that may exhibit temperature-dependent DNA binding.

2 Materials

Freshly prepare all solutions using ultrapure water and analytical-grade reagents. This protocol uses fluorescently labeled DNA for visualization of DNA-containing complexes. Radiolabeled DNA probes may be used in place of fluorescently labeled DNA with attention to all applicable safety and waste disposal regulations. The protocol specifies purified TFs and TF complexes and urea denaturation and refolding steps for EC formation; however, this may not be necessary depending on the TF or TF complex of interest. Binding buffers for TF-DNA complexes will vary depending on the TF of interest, and the buffer detailed in the protocol should be considered as a "starting point." Factors including salt concentration, pH, and additives (detergents, glycerol, BSA, competitor DNA) must be systematically varied and empirically tested in order to obtain an optimized TF-DNA binding buffer.

2.1 Protein Production and Purification

1. Bacterial strains for protein expression: Rosetta™ 2(DE3) pLysS Competent Cells or similar (*see* **Note 1**).
2. Expression media and antibiotics: LB-medium (Roth), for 1 L of LB, weigh out 20 g of LB power into a 2 L Erlenmeyer flask, add 1 L ultrapure water, and sterilize by autoclaving. Prepare 10 mL of 1000× antibiotic stock solutions: 50 mg/mL kanamycin stock solution in ultrapure water. 34 mg/mL chloramphenicol in 100% ethanol. Prepare 10 mL of a 1 M solution of IPTG in ultrapure water and filter-sterilize.
3. Protein purification buffers for ELF3-mCherry purification under denaturing conditions: Resuspension buffer: 200 mM N-cyclohexyl-3-aminopropanesulfonic acid (CAPS), pH to 9.7, 500 mM NaCl, 1 mM Tris(2-carboxyethyl) phosphine (TCEP) and 1× cOmplete™, EDTA-free protease inhibitors, or similar. Denaturing resuspension buffer: 8 M urea, 20 mM imidazole, 1 mM TCEP. Denaturing elution buffer: 8 M urea, 250 mM imidazole, 1 mM TCEP.
4. Protein purification buffers for LUX and ELF4 under nondenaturing conditions: Lysis buffer: 100 mM Bis-Tris-propane (BTP) pH 9.4, 500 mM NaCl, 20 mM imidazole, 1 mM TCEP and 1× cOmplete™, EDTA-free protease inhibitors, or similar. Wash buffer: 100 mM (BTP) pH 9.4, 500 mM NaCl, 20 mM imidazole, 1 mM TCEP. High-salt wash buffer: 100 mM BTP pH 9.4, 1 M NaCl, 20 mM imidazole and 1 mM TCEP. Elution buffer: 100 mM BTP pH 9.4, 500 mM NaCl, 300 mM imidazole, and 1 mM TCEP. Dialysis and SEC-buffer: 100 mM BTP pH 8.0, 500 mM NaCl, and 1 mM TCEP.
5. Size exclusion chromatography buffer: 50 mM BTP pH 8, 500 mM NaCl, and 1 mM TCEP.
6. Ni-NTA FastFlow resin (Qiagen) or similar.
7. 30 mL disposable column (BioRad) or similar.
8. Dialysis buffers for complex reconstitution: Buffer 1: 6 M urea, 1 mM TCEP. Buffer 2: 4 M urea, 1 mM TCEP. Buffer 3: 2 M urea, 50 mM sodium phosphate pH 7.3, 150 mM NaCl, and 1 mM TCEP. Buffer 4: 1 M urea, 50 mM sodium phosphate pH 7.3, 150 mM NaCl, and 1 mM TCEP.

2.2 Protein-DNA Binding Reaction and EMSAs

1. 10× binding buffer: 100 mM Tris, pH 7.5, 10 mM $MgCl_2$, 10 mM TCEP, 60% glycerol, 280 ng/μL herring sperm DNA, 200 μg/ml BSA, 25% CHAPS, and 12.5 mM spermidine.

2. 10× TBE: 121 g Tris base, 61.8 g boric acid, 7.4 g EDTA, add water to a final volume of 1 L. This gives a final concentration of 1.0 M borate, 1.0 M Tris base, and 20 mM EDTA.
3. 5% polyacrylamide nondenaturing gels: mix 5 mL of a 30% polyacrylamide (37.5:1) with 3 ml 10× TBE, 300 μL ammonium persulfate (10% stock solution), and 24 μL TEMED and 22 mL H_2O (sufficient for 4, mini protein gels 8.3 cm × 7.3 cm × 1 mm) in a 50 mL Falcon tube. Invert the tube 5× to mix and immediately pour the gel solution between the glass plates of the casting system to the top of the smaller plate and place a 10-well comb between the plates. Allow the gels to polymerize for at least 1 h before use. Gels may be stored at 4 °C by covering with a 1× TBE soaked paper towel and wrapped in plastic wrap to avoid evaporation and drying of the gel (*see* **Note 2**).
4. 2% nondenaturing agarose gels: Weigh out 2 g of molecular biology-grade agarose and add 100 mL of 1× TBE. Microwave to dissolve the agarose and immediately cast the gel. Let the gel cool for at least 1 h before use.

2.3 Additional Reagents and Equipment Required

1. Tobacco etch virus protease (TEV), 1 mg/mL stock.
2. Nuclease-free water (1 mL).
3. Three gels of 12%, 15%, and 18% polyacrylamide SDS or similar.
4. Instant Blue Coomassie Protein Stain or similar.
5. Precision Plus Protein dual color standard (BioRad) or similar.
6. Synthetic DNA oligomers labeled with an appropriate fluorophore (i.e., 5′- Cy5-ATGATGTCTTCTCA**AGATTCGA**TAAAAATGGTGTTG-3′) and (5′-CAACACACCATTTTTA**TCGAATCT**TGAGAAGACATCAT-3′). The LUX/EC binding site is shown in bold (*see* **Note 3**).
7. Recombinant plasmid DNA with the genes of interest (i.e., pESPRIT vectors encoding 6× His-ELF3-mCherry, 6× His-LUX and 6× His-ELF4).
8. Beckman Coulter Avanti J-26 XP centrifuge with the rotor JLA 8.1000 or similar.
9. Beckman Coulter Avanti J-26 XP centrifuge with the rotor JA-25.25 or similar.
10. Sonicator (Branson Sonifer 250) or similar.
11. Spin concentrators (Amicon Ultra 4, Ultracel-10 k, or similar).
12. Dialysis tubing MWCO 3500 Dalton (Spectra/Por™) or similar.
13. Superdex 200 Increase 10/300 GL or similar (Cytiva) or similar.

14. AktaPure or Biorad FPLC protein purification system or similar.
15. Polyacrylamide gel rig.
16. Agarose gel rig.
17. NanoDrop microvolume spectrophotometer (ThermoFisher) or similar.
18. ChemiDoc (ThermoFisher) or similar.

3 Methods

All protein-DNA binding reactions should be carried out using buffers equilibrated to the target temperature of interest unless otherwise specified. Pipette tips, Eppendorf tubes, and liquid media should be sterilized by autoclaving and all buffers and stock solutions filtered prior to use.

3.1 Protein Expression

1. Transform 50 μL of BL21(DE3) Rosetta cells with the genes of interest (*ELF3, LUX*, and *ELF4*) in the pESPRIT expression vector according to the manufacturer's instructions (i.e., gently thaw an aliquot of cells in a 1 mL Eppendorf tube on wet ice and add 1 μL of purified plasmid DNA at 20–50 ng/μL, tap the side of the tube to mix, and keep on ice for 15 min. Perform a heat shock at 42 °C for 30–45 s in a heating block or PCR machine followed by 2 min on ice and a 60 min incubation at 37 °C in 500 μL of LB medium. Plate the transformed cells onto an agar plate supplemented with 50 μg/mL kanamycin and 34 μg/mL chloramphenicol and incubate overnight at 37 °C (*see* **Note 2**).
2. In a 500 mL Erlenmeyer flask, prepare 100 mL of sterile LB media supplemented with 50 μg/mL kanamycin and 34 μg/mL chloramphenicol. Select a single colony from the agar plates of transformed cells (step 1) by gently touching with a sterile wand and swishing through the LB media. Perform this step for each individual construct to be expressed.
3. Grow the bacteria overnight at 37 °C in a shaker at 150–180 rpm.
4. Prepare 3× 1 L of LB media per target construct (1 L of LB in a 2 L or larger Erlenmeyer flask) and supplement with a final concentration of 50 μg/mL kanamycin and 34 μg/mL chloramphenicol. Inoculate each flask with 20 mL of the overnight culture from step 3.
5. Grow the bacteria at 37 °C, 150 rpm, until an optical density A600 = 0.6–0.8 is reached. Reduce the temperature to 18 °C

and induce protein expression by the addition of 1 mM IPTG (final concentration). Grow the culture overnight.

6. Transfer the liquid culture to a 1 L centrifuge container and pellet the cells at 4500 × g for 20 min.

3.2 Native Protein Purification (LUX and ELF4)

1. Resuspend the bacterial pellets in 50–100 mL lysis buffer.
2. Sonicate the sample in an ice bucket with an ice water mixture (1 s pulse, 1 s pause 2× 5 min total processing time, 70% power).
3. Spin the pellets at 45000 × g for 30 min.
4. Prepare the Ni-NTA column by adding 2 mL of Ni-NTA slurry (75% w/w Ni resin) for a final column volume of 1–1.5 mL to a 30 mL gravity column. Wash the resin with 50 column volumes (CV) of ultrapure water and equilibrate in 50 CV of wash buffer.
5. Load the supernatant onto the nickel affinity column pre-equilibrated with lysis buffer.
6. Wash with 50 CV wash buffer.
7. Wash with 50 CV high-salt wash buffer.
8. Elute the protein in 500 μL steps in elution buffer and collect each fraction in a 2 mL tube.
9. Run a denaturing SDS gel in Laemmli buffer and stain with Instant Blue or Coomassie stain or similar to visualize the eluted protein fractions.
10. Combine the fractions of interest and determine the concentration of protein using a NanoDrop spectrophotometer or Bradford assay.
11. Dialyze the fractions of interest against dialysis buffer.
12. Optional: Add 1:10 w/w TEV and incubate overnight to cleave the 6× His tag.
13. Optional: Perform a TEV-cleaved and uncleaved protein depletion by running the protein/TEV solution over a pre-equilibrated Ni-NTA column, collect the flow through, add 3 mL of dialysis buffer and collect in the same tube. This will be 6× His-tag cleaved protein depleted in TEV protease.
14. Concentrate the protein to 5–10 mg/mL using a spin concentrator.
15. Equilibrate a Superdex 200 in Dialysis SEC-buffer.
16. Inject the protein onto the column and collect 1 mL fractions using an automatic fraction collector.
17. Run a denaturing 15% SDS gel of the peak fractions in Laemmli buffer and stain with Instant Blue or Coomassie or similar to assess the purity of the sample.

18. Select the fractions of interest and concentrate the proteins to 5–10 mg/mL. The proteins can be stored at 4 °C overnight or at −80 °C for long-term storage.

3.3 ELF3-mCherry Purification from Inclusion Bodies

1. Resuspend the bacterial pellets in nondenaturing resuspension buffer.
2. Sonicate the sample in an ice bucket with an ice water mixture (1 s pulse, 1 s pause, 2× 5 min total processing time, 70% power).
3. Spin the pellets at 45000 × g for 30 min.
4. Decant the supernatant.
5. Resuspend the pellet in 30 mL lysis buffer.
6. Spin the pellets at 45000 × g for 30 min.
7. Decant the supernatant und resuspend the pellet in 20 mL of denaturing resuspension buffer.
8. Spin the pellets at 45000 × g for 30 min.
9. Load the supernatant onto a 1 mL nickel affinity column pre-equilibrated in denaturing resuspension buffer.
10. Wash with 50 CV denaturing resuspension buffer.
11. Elute the protein in 500 μL steps in denaturing elution buffer and collect each fraction in a 1.5 mL tube.
12. Run a denaturing 12% SDS gel in Laemmli buffer and stain with Instant Blue or Coomassie or similar to assess the purity of the sample.
13. Combine the ELF3-mCherry containing fractions (they should be pink) (*see* **Note 4**). Determine the concentration using a NanoDrop spectrophotometer or Bradford assay.

3.4 EC Reconstitution

1. Mix ELF3-mCherry: LUX: ELF4 in a ratio of 1:1:1.2 to a final volume of at least 100 μL (*see* **Notes 5** and **6**).
2. Dialyze the proteins against 250 ml complex reconstitution buffer 1 for at least 1 h at 4 °C, with slow stirring.
3. Dialyze the proteins against 250 mL complex reconstitution buffer 2 for at least 1 h at 4 °C, with slow stirring.
4. Dialyze the proteins against 250 mL complex reconstitution buffer 3 for at least 1 h at 4 °C, with slow stirring.
5. Dialyze the proteins against 250 mL complex reconstitution buffer 4 for at least 1 h at 4 °C, with slow stirring.
6. Pierce the dialysis membrane with a micropipette tip and pipette the protein solution into a fresh 1.5 mL reaction tube.

3.5 Annealing of Single-Stranded DNA Oligos

1. Resuspend Cy5-labeled single-stranded DNA oligos to a concentration of 200 μM in nuclease-free water.
2. Mix equimolar amounts (100 μM) of the forward and reverse oligomers in a reaction tube, seal the tube with parafilm, cover the sample with aluminum foil to reduce damage to the fluorophore by ambient light.
3. Boil 500 mL water in a 1 L glass beaker in the microwave, and place it onto a heating plate. Place the DNA sample into a floating microtube rack and heat it in the beaker for 5 min.
4. Switch of the heating plate and allow the 500 mL of water to cool to room temperature on the benchtop. The double-stranded DNA (dsDNA) stock solution can be stored at − 20 °C.
5. Dilute the dsDNA to 10 μM. This will be used for preparing the EMSA reaction.

3.6 Protein-DNA Complex Binding Reaction

1. Prepare 200 μL of 2× EMSA binding reaction buffer by combining 40 μL of 10× binding buffer with 0.8 μL of dsDNA probe and 159.2 μL of nuclease-free water.
2. Mix 10 μL of the 2× EMSA binding reaction buffer mix with varying amounts of protein or protein complex (i.e., 0.5 μM, 1 μM, 2 μM) to a final volume of 20 μL (sufficient for running 2 nondenaturing gels in parallel).
3. Incubate 1 h at the desired temperatures (i.e., 4 °C and 20 °C).

3.7 Electrophoretic Mobility Shift Assay Using Nondenaturing Polyacrylamide Gels

1. Rinse the wells of the 5% nondenaturing polyacrylamide gel with 1× TBE.
2. Pre-run the gel in 1× TBE at 30 V for at least 30 min (up to 1.5 h) at the desired temperature with buffer at the same experimental temperature (i.e., 4 °C and 20 °C).
3. Load 9 μL of the reaction mix into each gel (i.e., if running 2 gels in parallel at 2 different temperatures, divide the reaction mix equally between the gels for comparison) (*see* **Note 7**).
4. Run the gel at 100 V for 5 min to ensure the samples enter the gel (Fig. 1).
5. Run the gel at 75 V for 75 min at 4 °C or 55 min at 20 °C. Adjust times to result in similar migration distances (*see* **Note 8**).

3.8 Electrophoretic Mobility Shift Assays Using Nondenaturing Agarose Gels

1. Load 9 μL of the reaction mix into each well of the nondenaturing agarose gel.
2. Run the gel at 50 V for 60 min at 4 °C or 30 min at 20 °C. Adjust times to result in similar migration distances (Fig. 2).

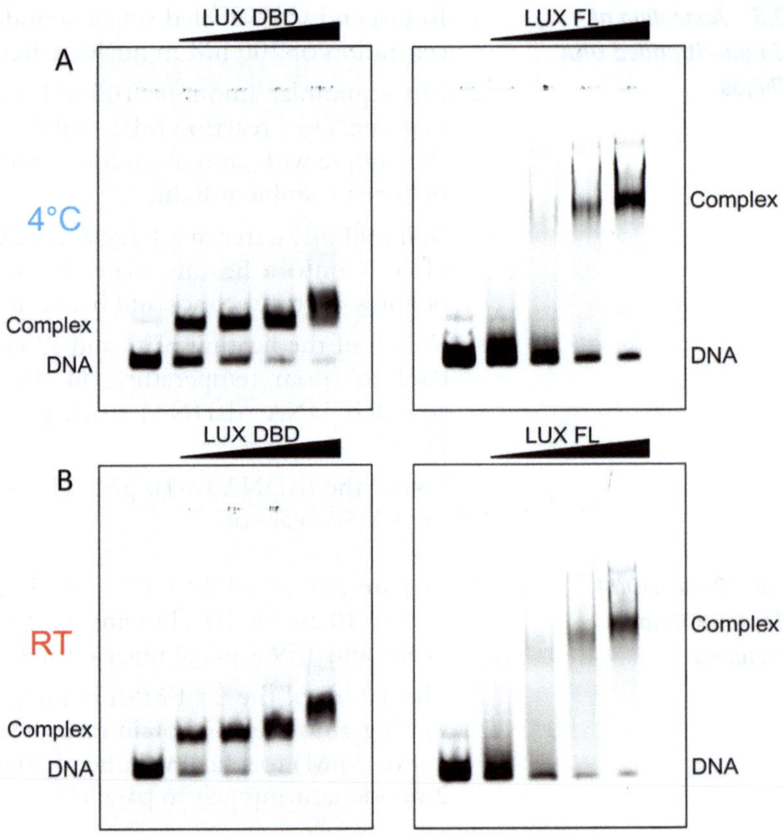

Fig. 1 Polyacrylamide EMSA of LUX DNA binding domain (DBD) and LUX full length (FL) protein. (**a**) Left, EMSA run at 4 °C with increasing concentrations of LUX DBD from 0 to 1 μM. Right, EMSA run at 4 °C with increasing concentrations of LUX FL from 0 to 1 μM nM. (**b**) Reactions as per A run at room temperature (RT). No temperature responsiveness in DNA binding is detected between the two experiments for LUX DBD and LUX FL. Free DNA and protein bound DNA are indicated

3.9 Visualization and Quantification of DNA Binding

1. Place the gel directly on the tray of the ChemiDoc or similar fluorescence imager and image the gel at the appropriate excitation wavelength (i.e., 650 nm for Cy5) (*see* **Note 9**).
2. Optional: Built-in ChemiDoc software such as Image Lab software may be used for quantification. Alternatively, ImageJ may be used for band quantification and K_d calculation provided the protein concentrations were determined for each reaction (*see* **Note 10**).

4 Notes

1. Choice of cell line for bacterial expression is important depending on the presence of rare codons. We have found that using cell lines with additional tRNAs such as Rosetta(DE3)

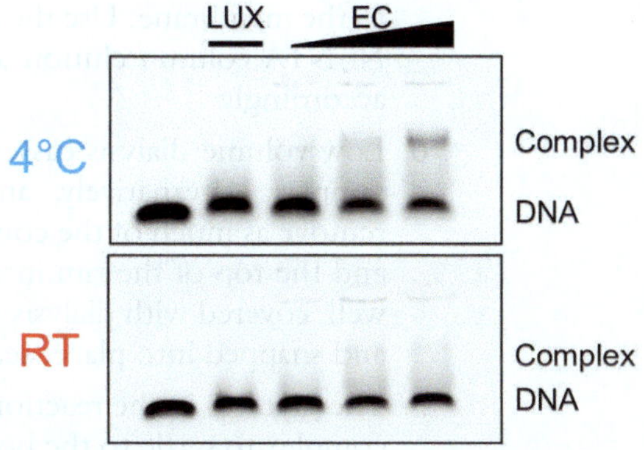

Fig. 2 Agarose EMSA of EC performed at two different temperatures. Free DNA, LUX full length, and the reconstituted EC (lanes 4–6) at increasing concentrations were run on a 2% agarose gel. At 4 °C, a band corresponding to EC-DNA is visible, whereas at room temperature (RT), this band no longer appears. All reactions were from the same master mix, equally divided and incubated at the gel running temperature for 1 h. LUX-DNA appears as a smear due to the agarose gel which allows diffusion of the small complex. EC-DNA appears as a distinct band

(Novagen) and BL-21 Codon Plus (Agilent) can be helpful in the expression of plant proteins. Alternatively, synthetic genes encoding the TF of interest with codons optimized for bacterial expression may be used.

2. When preparing nondenaturing gels, it is important to use an equipment that has not been in contact with SDS, since small amounts of SDS carryover from denaturing SDS-PAGE can interfere with the EMSA.
3. HPLC-purified DNA oligomers may be ordered with a covalently attached fluorophore directly from the manufacturer. These are generally of high quality and may be used directly without further purification. Alternatively, DNAs may be ordered unlabeled and derivatization with a fluorophore of interest. Generally, this requires a separate HPLC purification step prior to use to remove excess free dye.
4. mCherry is not fully denatured in 8 M urea and the mCherry tagged protein will generally be pink during purification. Longer-term storage at 4 °C in 8 M urea will result in gradual reduction of the pink color due to unfolding of the mCherry. This is also observable for other fluorescent protein tags such as GFP.
5. The concentration of EC will be limited by the concentration of ELF3-mCherry. We have found that ELF3-mCherry does not concentrate well using spin concentrators due to adherence

to the membrane. Use the most concentrated fraction from the Ni-NTA column elution and calculate LUX and ELF4 ratios accordingly.

6. Low volume dialysis cassettes may be used to dialyze 100 μL samples. Alternatively, an Eppendorf tube may be cut to remove as much of the conical tube as possible, leaving the cap and the top of the rim intact. The sample is applied to the lid well, covered with dialysis tubing and the top of the rim closed and snapped into place, sealing the dialysis membrane.
7. The glycerol in the reaction buffer will allow the protein-DNA complex to settle to the bottom of the gel well. Loading gels is difficult as no colored stain is added. Pipette slowly to minimize sample diffusion from the well.
8. Samples will migrate faster with increasing temperature. If similar migration distances are desired for the same samples run at different temperatures, running times will need to be adjusted with longer times required for low temperature experiments. Generally, for 20 °C differences in running temperature, times will need to be increased 1.5× for the lower temperature experiment at the same running voltage.
9. If the migration of the protein-DNA complex is not sufficient, the gel may be returned to the gel rig and a voltage applied for a longer running time.
10. Due to the denaturation and refolding steps for EC formation and difficulties in quantifying complex and free LUX concentrations, EC-DNA experiments are performed using a master mix of all components divided into separate reactions and incubated at the desired temperatures with all gels run at the same time and in parallel to allow for binding comparisons at different temperatures.

Acknowledgments

SH and CZ received support from the ANR (ANR-19-CE20-0021 and ANR-21-CE11-0037) and GRAL, a program from the Chemistry and Biology Health Graduate School of the University Grenoble Alpes (ANR-17-EURE-0003).

References

1. Babu MM, Luscombe NM, Aravind L et al (2004) Structure and evolution of transcriptional regulatory networks. Curr Opin Struct Biol 14:283–291
2. Robertson G, Hirst M, Bainbridge M et al (2007) Genome-wide profiles of STAT1 DNA association using chromatin immunoprecipitation and massively parallel sequencing. Nat Methods 4:651–657
3. Mukhopadhyay A, Deplancke B, Walhout AJM et al (2008) Chromatin immunoprecipitation (ChIP) coupled to detection by quantitative

real-time PCR to study transcription factor binding to DNA in Caenorhabditis elegans. Nat Protoc 3:698–709

4. Meers MP, Bryson TD, Henikoff JG et al (2019) Improved CUT&RUN chromatin profiling tools. Elife 8:e46314
5. Skene PJ, Henikoff S (2017) An efficient targeted nuclease strategy for high-resolution mapping of DNA binding sites. elife 6:e21856
6. Skene PJ, Henikoff JG, Henikoff S (2018) Targeted in situ genome-wide profiling with high efficiency for low cell numbers. Nat Protoc 13:1006–1019
7. Kaya-Okur HS, Wu SJ, Codomo CA et al (2019) CUT&tag for efficient epigenomic profiling of small samples and single cells. Nat Commun 10:1930
8. O'Malley RC, Huang SC, Song L et al (2016) Cistrome and epicistrome features shape the regulatory DNA landscape. Cell 165:1280–1292
9. Bartlett A, O'Malley RC, Huang SC et al (2017) Mapping genome-wide transcription-factor binding sites using DAP-seq. Nat Protoc 12:1659–1672
10. Li, M., Lin, W., Hinckley, W., et al. (2022). Uncovering the "ZIP code" for bZIP dimers reveals novel motifs, regulatory rules and one billion years of cis-element evolution, https://www.biorxiv.org/content/10.1101/2022.04.17.488518v1
11. Lai X, Vega-Léon R, Hugouvieux V et al (2021) The intervening domain is required for DNA-binding and functional identity of plant MADS transcription factors. Nat Commun 12:4760
12. Lai X, Stigliani A, Lucas J et al (2020) Genome-wide binding of SEPALLATA3 and AGAMOUS complexes determined by sequential DNA-affinity purification sequencing. Nucleic Acids Res 48:9637–9648
13. Hutin S, Blanc-Mathieu R, Rieu P et al (2023) Identification of plant transcription factor DNA-binding sites using seq-DAP-seq. In: Kaufmann K, Vandepoele K (eds) Plant gene regulatory networks: methods and protocols. Springer, New York, pp 119–145
14. Smaczniak C, Angenent GC, Kaufmann K (2017) SELEX-Seq: A method to determine DNA binding specificities of plant. Transcription Factors 1629:67–82
15. Hadži S, Lah J (2022) Analysis of protein-DNA interactions using isothermal titration calorimetry: successes and failures. Methods Mol Biol 2516:239–257
16. Layton CJ, Hellinga HW (2011) Quantitation of protein–protein interactions by thermal stability shift analysis. Protein Sci 20:1439–1450
17. Mueller AM, Breitsprecher D, Duhr S et al (2017) MicroScale thermophoresis: a rapid and precise method to quantify protein-nucleic acid interactions in solution. Methods Mol Biol 1654:151–164
18. Folta-Stogniew E (2021) Characterization of protein-nucleic acid complexes by size-exclusion chromatography coupled with light scattering, absorbance, and refractive index detectors. Methods Mol Biol 2263:381–395
19. Anderson BJ, Larkin C, Guja K et al (2008) Using fluorophore-labeled oligonucleotides to measure affinities of protein-DNA interactions. Methods Enzymol 450:253–272
20. Hellman LM, Fried MG (2007) Electrophoretic mobility shift assay (EMSA) for detecting protein-nucleic acid interactions. Nat Protoc 2: 1849–1861
21. Ezer D, Jung JH, Lan H et al (2017) The evening complex coordinates environmental and endogenous signals in Arabidopsis. Nature Plants 3:17087
22. Silva CS, Nayak A, Lai X et al (2020) Molecular mechanisms of evening complex activity in Arabidopsis. Proc Natl Acad Sci USA 117:6901–6909
23. Nusinow DA, Helfer A, Hamilton EE et al (2011) The ELF4-ELF3-LUX complex links the circadian clock to diurnal control of hypocotyl growth. Nature 475:398–402
24. Choi H, Oh E (2016) PIF4 integrates multiple environmental and hormonal signals for plant growth regulation in Arabidopsis. Mol Cells 39:587–593
25. Jung J-H, Barbosa AD, Hutin S et al (2020) A prion-like domain in ELF3 functions as a thermosensor in Arabidopsis. Nature 585:256–260
26. Hutin S, Kumita JR, Strotmann VI et al (2023) Phase separation and molecular ordering of the prion-like domain of the Arabidopsis thermosensory protein Early Flowering 3. Proc Natl Acad Sci USA 120:e2304714120

Chapter 15

Monitoring Real-Time Temperature Dynamics of a Short RNA Hairpin Using Förster Resonance Energy Transfer and Circular Dichroism

Martin Balcerowicz, Philip A. Wigge, Marco Di Antonio, and Betty Chung

Abstract

RNA molecules play crucial roles in gene expression regulation and cellular signaling, and these functions are governed by the formation of RNA secondary and tertiary structures. These structures are highly dynamic and subject to rapid changes in response to environmental cues, temperature in particular. Thermosensitive RNA secondary structures have been harnessed by multiple organisms to survey their temperature environment and to adjust gene expression accordingly. It is thus highly desirable to observe RNA structural changes in real time over a range of temperatures. Multiple approaches have been developed to study structural dynamics, but many of these require extensive processing of the RNA, large amounts of RNA input, and/or cannot be applied under physiological conditions. Here, we describe the use of a dually fluorescently labeled RNA oligonucleotide (containing a predicted hairpin structure) to monitor subtle RNA structural dynamics in vitro by Förster resonance energy transfer (FRET) and circular dichroism (CD) spectroscopy. These approaches can be employed under physiologically relevant conditions over a range of temperatures and with RNA concentrations as low as 200 nM; they enable us to observe RNA structural dynamics in real time and to correlate these dynamics with changes in biological processes such as translation.

Key words RNA structures, Hairpins, RNA thermometer, Circular dichroism, FRET

1 Introduction

RNA is a fundamental building block of life and plays a key role in the process of protein synthesis: messenger RNAs (mRNAs) provide the genetic information to be translated into proteins, transfer RNAs (tRNAs) deliver amino acids for protein assembly, and ribosomal RNAs (rRNAs) form an integral part of the ribosome that catalyzes this reaction. Beyond their direct involvement in translation, RNAs exert diverse regulatory functions, e.g., in the context of epigenetic regulation, splicing, and RNA interference. The functional diversity of RNA is ultimately based upon its ability to form a

Meng Chen (ed.), *Thermomorphogenesis: Methods and Protocols*, Methods in Molecular Biology, vol. 2795,
https://doi.org/10.1007/978-1-0716-3814-9_15,

plethora of sophisticated and dynamic secondary and tertiary structures such as hairpins, apical and internal loops, 3-way-junctions, or pseudoknots [1]. These structures can convey catalytic activity, serve as molecular scaffolds, and provide or occlude binding sites for RNA-binding proteins (RBPs) and other molecules. In mRNA, the formation of such structures is known to affect RNA stability and processing [2], alternative splicing [3], subcellular localization [4], and translation regulation [5].

The formation of secondary and tertiary structures is highly dependent on the thermodynamic landscape of a specific RNA sequence: while each molecule has the potential to adopt a multitude of distinct conformations, only the most thermodynamically and kinetically favorable will be populated under any given condition. Some highly structured RNA molecules, such as tRNAs or RNAs in the ribosome and spliceosome, show little change in their structure across cellular conditions. Other RNAs however will undergo substantial structural rearrangements as their cellular environment changes, and these rearrangements can confer specific functionality [6, 7]. Rearrangements can be brought about by the binding of additional molecules: riboswitches for instance change their conformation and exert regulatory function upon binding of small ligands. A well-studied example is the *E. coli* TPP riboswitch, a secondary structure in the 5′-UTR of thiamin biosynthesis genes. This structure has the capacity to bind thiamine pyrophosphate (TPP), which triggers a structural rearrangement that occludes the ribosome binding site and thereby reduces translation in a negative feedback mechanism [8].

RNA secondary structures are also highly sensitive to physical parameters such as temperature and osmolarity. Given that RNA folding is directly dependent on a structure's thermodynamic free energy, temperature is perhaps the most immediate factor to affect structure formation. Consequently, changes in temperature can be sensed by thermosensitive RNA secondary structures, which can in turn affect gene expression in a temperature-dependent manner. Such structures—referred to as RNA thermometers—are commonly found in bacteria; they often consist of one or multiple hairpins that melt or change conformation as temperature rises, altering access to the ribosome-binding site [9, 10]. A classic example of an RNA thermometer is found in the *prfA* transcript of the pathogenic bacteria *Listeria monocytogenes*, where it controls virulence in response to temperature: at temperatures up to 30 °C, a hairpin structure forms and prevents access to the ribosome binding site; when temperature rises to 37 °C (the body temperature of the human host), this structure partially melts, thereby exposing the ribosome binding site to enable translation [11]. Similar zipper-like RNA thermometers have been found to control translation in various prokaryotes, but thermosensitive RNA structures have recently also been discovered in eukaryotic systems. In the model flowering plant *Arabidopsis thaliana*, a rise in temperature from 22 to 27 °C triggers a conformational change in the 5′-UTR of the transcript encoding *PHYTOCHROME INTERACTING FACTOR 7 (PIF7)*; a more relaxed RNA conformation is adopted at higher temperatures, which enhances translation [12].

There can be no doubt that determining RNA structural dynamics is essential to understand RNA's biological function. Multiple biochemical and biophysical strategies have been developed to monitor such structural changes, and each comes with its own set of advantages and drawbacks. Chemicals or enzymes that react selectively with single- or double-stranded RNA can be employed to investigate secondary structure both in vitro and in vivo. In combination with next-generation sequencing, these methods now allow for probing RNA structure transcriptome-wide, determining the global RNA structure [13, 14]. However, these approaches are irreversible, leaving the RNA molecules permanently modified and preventing real-time observation of the same molecules' dynamics over a range of conditions. Biophyiscal methods such as nuclear magnetic resonance (NMR) spectroscopy and X-ray crystallography on the other hand require substantial amounts of RNA input and preclude observation under physiological conditions, while cryo electron microscopy still suffers from limited resolution and is thus currently restricted to larger structures [13]. In our characterization of plant RNA thermoswitches [12], we employed a combination of Förster resonance energy transfer (FRET) measurements and circular dichroism (CD) spectroscopy to observe RNA hairpin dynamics using a dually labeled RNA oligonucleotide (containing the predicted hairpin structure). FRET efficiency varies as a function of the fluorophores' distance; high efficiency can thus be correlated to a stable hairpin structure, while a reduction in FRET efficiency reflects a partial opening of the hairpin or a destablization of this structure.

The same RNA sequence can also be used for CD spectroscopy to directly observe global changes of RNA secondary structure at a given temperature (i.e., hairpin folding/unfolding). The combination of these approaches allowed us to monitor RNA structural dynamics over a range of temperatures in real time. A detailed protocol of these methods is outlined below.

2 Materials

2.1 Consumables

1. 1.5 mL Eppendorf tubes (*see* **Note 1**).
2. Semi-micro quartz cell 108F-QS (Sigma-Aldrich, Hellma Analytics, catalog number: HL108-F-10-40).
3. 1 mm quartz cell 350 mL (Sigma-Aldrich, Hellma Analytics, catalog number: HL110-F-1-40).

2.2 Chemicals and Solutions

1. RNase-free water.
2. 1 M KOH solution: Dissolve 5.61 g of KOH in 100 mL of RNase-free water (*see* **Note 2**).
3. 60 mM cacodylate buffer pH 7.4: Dissolve 8.280 g of cacodylic acid in 500 mL of RNase-free water (*see* **Notes 3** and **4**). Adjust pH by dropwise addition of 1 M KOH while stirring the solution (*see* **Note 5**). Measure the pH after each drop addition until it stabilizes at 7.4, which is the final pH required for both CD and FRET experiments. Add RNase-free water to the cacodylate pH 7.4 buffer to a total volume of 1 L and store at 4 °C for long-term usage (*see* **Note 6**).
4. RNA oligonucleotide with 5′-FAM and 3′-TAMRA fluorescent probes (5′-FAM-AAG AGA GCU UAA UUG UCA GUU UAU UCU CUG-3′-TAMRA in case of the *PIF7* hairpin, *see* **Note 7**) (IBA Gmbh, HPLC Purified). Store at 4 °C.

2.3 Equipment

1. 2 L glass beaker.
2. Stirring bar.
3. SevenCompact pH meter S220 (Mettler Toledo, model: 30019029).
4. −80 °C freezer.
5. Varian Cary Eclipse Fluorescence Spectrometer fitted with a Varian temperature control unit (Agilent, Varian, Cary Eclipse Fluorimeter).
6. ThermoMixer C (Eppendorf, model: EP5382000015) fitted with SmartBlock, 1.5 mL (Eppendorf, model: EP5361000031).
7. Applied Photophysics Chirascan fitted with temperature control unit (Applied Photophysics, ChirascanTM) for CD spectroscopy.

2.4 Software

1. Cary Eclipse Kinetics software (Agilent, https://www.agilent.com/en/product/molecular-spectroscopy/fluorescence-spectroscopy/fluorescence-software/cary-eclipse-software).
2. GraphPad Prism 8 (GraphPad Software, https://www.graphpad.com/scientific-software/prism/).

3 Methods

3.1 RNA Annealing

1. Resuspend the dually labeled RNA oligonucleotide in the appropriate volume of RNase-free water to obtain a stock solution with the final concentration of 100 μM. To achieve this, add a volume of water that is 10× with respect to the total numbers of RNA nanomoles. For example, for 5 nmol of RNA,

add 50 μL of RNase free water. Store this solution at −80 °C for long-term usage.

2. Dilute RNA as follows:
   (a) For FRET measurements, dilute 1 μL of the 100 μM RNA stock solution in 499 μL (200 nM final concentration, *see* **Note 8**) of 60 mM pH 7.4 cacodylate buffer in a 1.5 mL Eppendorf tube.
   (b) For CD measurements, dilute 10 μL of the 100 μM RNA stock solution in 190 μL (5 μM final concentration) of 60 mM pH 7.4 cacodylate buffer in a 1.5 mL Eppendorf tube.
3. Anneal the RNA solution prepared in Subhdeading 3.1.1 by heating the Eppendorf tube containing the solution at 90 °C for 10 min using the Eppendorf Thermomixer equipped with the 1.5 mL SmartBlock (*see* **Note 9**). Prepare an ice box while the RNA solution is heating and immediately place the RNA solution in ice after the 10 min heating cycle is completed.
4. Store the RNA solution at 4 °C overnight before performing any FRET or CD measurements (*see* **Note 10**).

3.2 FRET Protocol

In this experiment, it is possible to gain insight into the distance between the 5′ and the 3′ ends of the RNA sequence using FRET. To achieve this, excite the FAM fluorophore (placed at the 5′-end of the sequence) at 488 nm and record the emission between 500 and 700 nm, covering the emission peaks of both the FAM and the TAMRA (placed at the 3′ end of the sequence) fluorophore (*see* **Note 11**).

1. Transfer 500 μL of the RNA solution prepared in Subheading 3.1, **step 2a** in the Semi-micro quartz cell 108F-QS cuvette and insert it in the Cary Eclipse Fluorimeter pre-equilibrated at a temperature of 17 °C (*see* **Note 12**).
2. After equilibrating the cuvette in the fluorimeter for 10 min, record the fluorescence emission using the following parameters:
   Excitation wavelength 488 nm.
   Emission range 500–700 nm.
   Excitation slit 5 nm.
   Emission slit 5 nm.
   Detector Voltage High
3. Increase the temperature of the Cary Variant heating block at 27 °C (*see* **Note 12**), and once the temperature has been reached, equilibrate the sample for at least 10 min.

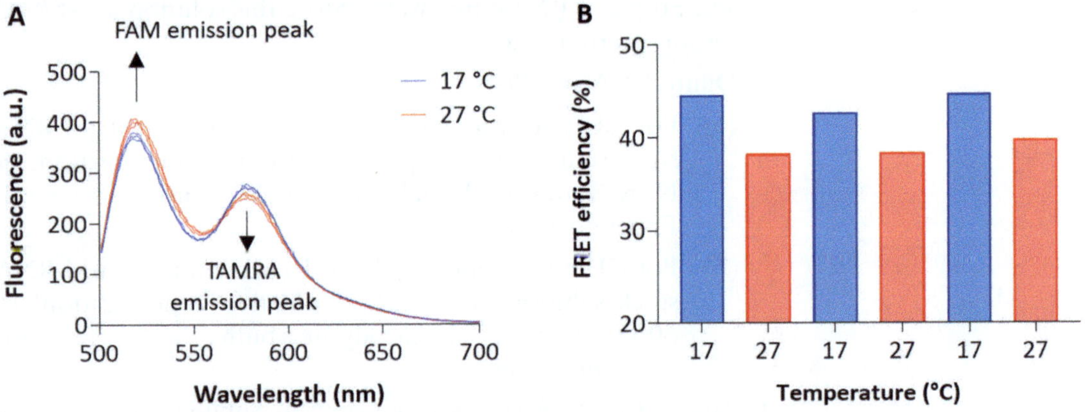

Fig. 1 Temperature-dependent hairpin dynamics measured as changes in FRET efficiency. A 5′-FAM and 3′-TAMRA fluorescently labeled 32 nt RNA oligonucleotide harboring the hairpin sequence was used for the experiment. Emission spectra between 500 and 700 nm (**a**) were recorded upon excitation at 488 nm over multiple shifts between 17 and 27 °C; peak emissions at 520 and 578 nm were used to calculate FRET efficiencies (**b**). FRET efficiency is calculated using the following equation: (Fa/(Fa + Fd)) × 100, where Fa is the emission of the acceptor (FAM) and Fd is the emission intensity of the donor (TAMRA)

4. Record the fluorescence emission using the following parameters:

   Excitation wavelength 488 nm.

   Emission range 500–700 nm.

   Excitation slit 5 nm.

   Emission slit 5 nm.

   Detector Voltage High

5. Repeat Subheading 3.2, **step 2–4** as appropriate. Due to the RNA's relatively low stability, this cycle can be repeated for a maximum of 12–14 times (*see* **Note 13**). Representative spectra are displayed in Fig. 1a.

3.3 CD Measurement

In this experiment, the overall secondary structure adopted by the RNA under different conditions will provide a characteristic CD signature that can be monitored in real time. Spectra with a maximum at ~270 nm and a minimum at ~210 nm are indicative of classical RNA hairpin structures.

1. Transfer 200 μL of the RNA prepared in Subheading 3.1, **step 2b** into a 1 mm quartz cell and place it in the ChirascanTM instrument pre-equilibrated at the temperature of 17 °C (*see* **Note12**).
2. After equilibrating the cuvette in the circular dichroism machine for at least 10 min, record the CD intensity (reported as molar ellipticity θ) between 300 and 180 nm using the following parameters:

   Record 1 point per nm.

   Bandwidth 0.5 nm.

   Time per point 1 s

3. Acquire the CD spectra at 17 °C for three times consecutively.
4. Raise the ChirascanTM to 27 °C (*see* **Note 12**) and equilibrate the sample at this temperature for at least 10 min.
5. Record the CD intensity between 300 and 180 nm using the following parameters:
   Record 1 point per nm.
   Bandwidth 0.5 nm.
   Time per point 1 s
6. Acquire the CD spectra at 27 °C for three times consecutively.
7. Repeat Subheading 3.3, **step 1–6** with 200 μL of 60 nM cacodylate buffer to record a baseline.

3.4 Data Analysis

1. For FRET measurements, divide the recorded emission at 578 nm (TAMRA emission peak) by the sum of the recorded emission at 578 nm and 520 nm (FAM emission peak) for each temperature (Fig. 1b); this ratio is inversely correlated with the distance between the two fluorophores (i.e., the ends of the RNA oligonucleotide). A reduction in FRET efficiency thus reflects a (partial) opening or destablization of the hairpin structure.
2. For CD spectroscopy, average the three replicate measurements for each temperature to reduce background noise and subtract the baseline recorded at the respective temperature. Plot the baseline-corrected CD intensity versus wavelength and overlay the two spectra obtained in Subheading 3.3, **step 3** and **6** to visualize structural differences between the RNA hairpin structures at 17 and 27 °C (Fig. 2). Hairpin-type characteristic signals display a detectable maximum at ~270 nm and a minimum ~210 nm, with a more pronounced peak ~210 nm suggesting an increased stability of the hairpin structure.

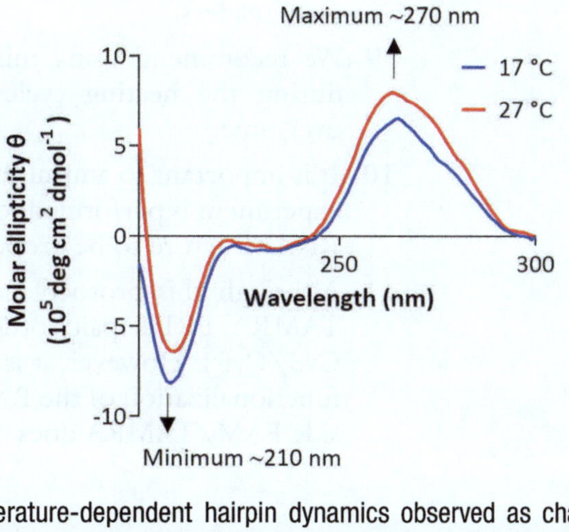

Fig. 2 Temperature-dependent hairpin dynamics observed as changes in CD spectra. CD spectra of the hairpin-containing RNA oligonucleotide were recorded at 17 and 27 °C

4 Notes

1. We recommend that RNase-free labware be used throughout the protocol.
2. Given the highly exothermic reaction caused by KOH dissolving in water, it is recommended to perform this operation in an ice bath.
3. Appropriate PPE should be worn while performing buffer preparation, including nitrile gloves, corrosive-resistant lab coat, and safety goggles.
4. It is important to ensure the powder is fully solubilized before proceeding with pH adjustment. If needed, use additional water up to a total volume of 700 mL to ensure full solubilization of the cacodylic acid.
5. A standard laboratory pH-meter can be used to measure pH. Calibration of the pH-meter should have been performed the same day.
6. Buffer storage at 4 °C should not exceed 3 months. After this time, preparation of fresh buffer is strongly recommended. Cacodylic acid contains arsenic and therefore is toxic and should be disposed in dedicated heavy metal waste collection.
7. The RNA sequence should cover the entire predicted hairpin-forming sequence but should not exceed more than three nucleotides beyond this sequence on either side to ensure efficient FRET.
8. The FRET assay is concentration independent; therefore, different concentrations of RNA can be used to perform this protocol. This might be needed to optimize the signal depending on the type of fluorimeter used or to adapt the protocol for plate readers.
9. We recommend using mineral oil to seal the RNA solution during the heating cycles as this will prevent evaporation over time.
10. It is important to anneal the RNA sequence the day before the experiment is performed to allow sufficient time for the correct RNA structure to be properly folded.
11. Although this protocol has been optimized for the FAM/-TAMRA FRET pair, other fluorophores can be used (i.e., Cy3/Cy5). However, it is important to verify by CD that the functionalization of the RNA sequence with fluorophores outside FAM/TAMRA does not affect hairpin formation.

12. The assay temperatures of 17 and 27 °C used in this protocol may be adjusted to the experimenter's temperature of interest.
13. The number of FRET cycles is also limited by the fluorophores getting bleached; it is important to use fresh batches of RNA for each FRET experiment.

Acknowledgments

MB and his group are supported by the Royal Society (URF\R1\211672). PW is supported by the Leibniz Association and the European Research Council through an Advanced Grant (TIP-TOP, ERC Adv 101021246). MDA is supported by Biotechnology and Biological Sciences Research Council (BBSRC) David Phillips Fellowship [BB/R011605/1]; MDA is a Lister Institute Fellow and is supported by a Lister Institute Research Prize. BYWC is supported by grants from the Medical Research Council [MR/R021821/1], BBSRC [BB/X001261/1, BB/V017780/1 and BB/V006096/1], and the Royal Society [RGS\R2\192222]. This protocol was initially published in bio-protocols (https://doi.org/10.21769/BioProtoc.3950).

References

1. Cruz JA, Westhof E (2009) The dynamic landscapes of RNA architecture. Cell 136:604–609. https://doi.org/10.1016/j.cell.2009.02.003
2. Soemedi R, Cygan KJ, Rhine CL et al (2017) The effects of structure on pre-mRNA processing and stability. Methods 125:36–44. https://doi.org/10.1016/j.ymeth.2017.06.001
3. Bartys N, Kierzek R, Lisowiec-Wachnicka J (2019) The regulation properties of RNA secondary structure in alternative splicing. Biochim Biophys Acta (BBA) 1862:194401. https://doi.org/10.1016/j.bbagrm.2019.07.002
4. Lazzaretti D, Bono F (2017) mRNA localization in metazoans: A structural perspective. RNA Biol 14:1473–1484. https://doi.org/10.1080/15476286.2017.1338231
5. Jacobs E, Mills JD, Janitz M (2012) The role of RNA structure in posttranscriptional regulation of gene expression. J Genet Genomics 39:535–543. https://doi.org/10.1016/j.jgg.2012.08.002
6. Ganser LR, Kelly ML, Herschlag D, Al-Hashimi HM (2019) The roles of structural dynamics in the cellular functions of RNAs. Nat Rev Mol Cell Biol 20:474–489. https://doi.org/10.1038/s41580-019-0136-0
7. Mustoe AM, Brooks CL, Al-Hashimi HM (2014) Hierarchy of RNA functional dynamics. Annu Rev Biochem 83:441–466. https://doi.org/10.1146/annurev-biochem-060713-035524
8. Winkler W, Nahvi A, Breaker RR (2002) Thiamine derivatives bind messenger RNAs directly to regulate bacterial gene expression. Nature 419:952–956. https://doi.org/10.1038/nature01145
9. Kortmann J, Narberhaus F (2012) Bacterial RNA thermometers: molecular zippers and switches. Nat Rev Microbiol 10:255–265. https://doi.org/10.1038/nrmicro2730
10. Krajewski SS, Narberhaus F (2014) Temperature-driven differential gene expression by RNA thermosensors. Biochim Biophys Acta (BBA) 1839:978–988. https://doi.org/10.1016/j.bbagrm.2014.03.006
11. Johansson J, Mandin P, Renzoni A et al (2002) An RNA Thermosensor controls expression of virulence genes in listeria monocytogenes. Cell

110:551–561. https://doi.org/10.1016/S0092-8674(02)00905-4
12. Chung BYW, Balcerowicz M, Di Antonio M et al (2020) An RNA thermoswitch regulates daytime growth in Arabidopsis. Nat Plants 6: 522–532. https://doi.org/10.1038/s41477-020-0633-3
13. Mailler E, Paillart J-C, Marquet R et al (2019) The evolution of RNA structural probing methods: from gels to next-generation sequencing. WIREs RNA 10:e1518. https://doi.org/10.1002/wrna.1518
14. Spitale RC, Incarnato D (2023) Probing the dynamic RNA structurome and its functions. Nat Rev Genet 24:178–196. https://doi.org/10.1038/s41576-022-00546-w

Part III

Temperature Signaling in Thermomorphogenesis

Chapter 16

Characterization of PIF4 Phosphorylation by SPA1

Sanghwa Lee and Enamul Huq

Abstract

The PHYTOCHROME INTERACTING FACTORs (PIFs) play pivotal roles in regulating thermo- and photo-morphogenesis in *Arabidopsis*. One of the main hubs in thermomorphogenesis is PIF4, which regulates plant development under high ambient temperature along with other PIFs. PIF4 enhances its own transcription and PIF4 protein is stabilized under high ambient temperature. However, the mechanisms of thermo-stabilization of PIF4 are less understood. Recently, it was shown that SUPPRESSOR OF PHYA-105 1 (SPA1) can function as a serine/threonine kinase to phosphorylate PIF4 in vitro, and the phosphorylated form of PIF4 is more stable under high ambient temperature conditions. In this chapter, we describe the in vitro kinase assay of PIF4 by SPA1. In principle, this protocol can be applied for other putative substrates and kinases.

Key words Phytochrome interacting factor 4, In vitro kinase assay, SUPPRESSOR OF PHYA-105 1 (SPA1), Protein purification, *Arabidopsis thaliana*

1 Introduction

The reversible protein phosphorylation regulates almost all aspects of cellular physiology and has been implicated in most, if not all, signaling pathways [1]. An addition of a phosphate by a kinase and removal of the phosphate by a phosphatase provide a reversible mechanism in regulating the activity, subcellular localization, and/or abundance of a protein. Mutation in a kinase or a phosphatase or the phosphorylation site in a target protein can abrogate cellular signaling pathways. Plants encode a large number of kinase and phosphatase genes in their genome [2, 3]; yet, the biological and molecular functions and the cellular targets of many of these crucial enzymes are still unknown.

Plants are acutely sensitive to their surrounding environment and have evolved intricate mechanisms to perceive and respond to multiple cues [4]. Two of the most important cues are light and

Meng Chen (ed.), *Thermomorphogenesis: Methods and Protocols*, Methods in Molecular Biology, vol. 2795,
https://doi.org/10.1007/978-1-0716-3814-9_16,

ambient temperature. Plants perceive and respond to both signals through a sensory photoreceptor system called phytochromes (phy). In the phy signaling pathways, PHYTOCHROME INTERACTING FACTORs (PIFs) were first identified as downstream proteins that directly interact with phyA and phyB. PIFs are degraded by the red/far-red light receptors phyA and phyB upon direct protein-protein interactions to accelerate photomorphogenesis under red light [4]. Eight PIF homologs were found in *Arabidopsis* so far, most of them function not only in an overlapping manner, but also they have their own distinct roles despite high degree of sequence homology among them [5, 6].

Although most PIFs contribute to regulating thermomorphogenesis in an overlapping manner [7–9], PIF4 has been considered as a main hub in thermomorphogenesis for more than a decade [10]. Majority of the thermomorphogenesis phenotypes are regulated by PIF4 including hypocotyl and petiole lengths, early flowering, stomatal development, and hypocotyl thickening [11–13]. PIF4 is transcriptionally activated and PIF4 protein is stabilized in response to high ambient temperature in a manner similar to that of other PIFs [7, 13–15]. Since the stabilization of PIF4 at high ambient temperature is a typical molecular indicator of thermomorphogenesis, finding components that regulate PIF4 stability is critical for better understanding thermomorphogenesis.

Recently, SUPPRESSOR OF PHYA-105 1 (SPA1) has been shown to have kinase activity toward PIF1 and PIF4 [16, 17]. Mutation in a conserved amino acid residue in the kinase domain (R517E; Fig. 1a) that drastically reduced and/or abolished the phosphorylation of PIF1 and PIF4 in vitro, suggesting that SPA1 is required to phosphorylate PIFs in vitro. Further analysis in *Arabidopsis* showed that wild-type SPA1 overexpressed in *spaQ* mutant was able to stabilize PIF4 protein due to PIF4 phosphorylation. However, mutated $SPA1^{R517E}$ overexpressed in *spaQ* background failed to complement the *spaQ* phenotypes under high ambient temperature [16]. These data suggest that phosphorylation of PIF4 by SPA1 kinase plays a crucial role in regulating thermomorphogenesis.

Here, we describe the method for the in vitro kinase assay of PIF4 by SPA1 [16]. The substrate is not limited to PIFs, any candidate target of SPA1 such as HY5 could be tested as recently described [18]. Although, the in vitro kinase assay method provides direct evidence of phosphorylation, the in vitro data still needs to be validated in vivo in combination with phenotypic analysis to provide significance of phosphorylation.

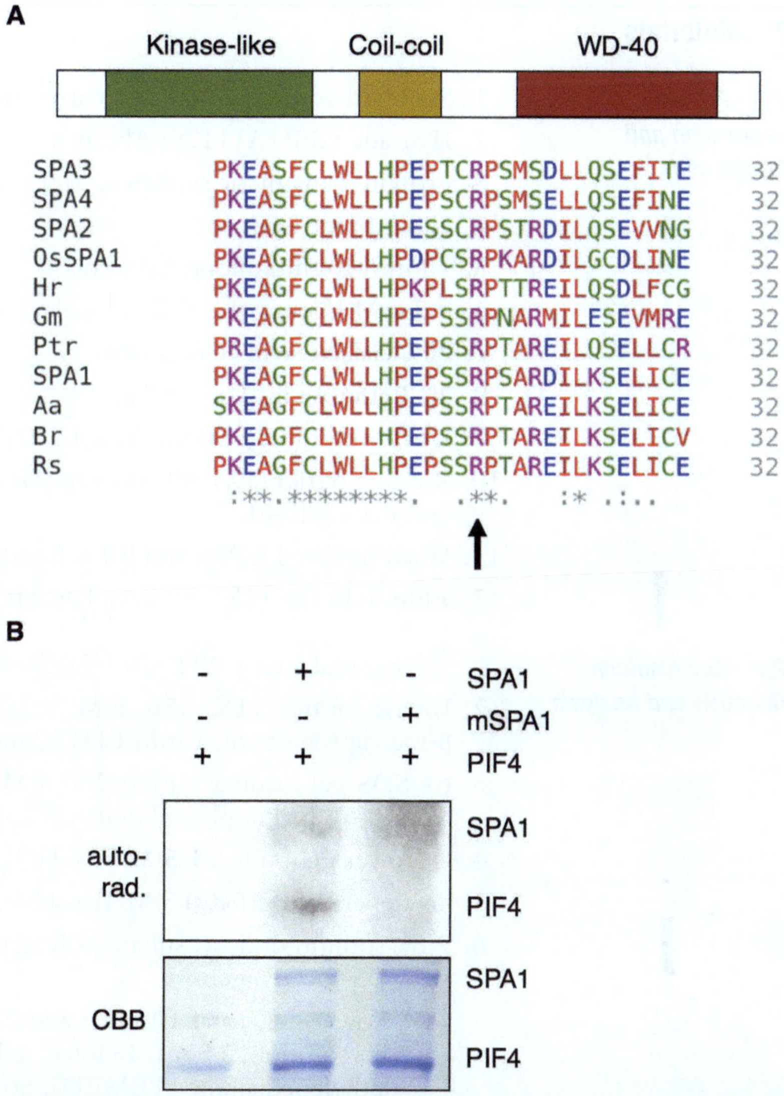

Fig. 1 In vitro kinase assay between SPA1 and PIF4. (**a**) Schematic diagram showing protein domains in SPA proteins (top panel). (Bottom panel) Sequence alignment of a kinase domain showing an arginine residue (arrow) which is crucial for kinase activity of SPA1 and is conserved among multiple plant species. *Os* Oryza sativa, *Gm* Glycine max*; Aa* Arabis alpine, *Br* Brassica rapa, *Ptr* Populus trichocarpa, *Hu* Herrania umbratical, *Rs* Raphanus sativus. Reproduced in part from Ref. (17). (**b**) The autoradiograph (top panel) shows that SPA1 could phosphorylate PIF4, while mutated SPA1 could not phosphorylate PIF4 using the in vitro kinase assay. The amount of the protein used in the assay is shown by Coomassie Blue-stained gel (bottom panel). Reproduced in part from Ref. (16)

2 Materials

2.1 Protein Expression and Purification

1. Standard equipment for incubator and centrifugation system.
2. IP grade GFP (A11120; Abcam).
3. Protein A magnetic sepharose xtra (GE).
4. Glutathione agarose.
5. 10 mM L-glutathione reduced.
6. 0.5 mM Isopropyl ß-D-1-thiogalactopyranoside (IPTG).
7. 20% Galactose in -Ura media.
8. 2% Raffinose in -Ura media.
9. Lysis buffer (1X): 50 mL Tris–HCl, pH 7.5, 0.1% NP-40.
10. Binding buffer (1X): 50 mM Tris–HCl pH 7.5, 200 mM NaCl, and 0.1% NP-40.
11. Wash buffer: 1X PBS and 0.1% Tween 20.
12. Elution buffer (1X): 50 mM Tris–HCl pH 7.5, EDTA 1 mM.

2.2 Radioisotope Reaction and Analysis

1. ^{32}P radiolabeled γ-ATP (BLU502A; Perkin Elmer).
2. Kinase buffer (1X): 50 mM Tris–HCl, pH 7.5, 4 mM β-mercaptoethanol, 1 mM EDTA, and 10 mM $MgCl_2$.
3. 6× SDS gel loading buffer: 300 mM Tris–HCl, pH 6.8, 12% SDS, 0.6% bromophenol blue, 60% glycerol, 600 mM DTT.
4. 4× lower gel buffer: 1.5 M Tris–HCl, pH 8.8, 0.4% SDS.
5. 4× upper gel buffer: 0.5 M Tris–HCl, pH 6.8, 0.4% SDS.
6. 7.8% sodium dodecyl sulfate-polyacrylamide gel electrophoresis (SDS-PAGE) gel:
   (a) Lower gel (total 10 mL): add 2.65 mL 30% acrylamide-Bis solution, 2.5 mL 4× lower gel buffer, 5 μL tetramethylethylenediamine (TEMED), 50 μL 10% ammonium persulfate (APS), 4.8 mL H_2O.
   (b) Upper gel (total 5 mL): add 0.83 mL 30% acrylamide-Bis solution, 1.25 mL 4× lower gel buffer, 5 μL TEMED, 50 μL 10% APS, 2.87 mL H_2O.
7. Standard equipment to run and dry SDS-PAGE gels.
8. Typhoon imaging system (GE Healthcare Life Sciences).
9. Coomassie Blue staining buffer: 0.1% Coomassie R-250 in 40% ethanol, 10% acetic acid.
10. Destaining buffer: 10% acetic acid, 50% methanol, and 40% H_2O.

3 Methods

3.1 PIF4 Protein Expression and Purification

1. Inoculate 10 μL of GST-PIF4 containing *E.coli* strain cell stock to 5 mL LB media in 14 mL culture tube at 37 °C overnight (*see* **Note 1**).
2. Put the seed culture solution into 200 mL LB media in 1 L flask at 37 °C.
3. Put 0.5 mM isopropyl ß-D-1-thiogalactopyranoside (IPTG) to the media at OD 0.6 ~ 1.
4. Incubate the flask at 28 °C overnight.
5. Pour the media from the flask to Nalgene bottle.
6. Centrifuge at 15,000× *g* for 30 min.
7. Discard the supernatant.
8. Gently dissolve the pellet with lysis buffer and pour the solution to 50 mL falcon tube.
9. Sonicate the solution using sonicator according to manufacturer's protocol.
10. Place the glutathione resin to the open column and wash twice with binding buffer.
11. The bacterial extract expressing GST-PIF4 was incubated with glutathione resin in the binding buffer for 2 h.
12. Wash glutathione resin for more than five times with wash buffer.
13. Elute samples were collected using elution buffer with 10 mM reduced glutathione.
14. Aliquot the samples and store in −80 °C deep freezer.

3.2 SPA1 Protein Expression and Purification

1. Inoculate 10 μL of SPA1-GFP and mSPA1-GFP containing *Saccharomyces* cell stock to 5 mL -Ura media with 2% raffinose in 14 mL culture tube at 30 °C overnight (*see* **Note 2**).
2. Put the seed culture solution into 100 mL -Ura media with 2% raffinose in 500 mL flask at 30 °C.
3. When OD reaches 0.6~1, centrifuge at 4000 rpm for 10 min.
4. Dissolve the pellet with 20% galactose solution.
5. Incubate at 30 °C overnight.
6. Centrifuge at 2000× *g* for 20 min.
7. Using liquid nitrogen, grind the pellet thoroughly.
8. Put the ground pellet into the 1.5 mL tube.
9. Add 10 μL of GFP antibody with 50 uL of Dynabead and rotate for 1 h.
10. Place the tube into magnetic plate and wait for 1 min.

11. Add wash buffer and wash the Dynabead for three times.
12. Remove the supernatant completely.
13. Use the Dynabead containing tube for further analysis (*see* **Note 3**).

3.3 Radioisotope Labeling

1. Put 15 μL of kinase buffer into the Dynabead-containing tube.
2. Put 5 μL of GST-PIF4 to the kinase buffer.
3. Add 5 μL of 32P radiolabeled γ-ATP to the sample and incubate at 30 °C for 30 min.
4. Add 8 μL of 6X SDS sample buffer and boil it for 5 min.

3.4 Analysis with SDS-PAGE

1. Samples were loaded onto a 7.8% SDS-PAGE gel depending on the size of the protein.
2. Use 100 V/20 mA to run the gel for typically 2 h to make the blue dye reach at the end of the gel.
3. Dry the gel on a gel dryer for 2 h at 80 °C (*see* **Note 4**).
4. Place the gel into the PhosphorImaging overnight.
5. The next day, scan the image using a phosphorimager (*see* **Note 5**). An example result is shown in Fig. 1b.
6. Add Coomassie Blue staining buffer to the gel for 1 h.
7. Wash the gel with destaining buffer three times for every 10 min.
8. Take the image for the loading control.

4 Notes

1. For PIF4-GST expression, pGEX4T-1 vector was used. Any type of tag can be used for this experiment.
2. For SPA1/mSPA1-GFP cloning, pYES2 vector was used. Any type of yeast expression vector with a tag can be used for this experiment.
3. For the experiment, we recommend purifying the PIF4-GST protein prior to SPA1/mSPA1-GFP purification so that you could do the reaction by putting PIF4-GST in the SPA1/mSPA1-GFP containing Dynabead.
4. To prevent damaging the phosphoscreen, wrap the gel with a plastic wrap.
5. Typhoon FLA phosphorimager was used for the imaging. The typical scanning parameters are following: laser: 650 nm, filter: IP, PMT: 950 V, pixel size: 25 nm.

Acknowledgments

We thank members of the Huq laboratory for critical reading of the manuscript. This work was supported by a grant from the National Science Foundation (MCB-2014408) to E.H.

References

1. Stark M (2004) Protein phosphorylation and dephosphorylation. In: Richard J, Dickinson MS (eds) The metabolism and molecular physiology of saccharomyces cerevisiae. Taylor & Francis Group, pp 284–375
2. Hardie DG (1999) Plant protein serine/threonine kinases: classification and functions. Annu Rev Plant Physiol Plant Mol Biol 50(1): 97–131. https://doi.org/10.1146/annurev.arplant.50.1.97
3. Luan S (2003) Protein phosphatases in plants. Annu Rev Plant Biol 54(1):63–92. https://doi.org/10.1146/annurev.arplant.54.031902.134743
4. Paik I, Kathare PK, Kim J-I, Huq E (2017) Expanding roles of PIFs in signal integration from multiple processes. Mol Plant 10(8): 1035–1046
5. Pham VN, Kathare PK, Huq E (2018) Phytochromes and phytochrome interacting Factors. Plant Physiol 176:1025–1038. https://doi.org/10.1104/pp.17.01384
6. Oh J et al (2020) Phytochrome interacting factor 8 inhibits Phytochrome A-mediated far-red light responses in Arabidopsis. Plant Cell 32(1):186–205. https://doi.org/10.1105/tpc.19.00515
7. Lee S, Zhu L, Huq E (2021) An autoregulatory negative feedback loop controls thermomorphogenesis in Arabidopsis. PLoS Genet. https://doi.org/10.1101/2021.01.04.425188
8. Chung BYW et al (2020) An RNA thermoswitch regulates daytime growth in Arabidopsis. Nat Plants 6(5):522–532. https://doi.org/10.1038/s41477-020-0633-3
9. Fiorucci A-S et al (2020) Phytochrome interacting factor 7 is important for early responses to elevated temperature in Arabidopsis seedlings. New Phytol 226:50–58. https://doi.org/10.1111/nph.16316
10. Koini MA et al (2009) High temperature-mediated adaptations in plant architecture require the bHLH transcription factor PIF4. Curr Biol 19(5):408–413. https://doi.org/10.1016/j.cub.2009.01.046
11. Kumar SV et al (2012) Transcription factor PIF4 controls the thermosensory activation of flowering. Nature 484:242–245. http://www.nature.com/nature/journal/vaop/ncurrent/abs/nature10928.html#supplementary-information
12. Lau OS et al (2018) Direct control of SPEECHLESS by PIF4 in the high-temperature response of stomatal development. Curr Biol 28(8):1273–1280 e1273. https://doi.org/10.1016/j.cub.2018.02.054
13. Lee S, Wang W, Huq E (2021) Spatial regulation of thermomorphogenesis by HY5 and PIF4 in Arabidopsis. Nat Commun 12(1): 3656. https://doi.org/10.1038/s41467-021-24018-7
14. Oh E, Zhu J-Y, Wang Z-Y (2012) Interaction between BZR1 and PIF4 integrates brassinosteroid and environmental responses. Nat Cell Biol 14:802–809
15. Stavang JA et al (2009) Hormonal regulation of temperature-induced growth in Arabidopsis. Plant J 60(4):589–601. https://doi.org/10.1111/j.1365-313X.2009.03983.x
16. Lee S, Paik I, Huq E (2020) SPAs promote thermomorphogenesis by regulating the phyB-PIF4 module in *Arabidopsis*. Development 147(19):dev189233. https://doi.org/10.1242/dev.189233
17. Paik I et al (2019) A phyB-PIF1-SPA1 kinase regulatory complex promotes photomorphogenesis in Arabidopsis. Nat Commun 10(1): 4216. https://doi.org/10.1038/s41467-019-12110-y
18. Wang W et al (2021) Direct phosphorylation of HY5 by SPA kinases to regulate photomorphogenesis in Arabidopsis. New Phytol 230(6): 2311–2326. https://doi.org/10.1111/nph.17332

Chapter 17

Chromatin Immunoprecipitation to Investigate H2A.Z Dynamics in Response to Environmental Changes

J. Paola Saldierna Guzmán, Mark Zander, and Björn C. Willige

Abstract

DNA methylation and posttranslational modifications of histones instruct gene expression in eukaryotes. Besides canonical histones, histone variants also play a critical role in transcriptional regulation. One of the best studied histone variants in plants is H2A.Z whose removal from gene bodies correlates with increased transcriptional activity. The eviction of H2A.Z is regulated by environmental cues such as increased ambient temperatures, and current models suggest that H2A.Z functions as a transcriptional buffer preventing environmentally responsive genes from undesired activation. To monitor temperature-dependent H2A.Z dynamics, chromatin immunoprecipitation (ChIP) of H2A.Z-occupied DNA can be performed. The following protocol describes a quick and easy ChIP approach to study in vivo H2A.Z occupancy.

Key words Thermomorphogenesis, Chromatin, Histones, H2A.Z, Chromatin immunoprecipitation

1 Introduction

Like in all eukaryotes, a plant's genomic DNA exists in the form of chromatin, a complex of DNA and proteins. This complex is organized in nucleosomes, which are composed of an octamer of histone proteins and the DNA that is wrapped around it [1]. Four different histones form an octamer, including H2A that can be replaced by a histone variant called H2A.Z. This exchange of H2A by H2A.Z and vice versa allows the regulation of gene activity. Collectively, research on diverse environmental stimuli indicates that reduced H2A.Z levels within gene bodies of environmentally responsive genes correlate with increased transcription in *Arabidopsis thaliana* [2–8]. This also includes responses to increased ambient temperatures. It was first reported in 2010 that the

The original version of the chapter has been revised. A correction to this chapter can be found at https://doi.org/10.1007/978-1-0716-3814-9_24

Meng Chen (ed.), *Thermomorphogenesis: Methods and Protocols*, Methods in Molecular Biology, vol. 2795, https://doi.org/10.1007/978-1-0716-3814-9_17,

Arabidopsis H2A.Z landscape is highly responsive to elevated ambient temperatures and that *arp6* (*actin-related protein 6*) mutants phenocopy thermomorphogenic responses. These mutants are deficient in an integral subunit of a chromatin remodeling complex which leads to defects in H2A.Z incorporation [9]. Furthermore, the same authors and their colleagues identified temperature-dependent dynamics between H2A.Z occupancy and DNA-binding by PIF4 (PHYTOCHROME INTERACTING FACTOR 4). Here, PIF4 binds to the *FT* (*FLOWERING LOCUS T*) locus to promote temperature-induced flowering [10]. This model was subsequently extended to the elevated temperature-induced *YUCCA8* expression which promotes hypocotyl elongation [11].

In humans as well as in yeast, H2A.Z is evicted by the chromatin remodeling complex INO80 [12, 13]. This complex is also present in plants; however, it was unclear how the perception of elevated temperatures is translated into positional information to guide INO80 to specific genes. This information could be provided by guiding transcription factors. Interestingly, it was found that the HSFA1 (HEAT SHOCK TRANSCRIPTION FACTOR A1) clade of heat shock factors causes the eviction of H2A.Z at its target genes [14]. Furthermore, a PIF-INO80 module was identified that regulate elongation growth after the detection of competing plant neighbors. Plants sense these neighbors by perceiving a reduced red to far-red light ratio (low R:FR) [15]. The low R:FR response pathway shares many overlapping components with temperature signaling. One of these overlapping factors is EEN (EIN6 ENHANCER), a small subunit of the chromatin remodeling complex INO80 [8]. In low R:FR, PIF7 recruits the INO80 complex by binding to EEN. This allows the removal of H2A.Z from PIF7 target genes through INO80 [7]. This transcription factor-mediated chromatin remodeler recruitment is also operational in the thermomorphogenesis pathway, where PIF4 associates with the INO80 complex to evict H2A.Z in order to promote hypocotyl elongation [16]. Altogether, these studies show the importance of H2A-H2A.Z dynamics for thermomorphogenesis.

The following protocol describes a quick and easy approach to study H2A.Z dynamics via chromatin immunoprecipitation (ChIP). The immunoprecipitation of H2A.Z-chromatin complexes and subsequent DNA purification allows the genome-wide identification of in vivo H2A.Z levels. This protocol can be modified by using antibodies against other histones, their variants, and modifications. We recommend to assess histone H3 levels in parallel to distinguish changes in H2A.Z levels from entire nucleosome disassembly or nucleosome sliding.

2 Materials

2.1 Tissue Collection and Cross-Linking

1. *Arabidopsis thaliana* seeds (*see* **Note 1**).
2. Tea bags or tea balls.
3. 1% formaldehyde in water (*see* **Note 2**).
4. 200 mM glycine in water.
5. Desiccator.
6. Vacuum pump.

2.2 Preparation of Magnetic Beads—Step a

1. Tube rotator.
2. Magnetic tube rack.
3. Dynabeads Protein G from Invitrogen (30 mg/mL).
4. BSA in PBS: 75 μM bovine serum albumin in 1X phosphate-buffered saline (PBS).
5. Anti-Htz1/Histone H2A.Z antibody (Active Motif Cat# 39647, RRID: AB_2793289).
6. Rabbit IgG, Jackson ImmunoResearch Cat# 011–000-003, RRID: AB_2337118.

2.3 Isolation of Nuclei

1. Liquid nitrogen.
2. Mortars and pestles.
3. Miracloth.
4. Nuclei Buffer 1 (NB1): 0.4 M sucrose, 10 mM Tris–HCl pH 8, 10 mM $MgCl_2$, 5 mM β-mercaptoethanol, 1 mM PMSF, 1X cOmplete Protease Inhibitor Cocktail from Roche.
5. Nuclei Buffer 2 (NB2): 0.25 M sucrose, 10 mM Tris–HCl pH 8, 10 mM $MgCl_2$, 1% Triton X-100, 5 mM β-mercaptoethanol, 1 mM PMSF, 1X Protease Inhibitor Cocktail.
6. Nuclei Buffer 3 (NB3): 1.7 M sucrose, 10 mM Tris–HCl pH 8, 2 mM $MgCl_2$, 0.15% Triton X-100, 5 mM β-mercaptoethanol, 1 mM PMSF, 1X Protease Inhibitor Cocktail.

2.4 Chromatin Extraction and Sonication

1. Multi-sample sonicator like Bioruptor Pico sonication device from Diagenode.
2. Fluorometer like Qubit fluorometer from Invitrogen.
3. Sonication Buffer (SB): 50 mM Tris–HCl pH 8, 10 mM EDTA, 1% SDS, 1 mM PMSF, 1X Protease Inhibitor Cocktail.

2.5 Preparation of Magnetic Beads—Step b

1. Tube rotator.
2. Magnetic tube rack.
3. BSA in PBS: 75 μM bovine serum albumin in 1X phosphate-buffered saline.

2.6 Chromatin Immunoprecipitation

1. Tube rotator.
2. Sonication Buffer (SB): 50 mM Tris–HCl pH 8, 10 mM EDTA, 1% SDS, 1 mM PMSF, 1X Protease Inhibitor Cocktail (*see* **Note 3**).
3. Dilution Buffer (DB): 1.7 M sucrose, 10 mM Tris–HCl pH 8, 2 mM $MgCl_2$, 0.15% Triton X-100, 5 mM β-mercaptoethanol, 1 mM PMSF, 1X Protease Inhibitor Cocktail.

2.7 Reverse Cross-Linking

1. Magnetic tube rack.
2. Shaking heat block.
3. Wash Buffer 1 (WB1): 50 mM Tris–HCl pH 7.4, 150 mM NaCl, 2 mM EDTA, 0.5% Triton X100.
4. Wash Buffer 2 (WB2): 50 mM Tris–HCl pH 7.4, 500 mM NaCl, 2 mM EDTA, 0.5% Triton X100.
5. Wash Buffer 3 (WB3): 50 mM Tris–HCl pH 7.4, 50 mM NaCl, 2 mM EDTA.
6. Elution Buffer (EB): 50 mM Tris–HCl pH 8, 10 mM EDTA, 1% SDS.
7. Proteinase K solution (20 mg/mL).

2.8 DNA Purification

1. 2 mL Phase Lock Gel tubes.
2. Phenol:Chloroform:Isoamyl alcohol (25:24:1).
3. Glycogen (5 mg/mL) from Invitrogen.
4. 4 M NaCl in water.
5. Pure ethanol.
6. Nuclease-free water.

3 Methods

Two to three days should be scheduled for this protocol. It consists of five main steps: cross-linking, nuclei isolation, chromatin sonication, the actual ChIP, and purification of DNA fragments. Formaldehyde is used to cause stable, but reversible, cross-links between proteins and DNA. After cross-linking, the tissue will be shock-frozen and can be long-term stored at −80 °C. Due to plant cell walls, fast and high-yielding nuclei isolation requires tissue grinding with liquid nitrogen. The isolated nuclei will be subsequently sonicated to break the nuclear envelope and to shear the chromatin into the desired fragment sizes. Chromatin fragments that contain H2A.Z will be precipitated for a couple of hours or overnight with a specific anti-H2A.Z antibody that is coupled to magnetic Protein G beads. Temperature-dependent reverse cross-linking and purification will yield in DNA fragments that can be stored at −20 °C.

These fragments can be subsequently used to prepare sequencing libraries or utilized in ChIP-PCR studies.

Like for any type of biological study, negative controls are crucial for the interpretation of the experimental outcome. We recommend the use of IgGs to determine the extent of nonspecific binding of DNA and DNA-protein complexes to the magnetic beads that are used for immunoprecipitation.

3.1 Tissue Collection and Cross-Linking (See Fig. 1)

1. Plate sterilized and stratified seeds on half-strength Murashige and Skoog (or half-strength Linsmaier and Skoog) medium containing 0.7% agar. As control condition, we recommend 21 °C. After 6 days of growth, transfer a subset of plates to the growth condition of interest (e.g., elevated temperature or low R:FR), while leaving control samples inside the control condition (*see* **Note 4**).
2. Harvest the plant material as quickly as possible after seedlings were exposed to the growth condition of interest. Collection should not take more than 3 min per sample (*see* **Note 5**).
3. Enclose each sample into a tea bag or tea ball. All subsequent steps of this section should be performed under a fume hood since formaldehyde is used for cross-linking.
4. Place tea balls/bags together in a beaker with 1% formaldehyde. If necessary, place a weight on top of tea balls/bags (*see* **Note 6**).
5. Apply vacuum for 10 min, release vacuum for 5 min, reapply vacuum for 5 min, release vacuum for another 5 min.

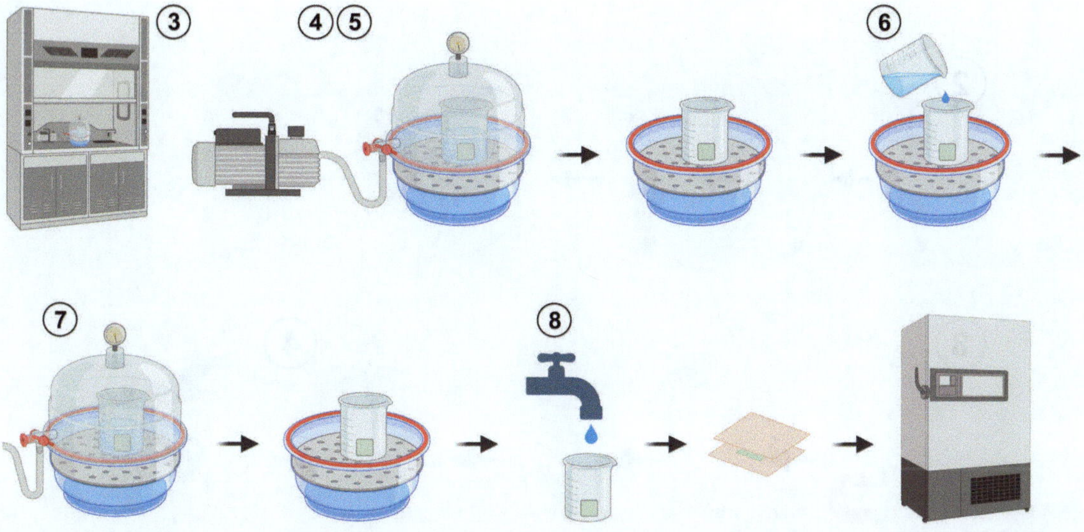

Fig. 1 Cross-linking. Numbers refer to the respective steps of this method section. (Figure was created with BioRender.com)

6. Stop cross-linking by exchanging the 1% formaldehyde solution for 200 mM glycine.
7. Apply vacuum for 10 min.
8. Rinse plant material thoroughly within the tea ball/bag with water. Take out the tissue and dry it between paper towels. Freeze plant material in liquid nitrogen. The cross-linked samples can be stored at −80 °C.

3.2 Preparation of Magnetic Beads—Step a (See Fig. 2)

1. Add 50 μL Protein G magnetic beads in a new 1.5 mL tube. This means, per biological sample and per antibody, one 1.5 mL tube needs to be prepared. Quickly wash beads twice with 800 μL BSA in PBS with the help of a magnetic tube rack.
2. Remove remaining solution with a pipette, add 800 μL BSA in PBS, and rotate tubes for 5 min.
3. Remove remaining solution with a pipette and add 300 μL BSA in PBS. Add 10 μL anti-H2A.Z antibody or 2 μL rabbit IgG, respectively, to each tube.
4. Leave tubes rotating at 4 °C until the start of Subheading 3.5. This allows the antibodies to bind to Protein G. Incubation with the antibodies should last at least for 2 h.

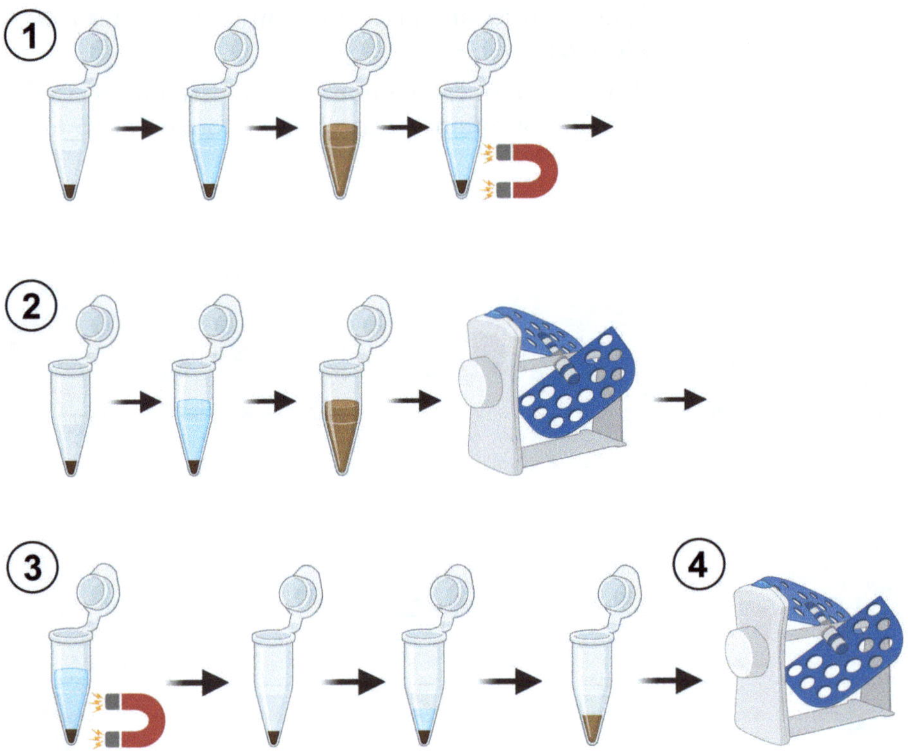

Fig. 2 Preparation of magnetic beads—step a. Numbers refer to the respective steps of this method section. (Figure was created with BioRender.com)

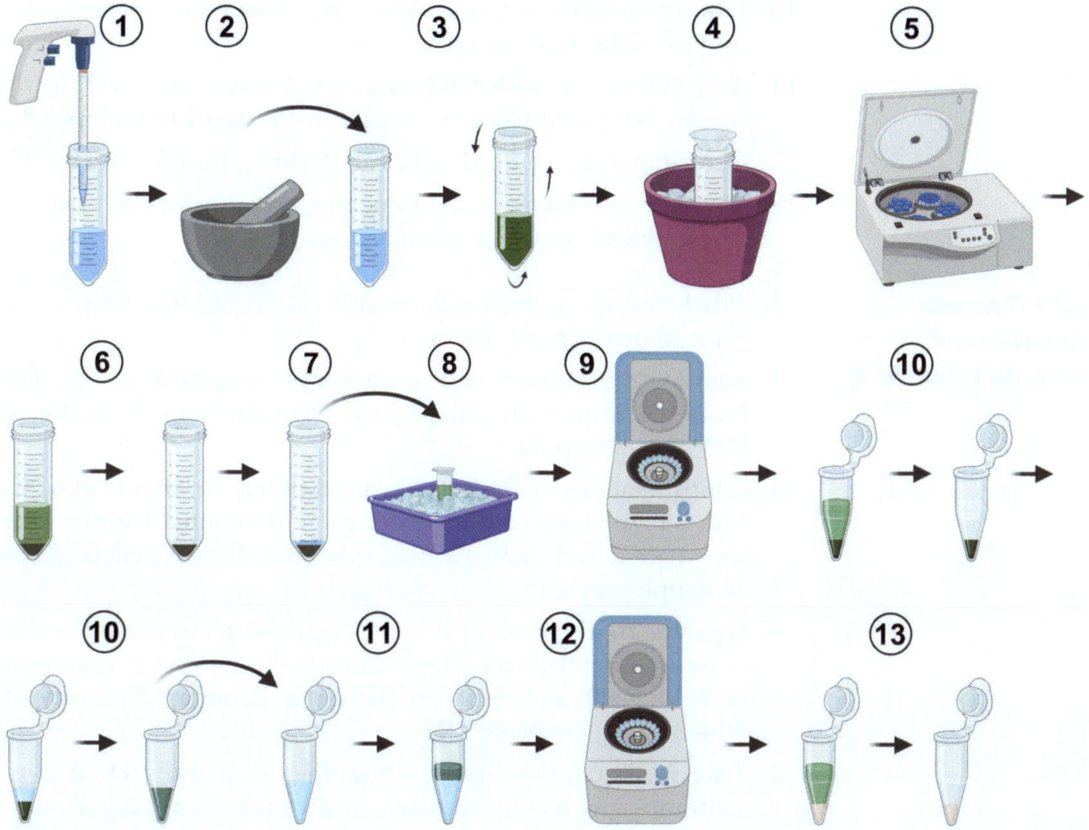

Fig. 3 Isolation of nuclei. Numbers refer to the respective steps of this method section. (Figure was created with BioRender.com)

3.3 Isolation of Nuclei (See Fig. 3)

1. Add 20 mL of cooled NB1 to a new 50 mL tube for each sample.
2. Grind the frozen tissue thoroughly in a mortar using liquid nitrogen.
3. Transfer powdered tissue to the 50 mL tube from **step 1**. Mix by inverting the tube slowly. Keep 50 mL tubes on ice. Repeat **steps 2** and **3** for each sample.
4. Filter each sample through a double layer of Miracloth and collect flow-through in a new 50 mL tube (*see* **Note 7**). Discard Miracloth. Keep all samples on ice.
5. Centrifugate the 50 mL tubes at 3,000 g for 20 min at 4 °C.
6. Discard supernatant, keep pellets in 50 mL tube on ice.
7. Slowly resuspend each pellet using 1 mL of cooled NB2.
8. Transfer suspension to a new 1.5 mL tube.
9. Centrifugate the 1.5 mL tubes at 18,000 g for 10 min at 4 °C.

10. Discard supernatant, and slowly resuspend each pellet using 500 μL NB3. Keep samples on ice.
11. Add 500 μL of cooled B3 to a new 1.5 mL tube and slowly overlay NB3 with the resuspension from **step 11** (*see* **Note 8**).
12. Centrifugate the 1.5 mL tubes at 18,000 g for 60 min at 4 °C.
13. Discard supernatant using a micropipette. Keep pellets on ice. These pellets contain the isolated nuclei.

3.4 Chromatin Extraction and Sonication (See Fig. 4)

1. Slowly resuspend each pellet in 300 μL SB and transfer suspensions to new 1.5 mL tubes.
2. Sonicate the samples in a multi-sample sonicator at 4 °C (*see* **Note 9**). Sonication should yield fragment sizes from 100 to 500 bp (*see* **step 4**).
3. Centrifugate the 1.5 mL tubes at 18,000 g for 30 min at 4 °C. The supernatants contain soluble chromatin. Transfer the supernatant to a new 1.5 mL tube and discard pellets. Keep all samples on ice.
4. Separate 20 μL per sample in a 1% agarose gel to assess the size of sonication products. The fragments should appear as a smear between 100 and 500 bp with the strongest intensity at 200–300 bp (*see* **Note 10**).
5. Take small aliquots from each sample to quantify DNA concentrations with a fluorometer. In the meantime, keep all samples on ice.

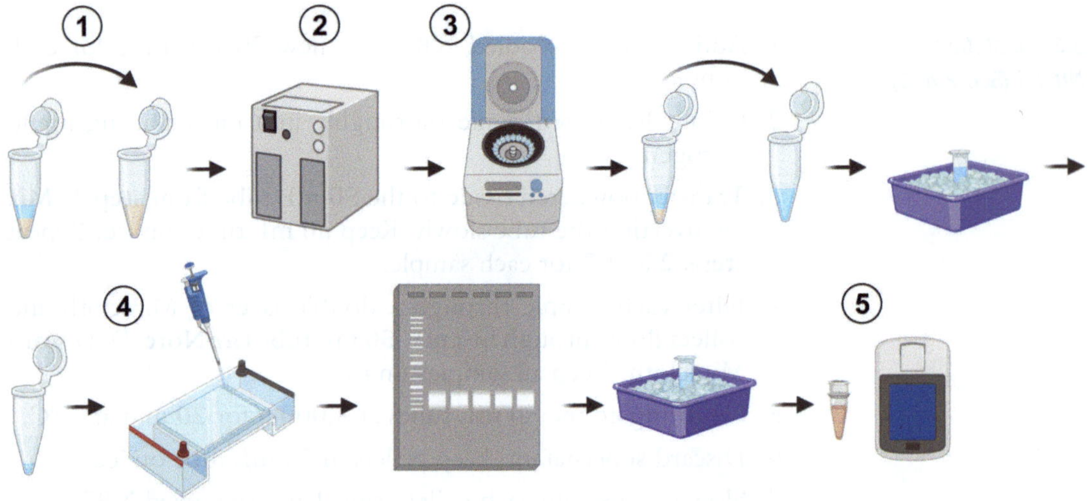

Fig. 4 Chromatin extraction and sonication. Numbers refer to the respective steps of this method section. (Figure was created with BioRender.com)

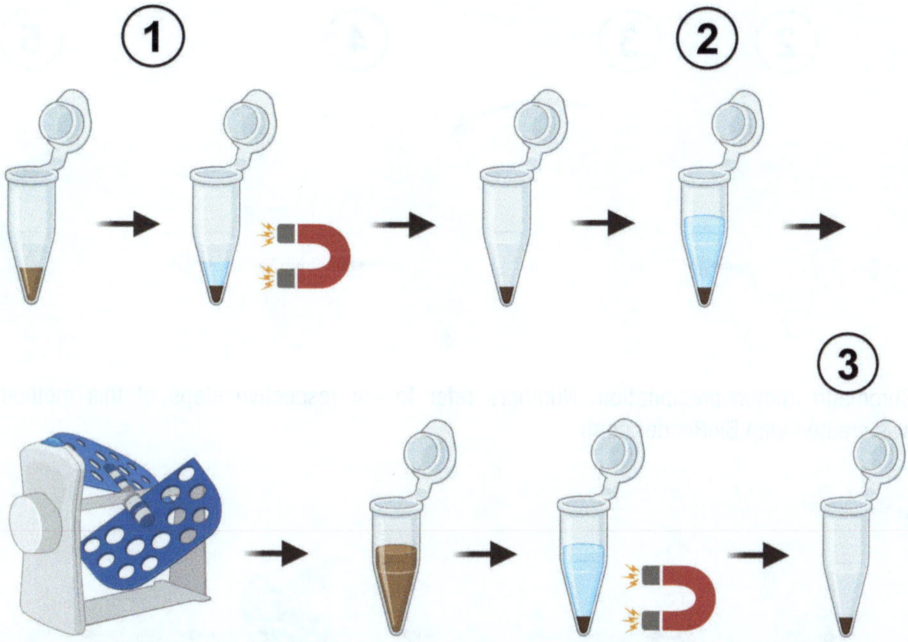

Fig. 5 Preparation of magnetic beads—step b. Numbers refer to the respective steps of this method section. (Figure was created with BioRender.com)

3.5 Preparation of Magnetic Beads—Step b (See Fig. 5)

1. Continue with tubes from Subheading 3.2: Remove supernatant with the help of a magnetic tube rack. Discard the supernatant.
2. Wash beads three times with 800 μL BSA in PBS with the help of a magnetic tube rack. For each step, rotate tubes for 5 min at 4 °C to wash magnetic beads thoroughly with BSA in PBS to remove any unbound antibody.
3. Remove and discard all supernatant after the last wash.

3.6 Chromatin Immunoprecipitation (See Fig. 6)

1. Depending on your DNA yield, use 4–8 μg of chromatin per sample for each ChIP. Add equal amounts of chromatin to new 1.5 mL tubes. This means, per biological sample and per antibody, one 1.5 mL tube needs to be prepared.
2. Fill each sample up to a total volume of 100 μL using SB to prepare samples with equal amounts of SDS.
3. Add 1 mL of cooled DB to each sample and mix to decrease the SDS concentration.
4. Add these 1.1 mL of each sample to the corresponding 1.5 mL tube from Subheading 3.5.
5. Leave all samples rotating at 4 °C for at least 4 h or overnight.

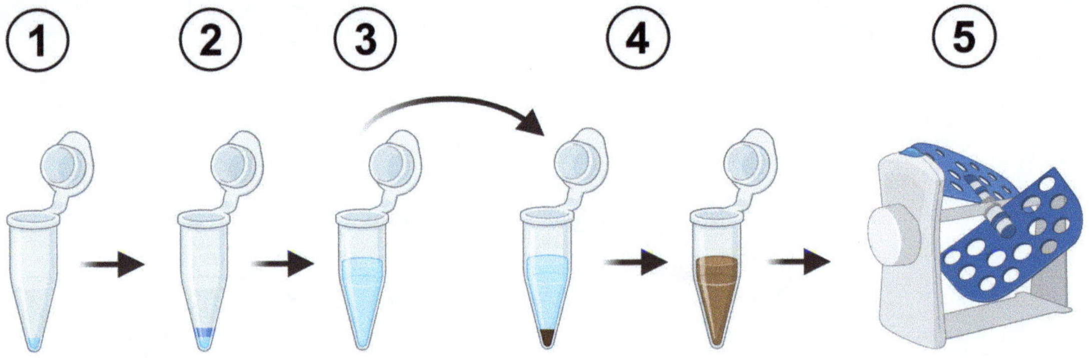

Fig. 6 Chromatin immunoprecipitation. Numbers refer to the respective steps of this method section. (Figure was created with BioRender.com)

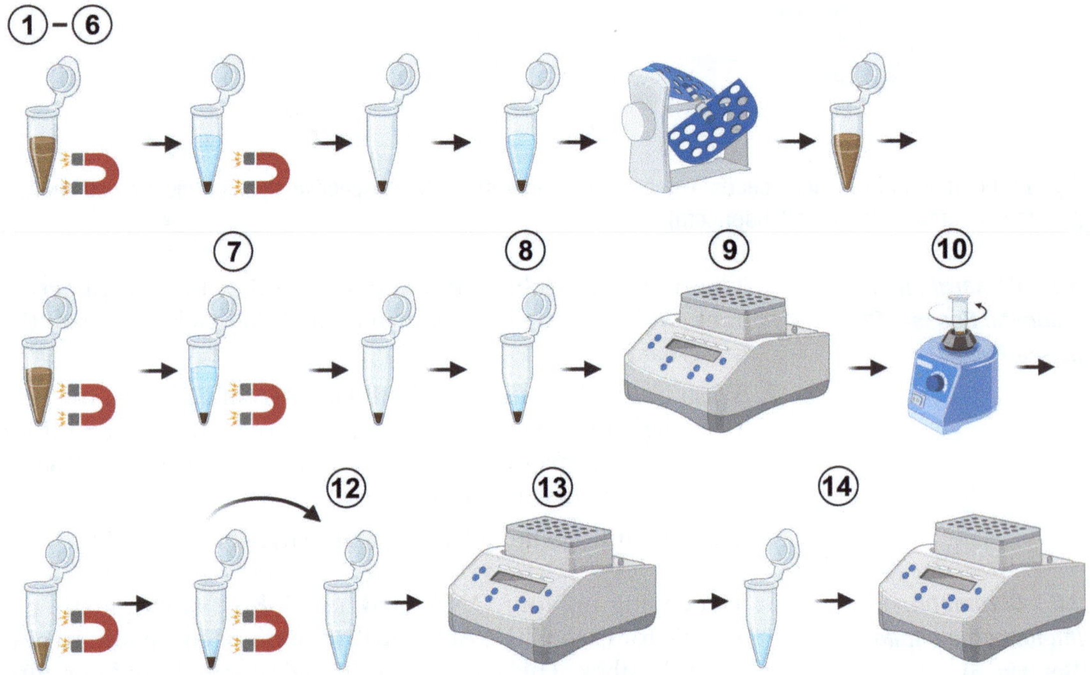

Fig. 7 Reverse cross-linking. Numbers refer to the respective steps of this method section. (Figure was created with BioRender.com)

3.7 Reverse Cross-Linking (See Fig. 7)

1. Place samples on a magnetic tube rack and remove and discard supernatant. The following wash steps serve to remove non-specifically bound DNA and proteins from the magnetic beads.
2. Wash twice with 1 mL of cooled WB1. Wash the magnetic beads quickly for the first wash and leave all samples rotating for 5 min at 4 °C for the second wash.
3. Place samples on a magnetic tube rack and remove and discard remaining WB1 after each wash step.

4. Wash twice with 1 mL of cooled WB2. Wash the magnetic beads quickly for the first wash and leave all samples rotating for 5 min at 4 °C for the second wash.
5. Place samples on a magnetic tube rack and remove and discard remaining WB2 after each wash step.
6. Wash twice with 1 mL of cooled WB3. Wash the magnetic beads quickly for the first wash and leave all samples rotating for 5 min at 4 °C for the second wash.
7. Place samples on a magnetic tube rack and remove and discard remaining WB3 after each wash step.
8. Add 200 μL EB to the magnetic beads in each tube.
9. Incubate tubes at 65 °C with agitation for 15 min to elute DNA-protein complexes.
10. Vortex and transfer supernatants to new 1.5 mL tubes with the help of a magnetic tube rack.
11. Repeat **steps 8** and **9**.
12. Vortex and combine both 200 μL elutes to yield 400 μL per sample. Discard the 1.5 mL tubes that contain the magnetic beads.
13. Incubate tubes at 65 °C with agitation for at least 4 h for reverse cross-linking of DNA-protein complexes.
14. Add 2 μL Proteinase K solution to each sample. Incubate tubes at 55 °C with slow agitation for a minimum of 2 h to digest proteins and peptides.

3.8 DNA Purification (See *Fig. 8*)

1. Prepare 2 mL Phase Lock Gel tubes and settle gels using a microcentrifuge.
2. Add each sample (approximately 400 μL) to a Phase Lock Gel tube.
3. Under a fume hood, add 400 μL Phenol:Chloroform:Isoamyl alcohol (25:24:1) to each tube.
4. Mix samples thoroughly by vortexing.
5. Centrifugate the Phase Lock Gel tubes at 18,000 g for 5 min at room temperature.
6. Under a fume hood, collect the upper aqueous phase (contains soluble DNA molecules) and transfer to a new 1.5 mL tube. Discard the lower organic phase (contains denatured proteins) (*see* **Note 11**).
7. Add 6 μL glycogen solution, 20 μL 4 M NaCl, and 875 μL pure ethanol to each tube and vortex.
8. Precipitate DNA molecules overnight at −80 °C.

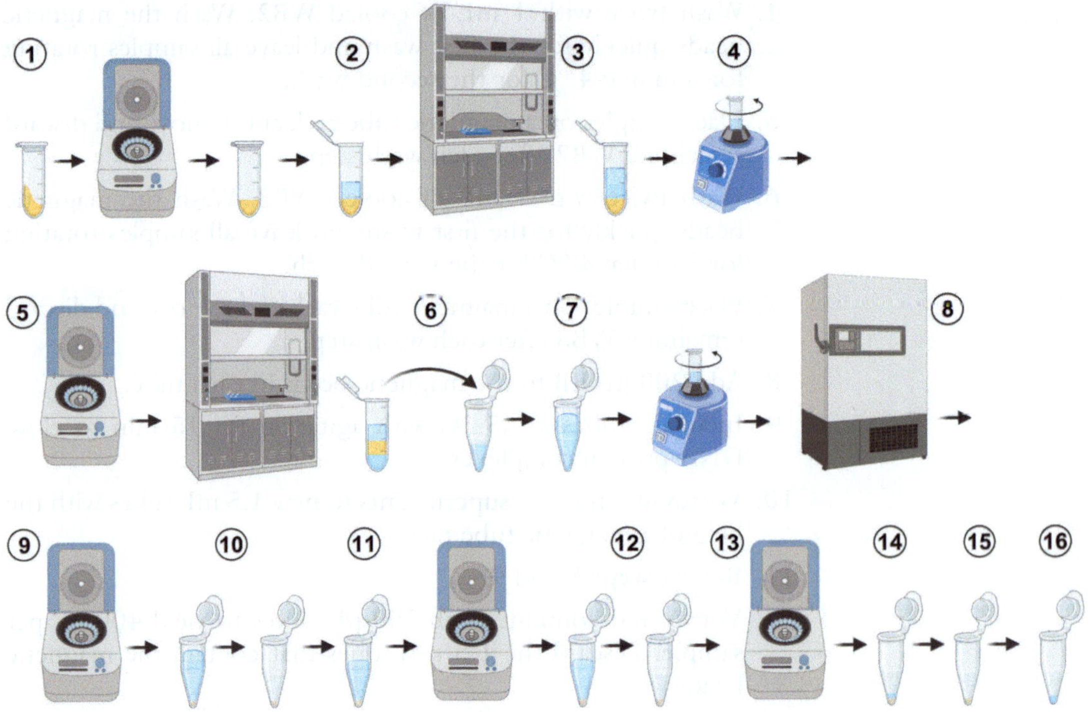

Fig. 8 DNA purification. Numbers refer to the respective steps of this method section. (Figure was created with BioRender.com)

9. Centrifugate samples at 18,000 g for 30 min at 4 °C.
10. Carefully remove supernatants without breaking or collecting the DNA/Glycogen pellets. Discard supernatants.
11. Add 1 mL of 70% ethanol. Centrifugate samples at 18,000 g for 10 min at 4 °C.
12. Carefully remove supernatants without breaking or collecting the pellets. Discard supernatants.
13. Centrifugate samples once more at 18,000 g for 5 min at 4 °C.
14. Remove all remaining 70% ethanol without breaking or collecting the pellets.
15. Allow pellets to dry to remove remaining ethanol.
16. Resuspend pellets in 50 μL nuclease-free water. The 50 μL samples contain immunoprecipitated and purified DNA fragments. They can be stored at −20 °C or be used for further analysis like sequencing library preparation or ChIP-PCR studies.

4 Notes

1. We found that 6-day-old seedlings from 80 mg of *Arabidopsis* seeds yield the right amount of plant material for one ChIP assay.
2. Be careful while handling formaldehyde and check its safety data sheet.
3. Do not keep SB too long on ice to prevent precipitation of SDS. In case of precipitation, keep SB at room temperature to dissolve the surfactant. A reduced SDS concentration will lead to insufficient sonication.
4. H2A.Z responds to various environmental cues. Therefore, we recommend growing the controls and the plants that will be exposed to elevated temperatures in the same chamber. For subsequent exposures to elevated temperatures, one set of plants should be moved to an identical chamber with an increased temperature setting.
5. Fast tissue harvesting is recommended since H2A.Z responds very quickly to environmental changes. It has been demonstrated that H2A.Z eviction due to an environmental stimulus can already be observed after 5 min [7].
6. It is important that the entire plant tissue is submerged. Therefore, we recommend using a weight to prevent containers from floating at the buffer surface.
7. Before filtering the lysate, make sure that there are no remaining ice crystals present.
8. Avoid mixing the two phases during the overlay process.
9. Needed sonication settings and length will depend on used instrument and amount of chromatin per sample and should take about 40–60 min. Choose settings with pulses and pauses to avoid overheating of samples.
10. In case the smear consists of larger fragments, the samples should be sonicated for an extended time until the appropriate fragment sizes are obtained.
11. Take special care not to mix the aqueous phase and the lower organic phases during the collection process.

References

1. Luger K, Mäder AW, Richmond RK et al (1997) Crystal structure of the nucleosome core particle at 2.8 A resolution. Nature 389(6648):251–260
2. Carter B, Bishop B, Ho KK et al (2018) The chromatin remodelers PKL and PIE1 act in an epigenetic pathway that determines H3K27me3 homeostasis in Arabidopsis. Plant Cell 30(6):1337–1352
3. Coleman-Derr D, Zilberman D (2012) Deposition of histone variant H2A.Z within gene

bodies regulates responsive genes. PLoS Genet 8(10):e1002988

4. Sura W, Kabza M, Karlowski WM et al (2017) Dual role of the histone variant H2A.Z in transcriptional regulation of stress-response genes. Plant Cell 29(4):791–807
5. Yelagandula R, Stroud H, Holec S et al (2014) The histone variant H2A.W defines heterochromatin and promotes chromatin condensation in Arabidopsis. Cell 158(1):98–109
6. Smith AP, Jain A, Deal RB et al (2010) Histone H2A.Z regulates the expression of several classes of phosphate starvation response genes but not as a transcriptional activator. Plant Physiol 152(1):217–225
7. Willige BC, Zander M, Yoo CY et al (2021) PHYTOCHROME-INTERACTING FACTORs trigger environmentally responsive chromatin dynamics in plants. Nat Genet 53(7): 955–961
8. Zander M, Willige BC, He Y et al (2019) Epigenetic silencing of a multifunctional plant stress regulator. elife 8:e47835
9. Kumar SV, Wigge PA (2010) H2A.Z-containing nucleosomes mediate the thermosensory response in Arabidopsis. Cell 140(1): 136–147
10. Kumar SV, Lucyshyn D, Jaeger KE et al (2012) Transcription factor PIF4 controls the thermosensory activation of flowering. Nature 484(7393):242–245
11. Van Der Woude LC, Perrella G, Snoek BL et al (2019) HISTONE DEACETYLASE 9 stimulates auxin-dependent thermomorphogenesis in Arabidopsis thaliana by mediating H2A.Z depletion. Proc Natl Acad Sci USA 116(50): 25343–25354
12. Papamichos-Chronakis M, Watanabe S, Rando OJ et al (2011) Global regulation of H2A.Z localization by the INO80 chromatin-remodeling enzyme is essential for genome integrity. Cell 144(2):200–213
13. Alatwi HE, Downs JA (2015) Removal of H2A.Z by INO80 promotes homologous recombination. EMBO Rep 16(8):986–994
14. Cortijo S, Charoensawan V, Brestovitsky A et al (2017) Transcriptional regulation of the ambient temperature response by H2A.Z nucleosomes and HSF1 transcription factors in Arabidopsis. Mol Plant 10(10):1258–1273
15. Casal JJ, Fankhauser C (2023) Shade avoidance in the context of climate change. Plant Physiol 191(3):1475–1491
16. Xue M, Zhang H, Zhao F et al (2021) The INO80 chromatin remodeling complex promotes thermomorphogenesis by connecting H2A.Z eviction and active transcription in Arabidopsis. Mol Plant 14(11):1799–1813

Chapter 18

Assessing the Function of CBF1 in Modulating the Interaction Between Phytochrome B and PIF4

Jing Peng, Xiaojing Dong, Shuhua Yang, and Jigang Li

Abstract

Phytochromes are red (R) and far-red (FR) light photoreceptors in plants. Upon light exposure, photo-activated phytochromes translocate into the nucleus, where they interact with their partner proteins to transduce light signals. The yeast two-hybrid (Y2H) system is a powerful technique for rapidly identifying and verifying protein-protein interactions, and PHYTOCHROME-INTERACTING FACTOR3 (PIF3), the founding member of the PIF proteins, was initially identified in a Y2H screen for phytochrome B (phyB)-interacting proteins. Recently, we developed a yeast three-hybrid (Y3H) system by introducing an additional vector into this Y2H system, and thus a new regulator could be co-expressed and its role in modulating the interactions between phytochromes and their signaling partners could be examined. By employing this Y3H system, we recently showed that both MYB30 and CBF1, two negative regulators of seedlings photomorphogenesis, act to inhibit the interactions between phyB and PIF4/PIF5. In this chapter, we will use the CBF1-phyB-PIF4 module as an example and describe the detailed procedure for performing this Y3H assay. It will be intriguing and exciting to explore the potential usage of this Y3H system in future research.

Key words Yeast three-hybrid, Protein-protein interactions, Phytochrome, phyB, PIF4, CBF1

1 Introduction

The yeast two-hybrid (Y2H) system, firstly established in the study of transcriptional regulation in eukaryotes in 1989 [1], is a powerful tool to examine whether two proteins could interact in yeast cells. It is often seen that a transcriptional activator can bind to the upstream activating sequence (UAS) of a target gene and activate its transcription. This transcriptional activator usually contains a DNA-binding domain (BD) and an activation domain (AD), and BD or AD alone could not activate the transcription. Only when an AD is recruited to the promoter region of a target gene can it activate the transcription. The Y2H system was established by fusing two proteins to the BD and the AD of the yeast GAL4 transcriptional activator, respectively. When the two proteins

Meng Chen (ed.), *Thermomorphogenesis: Methods and Protocols*, Methods in Molecular Biology, vol. 2795, https://doi.org/10.1007/978-1-0716-3814-9_18,

physically interact in yeast cells, AD and BD move in close proximity to each another, and AD is recruited by BD to the promoter region of a reporter gene (usually *HIS* or *LacZ*). If the reporter gene is expressed, it could be inferred that the two proteins physically interacted in yeast cells.

Plants have evolved several different classes of photoreceptors to perceive different wavelengths of light, among which, phytochromes are responsible for perceiving the red (R) and far-red (FR) light spectrums [2–6]. The *Arabidopsis* genome encodes five phytochrome proteins (designated phyA to phyE), among which, phyA is the photoreceptor responsible for perceiving FR light, whereas phyB is the primary photoreceptor mediating photomorphogenic responses in R light [2, 4, 5]. Phytochromes exist in vivo in two interconvertible forms: the R light-absorbing Pr form and the FR light-absorbing Pfr form [5, 7]. It has long been believed that the Pfr form of phytochrome is physiologically active while the Pr form is inactive; however, it was recently proposed that the Pr form of phyA may also have some biological activity when it is localized in the nucleus [8]. In darkness, phytochromes are synthesized in the cytoplasm in the Pr form; in response to R light, phytochromes convert to the Pfr form and translocate into the nucleus [4, 9, 10]. A group of basic helix-loop-helix (bHLH) transcription factors, named PHYTOCHROME-INTERACTING FACTORS (PIFs), accumulate in etiolated seedlings and repress photomorphogenesis; upon R light exposure, nuclear-localized Pfr forms of phytochromes physically interact with PIFs, leading to the rapid phosphorylation and degradation of PIFs via the 26S proteasome pathway [5, 11–13]. Notably, PIF3, the founding member of the PIF subset, was initially identified in a Y2H screen for phyB-interacting proteins [14]. The AP2/ERF-family transcription factors C-REPEAT BINDING FACTORS (CBFs) play central roles in plant cold acclimation. Cold stress rapidly induces the expression of *CBF* genes, and CBFs were shown to activate the expression of *COLD REGULATED* (*COR*) genes to enhance the plant freezing tolerance [15–18]. Recently, it was shown that CBF1 promotes PIF4/PIF5 protein accumulation and *Arabidopsis* hypocotyl growth at ambient temperatures [19].

The phytochromes are chromoproteins composed of a polypeptide and a covalently linked tetrapyrrole chromophore, and only the phytochrome holoprotein could undergo the conformational changes between Pr and Pfr forms [5]. Notably, the N-terminus of the phytochrome needs to be free for proper folding; however, the commercial Y2H systems were all designed to fuse BD or AD to the N-terminus of a protein of interest. In addition, the yeast cells could not produce the chromophores of phytochromes, and therefore, the chromophores should be exogenously supplied to allow the assembly of the phytochrome holoprotein in yeast cells. In 2002, Shimizu-Sato et al. [20] reported a GAL4-based light-switchable Y2H system by using the D153 vector that allows the fusion of the GAL4 DNA-BD to the C-terminus of a phytochrome and by adding phycocyanobilin (PCB), the chromophore of cyanobacterial phytochromes extracted from *Spirulina* into the system. By employing this system, it was demonstrated that PIF3 specifically interacted with the Pfr form of full-length phyA and the photosensory N-terminal domain of phyB [20]. Recently, a LexA-based light-switchable Y2H system was developed and was successful in detecting the phyA-COP1 and phyA-TZP interactions that had been confirmed by other experimental systems but were undetectable in the GAL4-based Y2H system [21].

Recently, we developed a yeast three-hybrid (Y3H) system by introducing an additional vector (to express a protein of interest) into the GAL4-based light-switchable Y2H system, so that the regulatory role of this protein in modulating the interactions between phytochromes and their signaling partners could be determined. By employing this Y3H system, we successfully demonstrated the inhibitory roles of CBF1 and MYB30 in the interactions between the Pfr form of phyB and PIF4/PIF5 in yeast cells [19, 22]. In this chapter, we will describe the detailed procedure for performing this Y3H assay using the CBF1-phyB-PIF4 module as an example (Fig. 1). It will be exciting to explore the applications of this Y3H system in more scenarios in future research.

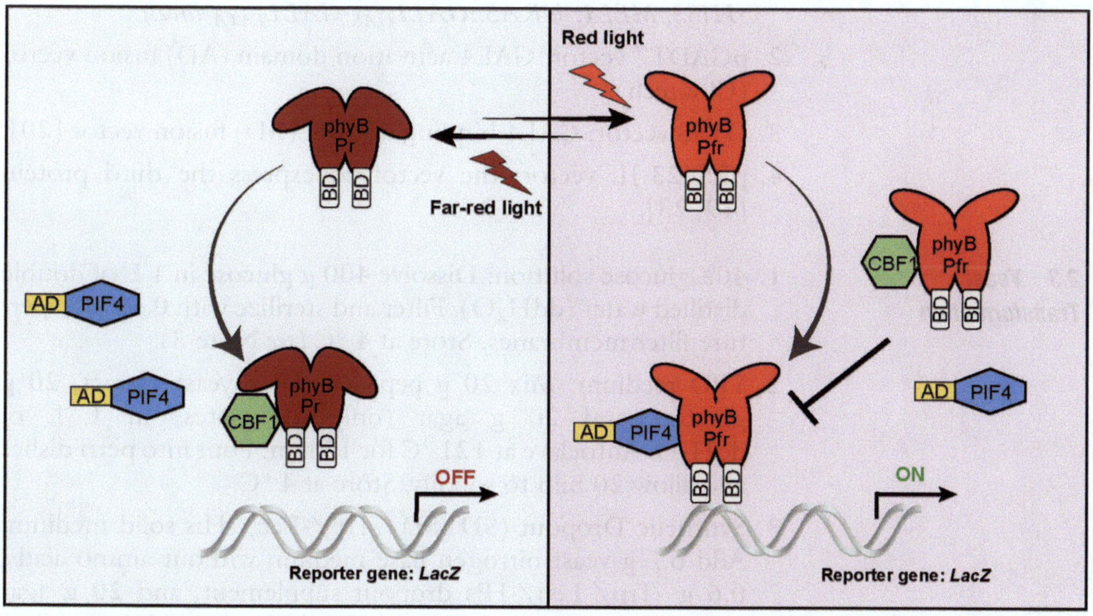

Fig. 1 Schematic illustration of the role of CBF1 in inhibiting the interaction between the Pfr form of phyB and PIF4 in yeast cells. Red light irradiation triggers the conversion of phyB-BD from the Pr form to the Pfr form, and the Pfr form of phyB-BD preferentially interacts with AD-PIF4 in yeast cells. When CBF1 is co-expressed in yeast cells, CBF1 acts to inhibit the interaction between AD-PIF4 and the Pfr form of phyB-BD

2 Materials

Prepare all solutions using ultrapure water (prepared by purifying deionized water to attain a sensitivity of 18 MΩ-cm at 25 °C) and analytical-grade reagents. Prepare and store all reagents at room temperature (unless indicated otherwise). Safety precautions must be observed when handling chloroform or β-mercaptoethanol; work within a fume hood. Do not mix waste solvents, and dispose of the waste solvents and β-mercaptoethanol-containing solutions according to local regulations.

2.1 Accessories and Supplies

1. A laminar flow bench.
2. An incubator and shaker set to 30 °C (*see* **Note 1**).
3. A temperature-controlled dark room with a weak green safety light (*see* **Note 2**).
4. The fluence rates of the light growth chambers (Percival Scientific) are 50 μmol m^{-2} s^{-1} for both continuous far-red and red light.
5. A spectrophotometer with disposable 1 mL cuvettes.
6. A water bath.

2.2 Yeast Strain and Cloning Vector

1. Y190 (*MATa*, *ura3-52*, *his3-200*, *ade2-101*, *lys2-801*, *trp1-901*, *leu2-3*, *112*, *gal4Δ*, *gal80Δ*, *cyh*^{r}*2*, *LYS2::GAL1*$_{UAS}$-*GAL1*$_{TATA}$ -*HIS3*, *MEL1*, *URA3::GAL1*$_{UAS}$-*GAL1*$_{TATA}$-*lacZ*).
2. pGADT7 vector: GAL4 activation domain (AD) fusion vector (Clontech).
3. D153 vector: GAL4 binding domain (BD) fusion vector [20].
4. pRS423-JL vector: the vector to express the third protein [19, 23].

2.3 Yeast Transformation

1. 40% glucose solution: Dissolve 400 g glucose in 1 L of double distilled water (ddH_2O). Filter and sterilize with 0.45 μm aperture filter membranes. Store at 4 °C (*see* **Note 3**).
2. YPD medium: Mix 20 g peptone, 10 g yeast extract, 20 g glucose, and 20 g agar (only for plates) in 1 L of ddH_2O. Autoclave at 121 °C for 15 min. Pour into petri dishes and allow 20 min to solidify. Store at 4 °C.
3. Synthetic Dropout (SD) SD/-Trp/-Leu/-His solid medium: Add 6.7 g yeast nitrogen base medium without amino acids, 0.6 g -Trp/-Leu/-His dropout supplement, and 20 g agar (only if solid medium is prepared) to 900 mL of ddH_2O, adjust pH to 5.8 with 2 M KOH, and add ddH_2O to a final volume of 950 mL. Autoclave at 121 °C for 15 min. Once cooled down to 55 °C, add 50 mL of sterile 40% glucose solution. Pour into petri dishes and allow 20 min to solidify. Store at 4 °C.

4. 10× TE buffer: Mix 10 mL 1 M Tris–HCl (pH 7.5) and 2 mL 0.5 M EDTA (pH 8.0) solutions and add ddH_2O to a final volume of 100 mL. Filter and sterilize with 0.45 μm aperture filter membranes. Store at room temperature.
5. 10× LiAc solution: Dissolve 10.2 g LiAc in ddH_2O, adjust pH to 7.5, and add ddH_2O to a final volume of 100 mL. Filter and sterilize with 0.45 μm aperture filter membranes. Store at room temperature.
6. 50% (w/v) PEG 3350: Dissolve 50 g polyethylene glycol MW 3350 (PEG 3350) in 100 mL of ddH_2O by heating up to 50 °C. Filter and sterilize with 0.45 μm aperture filter membranes. Store at room temperature.
7. Herring testes carrier DNA: Dissolve herring testes carrier DNA in ddH_2O to a final concentration of 10 mg/mL and store at −20 °C (*see* **Note 4**).
8. 1× TE buffer: Mix 3 mL 10× TE buffer and 27 mL sterile water.
9. 1× TE/1× LiAc solution: Mix 1 mL 10× TE, 1 mL 10× LiAc solution, and 8 mL sterile water.
10. PEG/LiAc solution: Mix 24 mL 50% PEG 3350, 3 mL 10× LiAc solution, and 3 mL 10× TE buffer.
11. Dimethyl sulfoxide (DMSO). Store at room temperature.

2.4 β-Galactosidase Assays

1. 40% galactose solution: Dissolve 400 g galactose in 1 L ddH_2O. Filter and sterilize with 0.45 μm aperture filter membranes. Store at 4 °C.
2. 40% raffinose solution: Dissolve 400 g raffinose in 1 L ddH_2O. Filter and sterilize with 0.45 μm aperture filter membranes. Store at 4 °C (*see* **Note 5**).
3. SD/-Trp/-Leu/-His liquid medium: Add 6.7 g yeast nitrogen base medium without amino acids and 0.6 g -Trp/-Leu/-His dropout supplement to ddH_2O. Adjust pH to 5.8 with 2 M KOH and add ddH_2O to a final volume of 1 L. Autoclave at 121 °C for 15 min. Store at 4 °C.
4. Z buffer: Dissolve 0.246 g $MgSO_4{\bullet}7H_2O$, 0.75 g KCl, 6.22 g $NaH_2PO_4{\bullet}2H_2O$, 21.5 g $Na_2HPO_4{\bullet}12H_2O$ in ddH_2O, adjust pH to 7.0, and add ddH_2O to a final volume of 1 L. Filter and sterilize with 0.45 μm aperture filter membranes. Store at room temperature.
5. 4 mg/mL ONPG: Dissolve 0.4 g ONPG in Z buffer, and then add ddH_2O to a final volume of 100 mL. Filter and sterilize with 0.45 μm aperture filter membranes. Store at −20 °C.
6. 1 M Na_2CO_3: Dissolve 106 g Na_2CO_3 in 1 L ddH_2O. Store at room temperature.

7. 2 mM PCB: Dissolve 5 mg PCB in 4.26 mL DMSO in the darkroom. Store at −20 °C in darkness (*see* **Note 6**).
8. β-Mercaptoethanol and chloroform (*see* **Note 7**).

2.5 Protein Extraction and Immunoblotting

1. 2× SDS loading buffer: Add 12.5 mL 1 M Tris–HCl pH 6.8, 4 g sodium dodecyl sulfate (SDS), 20 mL glycerol, 20 mg bromophenol blue (BPB), 3.1 g dithiothreitol (DTT), and add ddH_2O to a final volume of 100 mL.
2. Primary antibodies include anti-CBF1 [19], anti-phyB [19], anti-PIF4 (Agrisera), and anti-GADPH (ABclonal).

2.6 Primers

PHYB-Fr	TGGATCGCGGCCGCATGGTTTCCGGAGTCGGGGGT
PHYB-Rv	TGGATCGCGGCCGCCATATGGCATCATCAGCAT
PIF4-Fr	TGGATCGAATTCATGGAACACCAAGGTTGG
PIF4-Rv	TGGATCCTCGAGCTAGTGGTCCAAACGAGA
CBF1-Fr	TGGATCGGATCCATGAACTCATTTTCAGCT
CBF1-Rv	TGGATCCTCGAGTTAGTAACTCCAAAGCGA

3 Methods

3.1 Cloning and Construct Preparation

1. Amplify the full-length *PHYB* coding sequence via PCR using primers PHYB-Fr and PHYB-Rv, and then clone the PCR fragment into the *Not*I site of the D153 vector [20] to generate the PHYB-BD construct [24].
2. Amplify the full-length *PIF4* coding sequence via PCR using primers PIF4-Fr and PIF4-Rv, and then clone the PCR fragment into the *Eco*RI-*Xho*I sites of the pGADT7 vector (Clontech) to generate the AD-PIF4 construct.
3. Amplify the full-length *CBF1* coding sequence via PCR using primers CBF1-Fr and CBF1-Rv, and then clone the PCR fragment into *BamHI*-*Xho*I sites of the pRS423-JL vector [19].

3.2 LiAc/PEG-Mediated Yeast Transformation

The following steps should be carried out in a laminar flow hood to avoid contamination with other microorganisms. The stock solutions and media should be autoclaved, and solutions for each experiment should be prepared fresh to avoid contamination.

1. Inoculate 1 mL YPD with several colonies, 2–3 mm in diameter (*see* **Note 8**).

2. Vortex vigorously to disperse any clumps.
3. Transfer the mixture into a flask containing 50 mL of YPD medium.
4. Incubate at 30 °C for 16–18 h with shaking at 200 rpm to stationary phase (OD_{600} > 1.5).
5. Transfer 30 mL overnight culture to flask containing 300 mL of YPD. Check the OD_{600} of the diluted culture and, if necessary, add more overnight culture to bring the OD_{600} up to 0.2–0.3.
6. Incubate at 30 °C for 3 h with shaking at 200 rpm. At this point, the OD_{600} should be 0.4–0.6 (*see* **Note 9**).
7. Place cells in 50 mL tubes and centrifuge at 1000 *g* for 5 min at room temperature.
8. Discard the supernatants and thoroughly resuspend the cell pellets in ddH_2O. Collect the cells into one tube (final volume 25–50 mL).
9. Centrifuge at 1000 *g* for 5 min at room temperature.
10. Decant the supernatants.
11. Resuspend the cell pellet in 1.5 mL of freshly prepared, sterile 1× TE/1× LiAc.
12. Mix 0.1 μg plasmid DNA with 0.1 mg herring testes carrier DNA in a fresh 1.5 mL tube (*see* **Note 10**).
13. Add 0.1 mL yeast competent cells to each tube and mix well by vortexing.
14. Add 0.6 mL sterile PEG/LiAc solution to each tube and vortex at high speed for 10 s to mix.
15. Incubate at 30 °C for 30 min with shaking at 200 rpm.
16. Add 70 μL DMSO. Mix well by gentle inversion. Do not vortex (*see* **Note 11**).
17. Heat shock for 15 min in a 42 °C water bath.
18. Chill cells on ice for 1–2 min.
19. Centrifuge the cells for 5 s at 15,000 *g* at room temperature, and remove the supernatant.
20. Resuspend the cells in 0.1 mL of sterile 1× TE buffer.
21. Plate 60 μL of the cell suspension on SD/-His/-Leu/-Trp agar plate that will select for the desired transformants (*see* **Note 12**).
22. Incubate the plates at 30 °C for 3–5 day until colonies appear.

3.3 β-Galactosidase Assays

1. Pick yeast monoclonal and inoculate to the SD/-His/-Leu/-Trp agar plates, incubate at 30 °C for 1–2 day.

2. Inoculate 500 μL SD/-His/-Leu/-Trp liquid media (containing 2% glucose) in a 10 mL test tube with an inoculation line.
3. Incubate each culture at 30 °C with shaking at 200 rpm for 6–8 h.

Steps 4–12 should be performed in the darkroom.

4. Add 500 μL SD/-His/-Leu/-Trp liquid media (containing 1% raffinose, 2% galactose, 25 μM PCB) to test tubes and wrap the test tubes in foil paper in the darkroom, incubate at 30 °C and 200 rpm for the overnight culture.
5. Prepare and label new test tubes.
6. In the darkroom, divide the culture equally into two portions, each containing 500 μL (*see* **Note 13**).
7. Expose one of the cultures to red light (50 μmol m^{-2} s^{-1}) for 5 min; incubate the rest in red light (50 μmol m^{-2} s^{-1}) for 5 min followed by a 5-min far-red light (50 μmol m^{-2} s^{-1}) treatment (*see* **Note 14**).
8. Add another 1 mL SD/-His/-Leu/-Trp liquid medium (containing 1% raffinose, 2% galactose, 25 μM PCB) to each test tube and wrap in foil paper in the darkroom. Incubate at 30 °C and 200 rpm for 2 h.
9. Repeat the light irradiation in **step** 7, and then wrap in foil paper, incubate at 30 °C and 200 rpm for 1–2 h.
10. Transfer 1 mL culture into 1.5 mL microcentrifuge tubes in the darkroom. Centrifuge at 14,000 *g* for 5 min and discard the supernatant. Temporarily store the remaining culture at 4 °C for OD_{600} measurement and protein extraction (*see* **Note 15**).
11. Add 1 mL Z buffer to each microcentrifuge tube and vortex until cells are resuspended. Centrifuge cells again and remove the supernatant.
12. Resuspend each pellet in 150 μL of Z buffer containing 0.4 μL β-mercaptoethanol, 50 μL chloroform, and 20 μL 0.1% SDS. Set up a blank tube. Vortex until cells are resuspended (*see* **Note 16**).
13. Add 200 μL 4 mg/mL ONPG to each microcentrifuge tube and vortex to mix well.
14. Place the tubes in a 30 °C water bath. Start timer.
15. After the yellow color develops, add 500 μL 1 M Na_2CO_3 to the reaction and blank tubes. Record the elapsed time in minutes (*see* **Note 17**).
16. Centrifuge reaction tubes and blank tubes for 5 min at 14,000 *g* to pellet the cell debris.
17. Carefully transfer 800 μL supernatants to colorimetric dish. Calculate the spectrophotometer against the blank at A_{420}

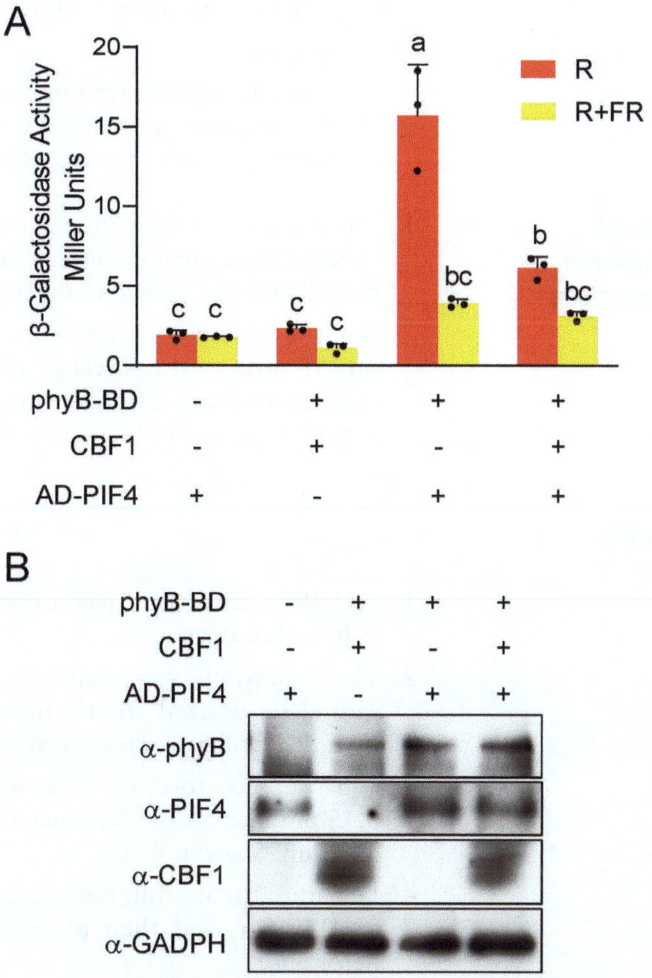

Fig. 2 Yeast three-hybrid assays showing that CBF1 inhibits the interaction between phyB-Pfr and PIF4 in yeast cells. (**a**) The β-galactosidase activities were measured by liquid culture assays using ONPG as the substrate. AD-PIF4, phyB-BD, and CBF1 were expressed in the yeast strain Y190 as indicated. Error bars represent SD of three independent yeast clones. Different letters represent significant differences by one-way ANOVA with Tukey' s post hoc test ($P < 0.05$). (**b**) Immunoblots showing the protein abundance of phyB-BD, CBF1, and AD-PIF4 in yeast cultures co-expressing the indicated combinations of proteins

and measure the OD_{420} of the samples relative to the blank. The ODs should be between 0.02 and 1.0 to be within liner range of the assay.

18. Dilute 200 μL the culture to 800 μL of SD/-His/-Leu/-Trp liquid medium containing PCB and measure the OD_{600} of the samples relative to the blank.
19. Calculate β-galactosidase units (Fig. 2A).

$$\beta - \text{galactosidase units} = 1000 \times OD_{420}/(t \times V \times OD_{600})$$

Where:
t = elapsed time (in min) of incubation
V = volume of culture
OD_{600} = culture concentration

3.4 Protein Extraction and Immunoblotting

1. The remaining yeast cultures in **step 10** of the above section are collected by centrifugation at 14,000 *g*, and then the pellets are mixed with 2× SDS loading buffer and vortexed vigorously.
2. Boil the mixtures for 10 min, and then perform immunoblotting to analyze the levels of phyB-BD, CBF1, and AD-PIF4 proteins in yeast cultures using antibodies against anti-CBF1, anti-phyB, anti-PIF4, and anti-GADPH, respectively (Fig. 2b).

4 Notes

1. The 30 °C incubator shaker should be carefully sealed to avoid light penetration.
2. The dark room, where all light-avoiding steps are performed, should be utilized strictly to avoid any light exposure other than the artificially dim green light.
3. Glucose, galactose, or raffinose should be sterilized through 0.45 μm filter. Autoclave may decrease the performance of the medium. Store at 4 °C.
4. Denature the herring testes carrier DNA in a 100 °C dry bath for 15 min, and then place on ice immediately for 5 min before use.
5. Galactose and raffinose are easily saturated and precipitated at low temperatures, so they should be melted before use.
6. PCB should be utilized strictly to avoid any light exposure other than the artificially dim green light.
7. β-Mercaptoethanol and chloroform are toxic! Work in a fume hood.
8. To improve the transformation efficiency, the fresh (one- to three-week-old) yeast colonies will yield best results for liquid culture inoculation.
9. For a successful transformation, the health and growth phase of cells for making competent cells are crucial. The cells in expansion medium should be in log-phase growth (i.e., OD_{600} between 0.4 and 0.6).
10. For simultaneous co-transformation (using two or three different plasmids), use 0.1 μg of each plasmid, in addition to the 0.1 mg carrier DNA.

11. Add DMSO and quickly invert the mixture to avoid precipitation.
12. Plate a moderate amount of the resuspended cells on SD agar plates to ensure that a plate with well-separated colonies will be obtained.
13. Vortex vigorously to disperse any clumps before dividing.
14. Transparent tubes should be used to better expose the culture to light.
15. Vortex the culture tube for 0.5–1 min to disperse cell clumps before checking the OD.
16. The thorough mixing is critical.
17. The length of required time will vary, and the yellow color will develop more intense with time. A new blank tube is required for every batch.

Acknowledgments

This work was supported by grants from the National Natural Science Foundation of China (32225006 to J.L.), the Beijing Natural Science Foundation (5232011 to J.L.), and the China Postdoctoral Science Foundation (2022M723410 to J.P.).

References

1. Fields S, Song O (1989) A novel genetic system to detect protein-protein interactions. Nature 340:245–246
2. Chen M, Chory J, Fankhauser C (2004) Light signal transduction in higher plants. Annu Rev Genet 38:87–117
3. Jiao Y, Lau OS, Deng XW (2007) Light-regulated transcriptional networks in higher plants. Nat Rev Genet 8:217–230
4. Franklin KA, Quail PH (2010) Phytochrome functions in Arabidopsis development. J Exp Bot 61:11–24
5. Li J, Li G, Wang H, Deng XW (2011) Phytochrome signaling mechanisms. Arabidopsis Book 9:e0148
6. Legris M, Ince Y, Fankhauser C (2019) Molecular mechanisms underlying phytochrome-controlled morphogenesis in plants. Nat Commun 10:5219
7. Bae G, Choi G (2008) Decoding of light signals by plant phytochromes and their interacting proteins. Annu Rev Plant Biol 59:281–311
8. Li J, Hiltbrunner A (2021) Is the Pr form of phytochrome biologically active in the nucleus? Mol Plant 14:535–537
9. Fankhauser C, Chen M (2008) Transposing phytochrome into the nucleus. Trends Plant Sci 13:596–601
10. Klose C, Viczián A, Kircher S, Schäfer E, Nagy F (2015) Molecular mechanisms for mediating light-dependent nucleo/cytoplasmic partitioning of phytochrome photoreceptors. New Phytol 206:965–971
11. Castillon A, Shen H, Huq E (2007) Phytochrome interacting factors: central players in phytochrome-mediated light signaling networks. Trends Plant Sci 12:514–521
12. Leivar P, Quail PH (2011) PIFs: pivotal components in a cellular signaling hub. Trends Plant Sci 16:19–28
13. Pham VN, Kathare PK, Huq E (2018) Phytochromes and Phytochrome interacting factors. Plant Physiol 176:1025–1038
14. Ni M, Tepperman JM, Quail PH (1998) PIF3, a phytochrome-interacting factor necessary for normal photoinduced signal transduction, is a novel basic helix-loop-helix protein. Cell 95: 657–667
15. Stockinger EJ, Gilmour SJ, Thomashow MF (1997) *Arabidopsis thaliana* CBF1 encodes an

AP2 domain-containing transcriptional activator that binds to the C-repeat/DRE, a cis-acting DNA regulatory element that stimulates transcription in response to low temperature and water deficit. Proc Natl Acad Sci USA 94:1035–1040

16. Liu Q, Kasuga M, Sakuma Y et al (1998) Two transcription factors, DREB1 and DREB2, with an EREBP/AP2 DNA binding domain separate two cellular signal transduction pathways in drought- and low-temperature-responsive gene expression, respectively, in Arabidopsis. Plant Cell 10:1391–1406
17. Thomashow MF (1999) Plant cold acclimation: freezing tolerance genes and regulatory mechanisms. Annu Rev Plant Physiol Plant Mol Biol 50:571–599
18. Chinnusamy V, Zhu J, Zhu JK (2007) Cold stress regulation of gene expression in plants. Trends Plant Sci 12:444–451
19. Dong X, Yan Y, Jiang B et al (2020) The cold response regulator CBF1 promotes Arabidopsis hypocotyl growth at ambient temperatures. EMBO J 39:e103630
20. Shimizu-Sato S, Huq E, Tepperman JM et al (2002) A light-switchable gene promoter system. Nat Biotechnol 20:1041–1044
21. Li H, Qin X, Song P et al (2021) A LexA-based yeast two-hybrid system for studying light-switchable interactions of phytochromes with their interacting partners. aBIOTECH 2:105–116
22. Yan Y, Li C, Dong X et al (2020) MYB30 is a key negative regulator of Arabidopsis photomorphogenic development that promotes PIF4 and PIF5 protein accumulation in the light. Plant Cell 32:2196–2215
23. Christianson TW, Sikorski RS, Dante M et al (1992) Multifunctional yeast high-copy-number shuttle vectors. Gene 110:119–122
24. Zhang S, Li C, Zhou Y et al (2018) TANDEM ZINC-FINGER/PLUS3 is a key component of phytochrome A signaling. Plant Cell 30: 835–852

Chapter 19

GST Pull-Down Assay to Study PIF4 Binding In Vitro

Abhishesh Bajracharya, Berry Dickey, and Yongjian Qiu

Abstract

The phytochrome-interacting factor 4 (PIF4) is a well-known transcription factor that plays a pivotal role in plant thermomorphogenesis, coordinating growth and development in response to temperature changes. As PIF4 functions by forming complexes with other proteins, determining its interacting partners is essential for understanding its diverse roles in plant thermal responses. The GST (glutathione-S-transferase) pull-down assay is a widely used biochemical technique that enables the investigation of protein-protein interactions in vitro. It is particularly useful for studying transient or weak interactions between proteins. In this chapter, we describe the GST pull-down approach to detect the interaction between PIF4 and a known or suspected interacting protein. We provide detailed step-by-step descriptions of the assay procedures, from the preparation of recombinant GST-PIF4 fusion protein to the binding and elution of interacting partners. Additionally, we provide guidelines for data interpretation, quantification, and statistical analysis to ensure robust and reliable results.

Key words PIF4, Thermomorphogenesis, Protein-protein interactions, GST pull-down, Immunoblots

1 Introduction

Temperature plays a pivotal role as a critical stimulus in regulating various plant processes. Plants possess the remarkable ability to sense and respond to temperature fluctuations, dynamically adjusting their growth, morphology, and physiology accordingly. Specifically, warm and non-stressful temperatures trigger a series of adaptive responses in plants, collectively known as thermomorphogenesis, which involves reorganizing and modifying plant structure and architecture [1]. At the heart of this regulatory network lies the basic Helix-Loop-Helix (bHLH) transcription factor—PHYTOCHROME-INTERACTING FACTOR 4 (PIF4), which plays a fundamental role in governing thermomorphogenesis. PIF4 is directly implicated in temperature-mediated shoot elongation [2, 3], flowering [4, 5], and stomatal development [6]. As a central thermomorphogenetic transcriptional regulator, PIF4 exhibits a

Meng Chen (ed.), *Thermomorphogenesis: Methods and Protocols*, Methods in Molecular Biology, vol. 2795,
https://doi.org/10.1007/978-1-0716-3814-9_19,

complex network of regulation, directly and indirectly interacting with various proteins and protein complexes to orchestrate a multitude of cellular responses [7].

As a well-established central regulator, PIF4 can be a valuable tool to explore protein-protein interactions and identify critical components that directly interact with PIF4. Notably, essential proteins involved in the regulation of thermomorphogenesis, such as SUPPRESSOR OF PHYA-105 (SPA) [8], EARLY FLOWERING 3 (ELF3) [9], and MEDIATOR SUBUNIT 14 (MED14) [10], have been confirmed as PIF4 interactors through immobilized protein-protein interaction approaches such as coimmunoprecipitation or glutathione-S-transferase (GST) pull-down assays. In this chapter, we specifically focus on the GST pull-down assay as an effective means to verify and demonstrate direct interactions between PIF4 and its potential binding partners, shedding light on the intricate molecular networks underpinning thermomorphogenetic responses. By employing the GST pull-down assay, we aim to contribute to the growing body of knowledge regarding thermomorphogenetic signaling, ultimately advancing our understanding of plant responses to temperature fluctuations and their potential applications in agriculture and climate resilience.

The GST pull-down assay takes advantage of the affinity of the GST protein (211 a.a., 26 kDa) [11] for glutathione, a tripeptide composed of glutamate, cysteine, and glycine. In this technique, the protein of interest, or "bait" protein, is fused to GST [12], and the potential interacting proteins, or "prey" proteins, are incubated with the GST-bait fusion protein. The bait-prey protein complexes are then captured using glutathione-coupled beads, and the interacting proteins can be identified and analyzed through subsequent steps such as immunoblots. Here, we demonstrate using GST pull-down assays to study PIF4's interaction with a known interactor MED14. Transcription factors like PIF4 induce gene expression by binding to their specific enhancer regions and initiating transcription by interacting with the Mediator co-activator complex, which bridges transcription factors and RNA Polymerase II [13]. PIF4 interacts with MED14, a tail component of the Mediator complex, to induce global transcriptomic changes during thermomorphogenesis [10]. The overall procedure testing the interaction between GST-PIF4 and HA (hemagglutinin)-tagged MED14 (MED14-HA) is illustrated in Fig. 1.

2 Materials

All buffers and media should be prepared using double-distilled Milli-Q water (ddH_2O). Heat-labile reagents should be filter-sterilized and added to the reagent after cooling to bearable

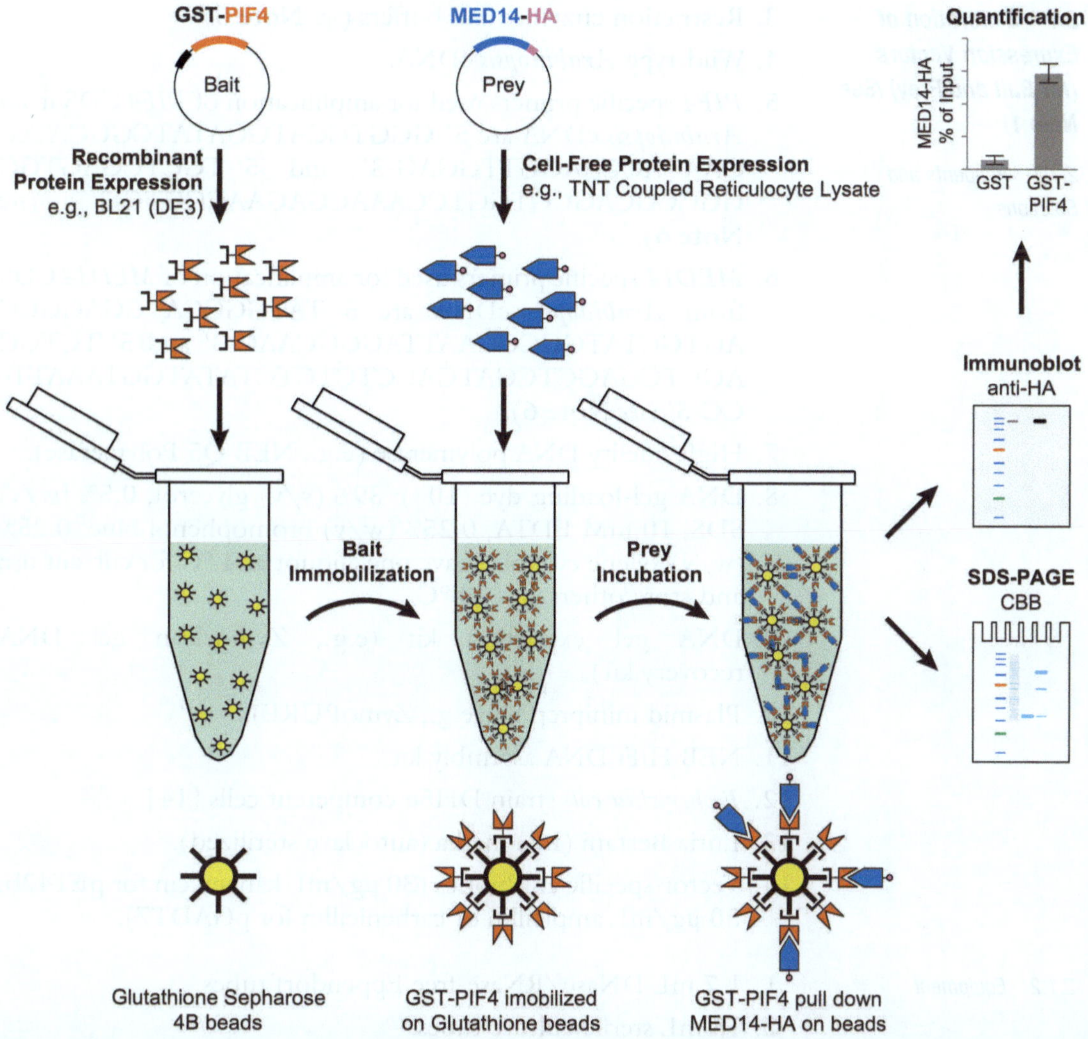

Fig. 1 Schematic outline of the GST pull-down pipeline. Once equilibrated in the EB buffer, glutathione sepharose 4B beads are mixed with recombinant GST-PIF4 proteins. Immobilized GST-PIF4 proteins are used to pull down in vitro transcribed/translated prey proteins (e.g., MED14-HA). The protein complex will be eluted and resolved in two SDS-PAGE gels—one for Coomassie Brilliant Blue R-250 (CBB) staining to visualize the bait proteins and the one for immunoblots to show the amount of precipitated prey protein, which is subsequently quantified using computer software such as ImageJ

temperatures. Nuclease-free water (nfH_2O) should be used as the solvent for all molecular biology reactions and dilutions.

1. Expression vector harboring a GST-tag (e.g., pET42b) (for the bait protein) (*see* **Note 2**).
2. T7-promoter-driven expression vector (e.g., pGADT7) harboring an epitope tag (for the prey protein) (*see* **Notes 3** and **4**).

2.1 Generation of Expression Vectors (for Bait and Prey) (See Note 1)

2.1.1 Reagents and Solutions

3. Restriction enzymes and buffers (*see* **Note 5**).
4. Wild-type *Arabidopsis* cDNA.
5. *PIF4*-specific primers used for amplification of *PIF4* CDS from *Arabidopsis* cDNA are 5′-GGGTCCATGGATATCGGGATG-GAACACCAAGGTTGGAG-3′ and 5′-TGCTCGAGTGC GGCCGCAGCGTGGTCCAAACGAGAACCGTC-3′ (*see* **Note 6**).
6. *MED14*-specific primers used for amplification of *MED14* CDS from *Arabidopsis* cDNA are 5′-TATGGCCATGGAGGCC AGTGCTATGGCGGAATTAGGGCAAC-3′ and 5′-TCTGC AGCTCGAGCTCGATGACCTCTCTCTATATGGTAAATT CC-3′ (*see* **Note 6**).
7. High-fidelity DNA polymerase (e.g., NEB Q5 Polymerase).
8. DNA gel-loading dye (10×): 39% (v/v) glycerol, 0.5% (w/v) SDS, 10 mM EDTA, 0.25% (w/v) bromophenol blue, 0.25% (w/v) xylene cyanol. Leave one aliquot at 4 °C for current use and store others at −20 °C.
9. DNA gel extraction kit (e.g., Zymoclean gel DNA recovery kit).
10. Plasmid miniprep kit (e.g., ZymoPURE).
11. NEB HiFi DNA assembly kit.
12. *Escherichia coli* strain DH5α competent cells [14].
13. Luria Bertani (LB) media (autoclave sterilized).
14. Vector-specific antibiotics (30 μg/mL kanamycin for pET42b, 50 μg/mL ampicillin or carbenicillin for pGADT7).

2.1.2 Equipment

1. 1.7 mL DNase/RNase-free Eppendorf tubes.
2. 15 mL sterile culture tubes.
3. PCR tubes.
4. DNA electrophoresis apparatus (e.g., Bio-Rad wide mini-sub cell GT system).
5. Gel documentation imaging system (e.g., Azure C600 advanced imaging system).
6. Thermal cycler.
7. Nanodrop.
8. Water bath.
9. Incubator with shaker (37 °C).
10. Inoculating stick/beads (autoclave sterilized).
11. Incubator (37 °C).
12. Centrifuge (able to hold 10 mL culture tubes).

2.2 Expression of the Bait Protein in E. Coli

2.2.1 Reagents and Solutions

1. Plasmid harboring GST-PIF4.
2. *E. coli* strain BL21 (DE3) (Agilent Technologies).
3. LB media (autoclave sterilized).
4. Antibiotics: 30 μg/mL kanamycin (for pET42b), 30 μg/mL chloramphenicol (for BL21 cells).
5. Glycerol stock for storing bacterial cultures: 52% (v/v) glycerol, 25 mM Tris-HCl (pH 7.5), 100 mM $MgSO_4$.
6. EB buffer: 50 mM Tris-HCl (pH 7.5), 100 mM NaCl, 1 mM EDTA, 1 mM EGTA, 1% DMSO.
7. Wash buffer: EB buffer, 2 mM DTT, 0.1% Nonidet P-40.
8. Protease inhibitor cocktail (e.g., Halt Protease Inhibitor Cocktail from Thermo Fisher Scientific).
9. Isopropyl β-D-1-thiogalactopyranoside (IPTG): 1 M solution in water.
10. 8% polyacrylamide gel: 4.50 mL of Milli-Q water, 2.86 mL of 1.25 M Bis-Tris-HCl (pH 6.7), 2.66 mL of 30% acrylamide and bis-acrylamide solution (37.5:1, stored in 4 °C), 70 μL of 10% (w/v) ammonium persulfate (APS) (stored in 4 °C), 20 μL of N,N,N′,N′-tetramethylethylenediamine (TEMED) (*see* **Note 7**).
11. MOPS buffer (10×): 250 mM MOPS, 250 mM Tris, 5 mM EDTA, 25 mL of 20% SDS. Before use, dilute 10× to 1× MOPS buffer and add 5 mL of 1 M sodium bisulfite per liter of MOPS buffer (*see* **Note 7**).
12. Prestained protein molecular weight markers.
13. Coomassie brilliant blue (CBB) staining buffer: 45% (v/v) methanol, 10% (v/v) glacial acetic acid, 0.25% (w/v) coomassie brilliant blue R-250.
14. CBB destaining buffer: 45% (v/v) methanol, 10% (v/v) glacial acetic acid.

2.2.2 Equipment

1. BioRad Micropulser electroporator.
2. Disposable cuvettes (for Micropulser).
3. 15 mL sterile culture tubes.
4. 50 mL sterile falcon tubes.
5. 1.7 mL DNase/RNase-free Eppendorf tubes.
6. 2 mL microcentrifuge tubes with screw caps.
7. 0.22 μm polyethersulfone (PES) syringe filter.
8. Conical flask (2 L).
9. Large volume centrifuge (500 mL) with temperature control (e.g., Thermo Fisher Scientific Sorvall LYNX 4000 superspeed centrifuge).

10. 500 mL centrifuge bottles (autoclave sterilized).
11. Sterile wooden applicator for streaking bacterial strains (autoclave sterilized).
12. Centrifuge (50 mL) with temperature control (e.g., Eppendorf 5810R benchtop refrigerated swing-bucket centrifuge).
13. Protein electrophoresis apparatus (e.g., Bio-Rad Mini-PROTEAN system).
14. Incubator with shaker.
15. Nanodrop.
16. −80 °C freezer.
17. Cell homogenizer (e.g., Benchmark Scientific Pulse 150 Ultrasonic Homogenizer).
18. Square Petri plate (100 mm × 100 mm).

2.3 HA-Tagged Prey Protein Production

2.3.1 Reagents and Solutions

1. pGADT7/MED14 plasmids.
2. TNT T7 Coupled Reticulocyte Lysate System (Promega, Madison, WI).
3. Ribonuclease Inhibitor (e.g., Thermo Fisher Scientific RNaseOUT).

2.3.2 Equipment

1. DNase/RNase-free PCR tubes.
2. Thermal cycler.
3. 1.7 mL DNase/RNase-free Eppendorf tubes.

2.4 GST Pull-Down Assay

1. *E. coli* expressed GST and GST-PIF4 lysates (filtered).
2. In vitro transcribed/translated MED14-HA protein.
3. Glutathione sepharose 4B beads (GE Healthcare, Chicago, IL).
4. EB Wash buffer.
5. 5× SDS sample buffer: 250 mM Tris-HCl (pH 6.8), 10% (w/v) SDS, 50% glycerol, 0.5 M DTT, 0.25% bromophenol blue. Leave one aliquot at RT for current use and store others at −20 °C.

2.5 Immunoblotting

2.5.1 Reagents and Solutions

1. Polyacrylamide gels.
2. MOPS buffer (1×) with freshly added sodium bisulfite.
3. Protein molecular weight markers.
4. Relevant primary and secondary antibodies (e.g., rabbit anti-HA primary antibody and goat-anti-rabbit secondary antibody conjugated with Horseradish Peroxidase (HRP)).
5. Tris-buffered saline (TBS; 10×): 1.5 M NaCl, 0.1 M Tris–HCl (pH 7.4).

6. TBST: TBS containing 0.05% Tween-20.
7. Enhanced chemiluminescence (ECL) western blotting substrate (e.g., Thermo Fisher Scientific SuperSignal West Dura Extended Duration Substrate).

2.5.2 Equipment

1. Protein electrophoresis apparatus.
2. Wet tank blotting apparatus (e.g., Bio-Rad Criterion Blotter).
3. Nitrocellulose membranes.
4. Square Petri plate (100 mm × 100 mm).
5. Horizontal plate shaker.
6. Chemiluminescence Imaging Systems (e.g., Azure C600 advanced imaging system).

2.6 Quantification and Statistical Analysis

1. Quantification of band intensity: the National Institutes of Health (NIH) ImageJ or equivalent software.
2. Statistical analysis: Graphpad Prism 10, Microsoft Excel, or equivalent software.

3 Methods

Carry out all procedures at room temperature unless otherwise specified. The generation of vectors (*see* Subheading 3.1) takes around 1–2 weeks to complete depending on the sequence integrity. Production of the recombinant GST-PIF4 protein (*see* Subheading 3.2) can be performed before the pull-down assays (*see* Subheading 3.4), and the protein lysates can be stored at −80 °C.

3.1 Generation of Expression Vectors for Bait and Prey

1. Design primers to clone *PIF4* into pET42b and *MED14* into pGADT7 (*see* **Note 6**).
2. Amplify the PIF4 and MED14 sequences from plant cDNA by PCR using high-fidelity Q5 DNA polymerase.
3. Digest the pET42b and pGADT7 plasmids using appropriate restriction enzymes and their respective buffers (*see* **Note 5**).
4. Ligate the amplified *PIF4* and *MED14* sequences into the linearized pET42b and pGADT7 vector, respectively (*see* **Note 8**).
5. Transform the ligated pET42b/PIF4 and pGADT7/MED14 into DH5α competent cells using the heat-shock method (*see* **Note 9**). Grow the transformed cells in LB broth in the incu-shaker at 37 °C for 1 h.
6. Streak the transformed cells to LB agar media containing appropriate antibiotics and incubate the Petri plates at 37 °C overnight.

7. After antibiotic screening, inoculate and grow the transformed *E.coli* cells in LB broth with antibiotics at 37 °C overnight. (*see* **Note 10**).
8. Perform plasmid extraction and purification using the Zymo-Pure Plasmid Miniprep Kit.
9. Verify sequence integrity by Sanger sequencing.

3.2 Expression of the Bait Protein in E. Coli (See Note 11)

1. Transform the plasmid pET42b/PIF4 into *E. coli* strain BL21 (DE3) using electroporation (*see* **Note 12**). Use the following electroporation settings for 1.0 mm cuvettes: capacitance, 10 μF; resistance, 600 Ω; voltage, 1.8 kV.
2. Grow transformed BL21 (DE3) cells in 2 mL of LB broth in a sterile 15 mL culture tube for 1 h using an incu-shaker at 37 ° C.
3. Plate 200 μl of BL21 (DE3) culture into an LB agar plate with 30 μg/mL kanamycin and 34 μg/mL chloramphenicol and grow overnight in a 37 °C incubator (*see* **Note 12**).
4. Pick a single colony and inoculate in 11 mL LB broth containing antibiotics in a 50 mL culture tube. Resuspend well by vortexing, and incubate overnight in an incu-shaker (37 °C).
5. The next day, dilute 10 mL overnight culture into 1 L of LB broth (autoclave sterilized, without antibiotics) in a 2-L conical flask. (Optional) Save the strain by mixing ~700 μL of the overnight culture with glycerol stock at 1:1 (v/v) and store in the −80 °C freezer.
6. Incubate the 1-L culture in an incu-shaker at 37 °C for 3–4 h (180–200 rpm) until the optical density (OD_{600}) is 0.5 (*see* **Note 13**).
7. Add 1 mL of the culture into a sterile 1.7-mL Eppendorf tube and label it as the uninduced fraction (−) (*see* **Note 14**).
8. Induce the expression of the recombinant fusion protein in the 1-L culture by adding IPTG to a final concentration of 0.5 mM.
9. Incubate the 1-L culture and the uninduced fraction (−) in an incu-shaker at 37 °C for 3–4 h (180–200 rpm) (*see* **Note 14**).
10. Place the culture and the (−) sample into an ice bath. Add 1 mL of the induced culture into a sterile 1.7-mL Eppendorf tube, and label it as the induced fraction (+) (*see* **Note 14**).
11. Spin the (−) and (+) fractions at 20,000 × g for 2 min. Discard the supernatant and resuspend the bacterial pellet in 100 μL of 1× SDS sample buffer. Store the sample at −20 °C for step 30 (*see* **Note 14**).
12. Pour the induced culture into centrifuge bottles and spin at 5000 × g at 4 °C for 15 min (*see* **Note 15**).

13. Resuspend the bacterial pellet in ~30 mL of EB buffer and transfer it into a 50 mL Falcon tube using a squeezable transfer pipette. Centrifuge at 5000 × g at 4 °C for 10 min (*see* **Note 16**).
14. Discard the supernatant, aspirate any excess liquid from the tube, and measure the weight of the bacterial pellet (*see* **Note 17**).
15. Label and save the pellet at −80 °C until use. If using the frozen pellet, thaw the cell pellet in cold tap water before adding EB buffer. Alternatively, use the fresh cell pellet for protein extraction by following the procedure below. Perform all the steps on ice or at 4 °C.
16. Add 500 μL of PMSF, 500 μL of Halt protease inhibitor cocktail (100×), and 100 μL of DTT to 50 mL of cold EB buffer.
17. Add 5 mL of inhibitor-containing EB buffer to each gram of pellet. Resuspend the pellet by using a squeezable plastic transfer pipette.
18. Lyse the cells in the resuspended pellet using the Benchmark Scientific Pulse 150 Ultrasonic Homogenizer with the following settings: 50% power output, 4 s ON/4 s OFF per cycle, 10 cycles (*see* **Note 18**).
19. Centrifuge at 20,000 × g at 4 °C for 20 min to pellet down cellular debris.
20. Transfer the lysate to a prechilled fresh 50 mL Falcon tube.
21. Dilute 20 μL of the lysate in 1× SDS sample buffer to make a final volume of 100 μL. Save in −20 °C as 'cell lysate' (CL) for step 30. Add an equal volume of EB buffer to the pellet (step 19) and resuspend. Dilute 20 μL of this mixture in 1× SDS sample buffer to make a final volume of 100 μL. Save at −20 °C as 'cell pellet' (CP) for step 30 (*see* **Note 14**).
22. Add 0.4368 g/mL of ammonium sulfate to the lysate (step 20). Rotate at 4 °C for 4 h to precipitate proteins.
23. Centrifuge at 10,000 × g at 4 °C for 30 min to pellet down proteins.
24. Discard the supernatant and resuspend the protein pellets in the same volume of cold EB buffer (with inhibitors).
25. Centrifuge at 20,000 × g at 4 °C for 30 min to remove debris.
26. Collect the supernatant and dialyze overnight against 500 mL EB buffer (with DTT) per 10 mL sample at 4 °C.
27. The next day, dialyze against 500 mL freshly prepared EB buffer (with DTT) per 10 mL sample at 4 °C for another 4 h.

28. Filter the dialyzed lysate to a fresh 50 mL Falcon tube using a 0.22 μm PES syringe filter.
29. Save the filtered lysate in 500-μL aliquots at −80 °C.
30. Boil the (−) and (+) samples (step 11) and CL and CP samples (step 21) for 10 min. After cooling them to room temperature, vortex and centrifuge the tubes at 20,000 × g for 2 min.
31. Load 10 μL of each sample and run an SDS-PAGE gel with a protein molecular weight marker.
32. Stain the gel with Coomassie brilliant blue R-250 stain to check the protein induction, expression, solubility, and quality in the lysate.

3.3 HA-Tagged Prey Protein Production in Vitro

1. Thaw the TNT RNA Polymerase on ice. Thaw all the other components in the TNT T7 Coupled Reticulocyte Lysate System at room temperature and immediately place them on ice.
2. Prepare the following reaction mixture for expressing MED14-HA on ice (*see* **Note 20**):

Reagents	For 50 μL
DNA template	2 μg
TNT rabbit reticulocyte lysate	25 μL
TNT reaction buffer	2 μL
TNT RNA polymerase (T7)	1 μL
Amino acid mixture, minus leucine	0.5 μL
Amino acid mixture, minus methionine	0.5 μL
Ribonuclease inhibitor (*see* **Note 19**)	1 μL
Nuclease-free water	To 50 μL

3. Incubate the reaction at 30 °C for 90 min using a thermal cycler.
4. Transfer 2.3 μL (10% of reaction volume) of each lysate into a 1.7-mL Eppendorf tube and mix with 60 μL of 1× SDS sample buffer. This will serve as the "input" sample.
5. Immediately use the lysate for GST pull-down assays or save the lysate and the "input" sample at −20 °C until the experiment.

3.4 GST Pull-Down Assay

1. If frozen, thaw GST and GST-PIF4 lysate produced in 3.2.
2. Resuspend the glutathione sepharose 4B beads in the original container until homogeneous.
3. Transfer 1 mL of the suspension into a 1.7-mL Eppendorf tube and centrifuge at 500 × g for 1 min.

4. Remove the supernatant and wash the beads with 1 mL of cold EB buffer. Centrifuge again at 500 × g for 1 min.
5. Repeat the wash (step 4) 3 times to equilibrate the beads.
6. Add one bed volume of cold EB buffer to the beads to make 50% slurry.
7. For each prey, use appropriate amounts of GST and GST-PIF4 lysates and dilute to 1 mL using cold EB buffer in a 1.7-mL Eppendorf tube.
8. Add 50 μL of the equilibrated beads (step 6) to the diluted lysate.
9. Incubate the mixtures at 4 °C for 1.5 h with end-to-end rotation.
10. Centrifuge the mixtures at 500 × g for 1 min and carefully remove the supernatant. Ensure to leave a cushion of liquid above the pellet to avoid loss of samples.
11. Add 1 mL cold Wash buffer and invert the tube several times to resuspend the pellet.
12. Repeat the wash (steps 10–11) 4 times.
13. After the fourth wash, add 500 μL of Wash buffer to each tube of the protein-bead mixture.
14. Add 23 μL of the MED14 lysate (produced in 3.3) to the GST and GST-PIF4 tubes and incubate the mixtures at 4 °C for 2 h with end-to-end rotation.
15. Wash the beads 4 times by following steps 10–11. For the last two washes, incubate the beads in 500 μL Wash buffer with end-to-end rotation for 10 min.
16. Remove as much Wash buffer as possible after the final wash.
17. Add 60 μL of 2× SDS sample buffer to each tube, mix, and store at −20 °C until use.

3.5 SDS-PAGE Gel Electrophoresis and Immunoblotting

1. Thaw the "input" (step 3.3) and "pull-down" samples (step 3.4) at room temperature and boil for 10 min. Centrifuge at 20,000 × g for 2 min and use the supernatant for SDS-PAGE.
2. For each pull-down assay, run two identical gels in the following order: protein marker (3 μL), input (30 μL), GST pull-down sample (30 μL), and GST-PIF4 pull-down sample (30 μL).
3. Electrophorese at 120 v until the dye front (from the bromophenol blue dye in the samples) reaches the bottom of the gel.
4. Use one SDS-PAGE gel for Coomassie brilliant blue R-250 staining to detect immobilized GST and GST-PIF4 proteins (Fig. 2).

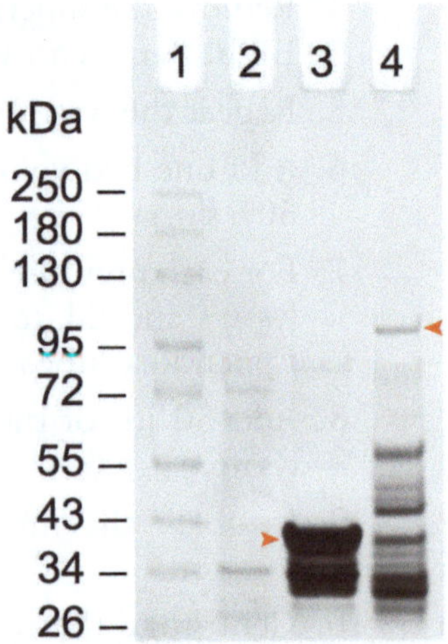

Fig. 2 Coomassie brilliant blue R-250 staining for bait protein visualization. *Lane* 1 shows prestained SDS-PAGE molecular weight standards. *Lane* 2 shows the input prey protein produced using the TNT T7 coupled reticulocyte lysate system. *Lane* 3 and *Lane* 4 show the GST and GST-PIF4 proteins immobilized on the glutathione sepharose 4B beads, respectively. The full-length protein band is indicated by an orange arrowhead. Multiple degradation bands may be visible under the full-length protein band

5. Use the other gel to perform immunoblot and blot into a nitrocellulose membrane using a wet tank blotting apparatus. Set the voltage to 70 v and transfer the proteins for 1.5 h at 4 ° C.
6. Briefly rinse the blotted nitrocellulose membrane with ddH_2O and block it with 15 mL TBS containing 2% non-fat milk for 1 h at room temperature.
7. Incubate the blocked membrane with anti-HA primary antibodies (diluted in 15 mL TBS containing 2% non-fat milk) overnight at 4 °C on an orbital shaker.
8. Wash the membrane 4× with 15 mL TBST at room temperature, 10 min each time.
9. Incubate the membrane with HRP-conjugated secondary antibodies (diluted in 15 mL TBS containing 2% non-fat milk) for 30–45 min at room temperature on an orbital shaker.
10. Wash the membrane 4× with 15 mL TBST at room temperature, 10 min each time.
11. Cut a piece of plastic wrap with a size larger than the membrane and lay it on an even surface.
12. Place the membrane on the plastic wrap and add the ECL reagent.

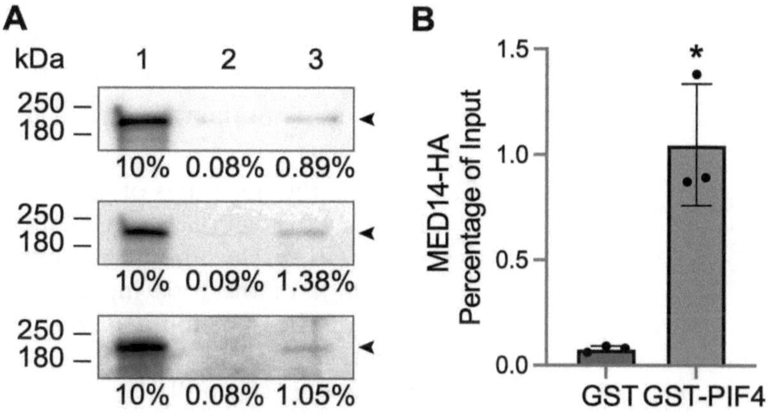

Fig. 3 Immunoblots for detecting the amount of prey proteins recovered from GST pull-down assays. (**a**) Three replicates of immunoblots in which anti-HA polyclonal antibodies are used to detect the amount of MED14-HA (black arrowheads) co-precipitated by GST-PIF4. *Lane* 1 shows 10% input of MED14-HA produced by the TNT T7 coupled reticulocyte lysate system. *Lane* 2 and *Lane* 3 show the amount of MED14-HA co-precipitated by GST and GST-PIF4 proteins, respectively. Numbers under the immunoblots indicate the percentages of MED14-HA input used in the GST pull-down assays. (**b**) The amounts of MED14-HA co-precipitated by GST and GST-PIF4 in (a) are quantified using NIH ImageJ and shown as a bar figure. An unpaired *t*-test with Welch's correction is used to demonstrate the statistical significance between the amounts of MED14-HA precipitated by GST-PIF4 and GST ($p = 0.028$, $n = 3$)

13. Incubate the membrane with the ECL reagent for 5 min at room temperature.
14. Visualize the input and pull-down fractions of MED14 using a chemiluminescence imaging system (Fig. 3a). Save the images for data analysis.

3.6 Quantification and Data Analysis

1. Open the saved immunoblot image in ImageJ (*see* **Note 21**).
2. Select an area in the background region using the "Rectangle" tool. Go to "Process" > "Subtract Background" to remove the background intensity.
3. Use the "Rectangle" tool to draw a frame around the protein band in *Lane* 1 (10% input). Press "1" or go to "Analyze" > "Gels" > "Select First Lane." Move the same frame to the following lanes—*Lane* 2 (GST) and *Lane* 3 (GST-PIF4)—press "2" or go to "Analyze" > "Gels" > "Select Next Lane" (*see* **Note 22**). Once all the protein bands have been selected, press "3" or go to "Analyze" > "Gels" > "Plot Lanes" to draw a profile plot of each lane.
4. On the profile plot, use the "Straight" tool to draw a line across the base of each peak to enclose the peak (*see* **Note 23**).

5. Utilize the "Wand" tool to click within each enclosed peak. Upon highlighting each peak, measurements will automatically appear in the "Results" window that emerges.
6. Copy and paste the values from the "Results" window into a spreadsheet.
7. Calculate the quantities of prey proteins (MED14-HA) precipitated by the bait (GST or GST-PIF4) using the values of peak density (PD). The amount of prey proteins will be shown as a percentage of the total input by comparing it with the PD values of the 10% input (*Lane* 1). For *Lane* 2 (GST), the percentage is calculated as $PD_{Lane2}/PD_{Lane1} \times 10\%$; similarly, for *Lane* 3 (GST-PIF4), the percentage is calculated as $PD_{Lane3}/PD_{Lane1} \times 10\%$. The calculated values are shown under the immunoblot images in Fig. 3a.
8. The same GST pull-down assay should be repeated at least two more times, and a minimum of three immunoblots should be generated. Repeat the band quantifications for all the replicates.
9. Using the data from all the replicates, create a bar graph depicting the quantities of precipitated prey proteins by the bait, presented as a percentage of the total input (Fig. 3b). Conduct statistical analysis, such as an unpaired t-test with Welch's correction, to assess the significant interaction between the bait protein and the prey.

4 Notes

1. It is generally a good idea to generate reciprocal bait/prey vectors. Even though only one way is shown here, performing the reciprocal GST pull-down assay is also recommended.
2. Generation of both N-terminal and C-terminal GST-tagged bait is recommended due to the probability of GST protein interference during protein-protein interaction. Adding a flexible polypeptide linker (e.g., GS linker) between GST and the bait protein is also recommended.
3. When using GST pull-down to test the interaction of two known proteins, if the prey protein-specific antibodies are unavailable, the gene encoding the prey protein can be cloned into an expression vector containing an epitope tag such as an HA-tag, which can be used to detect the presence of the prey protein during the subsequent immunoblotting steps.
4. Use an expression vector with a promoter corresponding to the in vitro transcription/translation system. For example, since we decided to use the TNT T7 Coupled Reticulocyte Lysate System (Promega, Madison, WI) for in vitro transcription/

translation, we chose an expression vector with the T7 promoter (e.g., pCMX or pGADT7) for the prey protein expression.

5. BamHI and HindIII were used to digest pET42b for the *PIF4* insertion, and EcoRI and BamHI were used to digest pGADT7 for *MED14* insertions.
6. Standard molecular techniques are used to construct both expression vectors. Gibson assembly was employed to ligate both the bait expression vector (pET42b containing the GST-PIF4) and the prey expression vector (pGADT7 containing the MED14-HA driven by the bacteriophage T7 promoter).
7. The percentage of polyacrylamide can be increased to 10–12% for smaller proteins for better band separation. The volume of MOPS buffer in the electrophoresis tank depends upon the number of gel cassettes being run. Use 500 mL of 1× MOPS buffer for one cassette (2 gels) and 1 L of 1× MOPS buffer for two cassettes (4 gels).
8. The vector:insert molar ratio during ligation should be between 1:2 and 1:10 for high efficiency.
9. Competent cells must be stored at −80 °C and thawed in ice for ~10 min before adding the construct. Gentle tapping can ensure a homogeneous mixture between the competent cells and the construct. The mixture should be incubated for ~30 min before the heat-shock treatment. Heat-shock treatment at 42 °C for 45 sec is optimal.
10. The plasmid selection markers for pET42b and pGADT7 are kanamycin (30 μg/mL) and ampicillin/carbenicillin (100 μg/mL), respectively.
11. *Escherichia coli* strains, such as BL21 (DE3), and the pET expression system are widely utilized for high-level recombinant protein expression. In these systems, the recombinant protein expression is driven by the T7 promoter, a powerful and tightly regulated promoter derived from the T7 bacteriophage. The gene of interest is cloned downstream of the T7 promoter in an expression vector. Expression from the T7 promoter is driven by the chromosomally encoded bacteriophage T7 RNA polymerase [15]. Upon the addition of IPTG, the lacUV5 promoter in the BL21 (DE3) strain is activated, producing T7 RNA polymerase that binds specifically to the T7 promoter on the expression vector and initiates transcription of the gene of interest, resulting in the synthesis of mRNA encoding the recombinant protein. Subsequent translation of the mRNA by ribosomes leads to the expression of the recombinant fusion protein. However, it is essential to note that prokaryotic systems like *E. coli* lack the machinery to

perform post-translational modifications (PTMs) and may not be suitable for expressing proteins requiring such PTMs or containing cysteine-rich motifs [16]. Researchers should consider alternative systems like yeast or insect cells for protein production to ensure accurate representation and functionality.

12. The amount of plasmids used for trasnsformation depends on the competent cells' transformation efficiency (TE). If using homemade BL21 (DE3) competent cells with a low TE (e.g., 1×10^5 cfu/μg) and aiming to obtain ~100 colonies, transform 10 ng of plasmid DNA into 50 μL of competent cells and plate 10% of the cell culture diluted in LB media. If using commercial competent cells with a high TE (e.g., 1×10^9 cfu/μg) and aiming to obtain ~100 colonies, transform 100 pg of plasmid DNA into 50 μL of competent cells, and plate 0.1% of the cell culture diluted in LB media.
13. The OD_{600} should be measured every 20 min after the first two hours to ensure bacterial cells are in the early logarithmic phase (0.4–0.6) before IPTG induction. To improve the yield of unstable fusion proteins, delay the addition of IPTG to about 1 h before cell harvest [12].
14. The induced (+) and uninduced (−) fractions are used later in step 3.2.4, along with the CL and CP fractions, to test whether the induction was successful. The Coomassie blue staining results should show a difference in band intensities between the (−) and (+) fractions. The band intensities in CL and CP fractions indicate whether the protein is soluble or insoluble.
15. The centrifuge bottles should only be 4/5 full, i.e., add a maximum of 400 mL bacterial culture into a 500 mL bottle.
16. The centrifuge bottle should be placed in an ice bath while the pellet is being resuspended.
17. Excess liquid can be drained from the tube by placing it upside down on a paper towel for a minute. The pellet weight can be calculated by subtracting the weight of an empty Falcon tube from that of the Falcon tube with the bacterial pellet.
18. The mixture should be sticky after the first few cycles of cell lysis using the homogenizer. This occurs due to the release of a large amount of DNA after cell lysis and can be used to verify the extent of cell lysis.
19. Ribonuclease inhibitors like RNasin or RNaseOUT can be used to limit RNA degradation.
20. Precise pipetting of the reaction reagents must be performed. Minimize bubble formation and reagent loss.
21. The immunoblot image needs to be a gray-scale image for this analysis in ImageJ. To convert the immunoblot image to grayscale, go to "Image" > "Type" > "8-bit".

22. When moving the frame drawn in *Lane* 1 to other Lanes, center the rectangle over the protein bands horizontally (from left to right) without the need for precise vertical alignment. ImageJ will automatically adjust the vertical alignment of the rectangle to match that of the first frame in the subsequent step.
23. The profile plot illustrates the relative density of frame contents across each lane. Since three lanes are selected, three profile plots will be generated. Despite background subtraction, peaks often do not extend to the profile plot's baseline. Closing off the peak is essential for size measurement. This step entails subjective judgment in determining the point where the peak concludes and background noise commences.

Acknowledgments

This work was supported by the National Science Foundation (NSF) grants IOS-2200200 and IOS-2239963 to Y.Q. The authors want to thank Anupa Wasti, Ranjeeta Odari, and other members of the Qiu Laboratory at the University of Mississippi for reading and discussing this book chapter during the manuscript preparation. This method is modified based on a GST pull-down protocol developed in Dr. Meng Chen's laboratory at the University of California, Riverside.

References

1. Casal JJ, Balasubramanian S (2019) Thermomorphogenesis. Annu Rev Plant Biol 70:321–346
2. Koini MA, Alvey L, Allen T et al (2009) High temperature-mediated adaptations in plant architecture require the bHLH transcription factor PIF4. Curr Biol 19:408–413
3. Sun J, Qi L, Li Y et al (2012) PIF4-mediated activation of YUCCA8 expression integrates temperature into the auxin pathway in regulating Arabidopsis hypocotyl growth. PLoS Genet 8:e1002594
4. Kumar SV, Lucyshyn D, Jaeger KE et al (2012) Transcription factor PIF4 controls the thermosensory activation of flowering. Nature 484: 242–245
5. Thines BC, Youn Y, Duarte MI et al (2014) The time of day effects of warm temperature on flowering time involve PIF4 and PIF5. J Exp Bot 65:1141–1151
6. Lau OS, Song Z, Zhou Z et al (2018) Direct control of SPEECHLESS by PIF4 in the high-temperature response of Stomatal development. Curr Biol 28:1273–1280.e3
7. Qiu Y (2020) Regulation of PIF4-mediated thermosensory growth. Plant Sci 297:110541
8. Lee S, Paik I, Huq E (2020) SPAs promote thermomorphogenesis by regulating the phyB-PIF4 module in Arabidopsis. Development 147. https://doi.org/10.1242/dev.189233
9. Nieto C, López-Salmerón V, Davière J-M et al (2015) ELF3-PIF4 interaction regulates plant growth independently of the evening complex. Curr Biol 25:187–193
10. Bajracharya A, Xi J, Grace KF et al (2022) Phytochrome-interacting factor 4/hemera-mediated thermosensory growth requires the Mediator subunit MED14. Plant Physiol. https://doi.org/10.1093/plphys/kiac412
11. Smith DB, Davern KM, Board PG et al (1986) Mr 26,000 antigen of Schistosoma japonicum recognized by resistant WEHI 129/J mice is a parasite glutathione S-transferase. Proc Natl Acad Sci USA 83:8703–8707

12. Smith DB, Johnson KS (1988) Single-step purification of polypeptides expressed in Escherichia coli as fusions with glutathione S-transferase. Gene 67:31–40
13. Richter WF, Nayak S, Iwasa J et al (2022) The Mediator complex as a master regulator of transcription by RNA polymerase II. Nat Rev Mol Cell Biol 23:732–749
14. Green MR, Sambrook J (2020) The Inoue method for preparation and transformation of competent Escherichia coli: "Ultracompetent" cells. Cold Spring Harb Protoc 2020:101196
15. Studier FW, Moffatt BA (1986) Use of bacteriophage T7 RNA polymerase to direct selective high-level expression of cloned genes. J Mol Biol 189:113–130
16. Demain AL, Vaishnav P (2009) Production of recombinant proteins by microbes and higher organisms. Biotechnol Adv 27:297–306

Chapter 20

Investigating Circadian Gating of Temperature Responsive Genes

Rachel I. Strout, Calum A. Graham, Antony N. Dodd, and Dawn H. Nagel

Abstract

Understanding gene expression dynamics in the context of the time of day and temperature response is an important part of understanding plant thermotolerance in a changing climate. Performing "gating" experiments under constant conditions and light-dark cycles allows users to identify and dissect the contribution of the time of day and circadian clock to the dynamic nature of stress-responsive genes. Here, we describe the design of specific laboratory experiments in plants (*Arabidopsis thaliana* and bread wheat, *Triticum aestivum*) to investigate temporal responses to heat (1 h at 37 °C) or cold (3 h at 4 °C), and we include known marker genes that have circadian-gated responses to temperature changes.

Key words Phase, Gating, Entrainment, Temperature stress, Transcript abundance

1 Introduction

A fundamental property of the circadian clock is to control the expression of a repertoire of genes involved in key biological processes [1, 2]. Modulation of the dynamics of gene expression responses to fluctuating daily environmental stimuli or stressors, according to the time of day, is referred to as circadian or time of day gating of the response [1]. Depending on the context of the organism and stimuli, the concept of circadian gating can differ. For example, there is circadian gating of the emergence of adult *Drosophila* to specific times of the day, and circadian control of cell division can be confined to specific periods in a day [3, 4]. Circadian gating also acts on entrainment pathways. For example, the clock can gate environmental inputs such as light and temperature to appropriately synchronize and calibrate the phase of the oscillator [5–8]. In another context and increasingly demonstrated in plants, circadian gating or time of day control often considers the application of a short treatment (1–3 h) at different times of the day, and interpreting the subsequent gene expression responses based on

Meng Chen (ed.), *Thermomorphogenesis: Methods and Protocols*, Methods in Molecular Biology, vol. 2795,
https://doi.org/10.1007/978-1-0716-3814-9_20,

changes in transcript abundance or other readouts such as luciferase reporters of promoter activity and fluorescent proteins [9–21]. Both transcripts that are cycling (rhythmic) and non-cycling (non-rhythmic) in the absence of the stimulus can have a gated response to a given stimulus, regardless of whether the plants are cultivated under 24 h cycles of light and dark (diel cycles) or under conditions of constant light and temperature (free running conditions) (Fig. 1). Whilst the gating of stimulus-induced responses under cycles of light and dark might be regulated by the circadian clock, experimentation under constant light conditions or with clock mutants becomes necessary to confidently assign the gating response to the circadian clock. Understanding the concept of gating as it relates to stress responses is critical (i) in the context of climate change, because temperature variations can have negative impacts on plant growth and yield depending on the time of day; (ii) in unraveling the dynamics of gene regulatory networks involved in stress response and tolerance; (iii) in helping to interpret datasets generated by different research groups for which stress was applied at different times of the day. In this methods chapter, we provide a framework for designing an experiment to investigate circadian gating of temperature responses in plants, using *Arabidopsis thaliana* (*Arabidopsis*) and *Triticum aestivum* (hexaploid bread wheat) as examples in relation to heat (37 °C) and cold stress (4 °C), respectively [18, 19]. In addition, we provide examples of genes that can aid with the validation of a successful gating experiment.

2 Materials

All Petri dishes and reagents are prepared using sterile deionized water and under sterile conditions where appropriate. Calculate carefully the number of plants that are needed for the experiment. We recommend harvesting at least one spare sample per time point and condition, in case of sample loss during extraction. With large experiments involving long time courses, there is always a risk of sampling the wrong material, especially where several genotypes are used. One solution can be to label the plates or plant pots with colored tape, using different colors per genotype, because this can be more instantly recognizable than text on labels.

2.1 *Arabidopsis* Plant Growth

1. 1 × MS medium plates: 4.3 g/L Murashige and Skoog Basal Salts (Caisson Labs MSP01-50LT), 1.5% (w/v) Sucrose (Fisher Scientific S5-500), pH to 5.7–5.9 using KOH, add 8 g/L agar (Sigma-Aldrich A1296). Autoclave and prepare circular petri dishes (100 × 15mm, VWR 25384-088) containing 25–30 mL of MS media supplemented with 1.5% (w/v) sucrose (*see* **Note 1**).

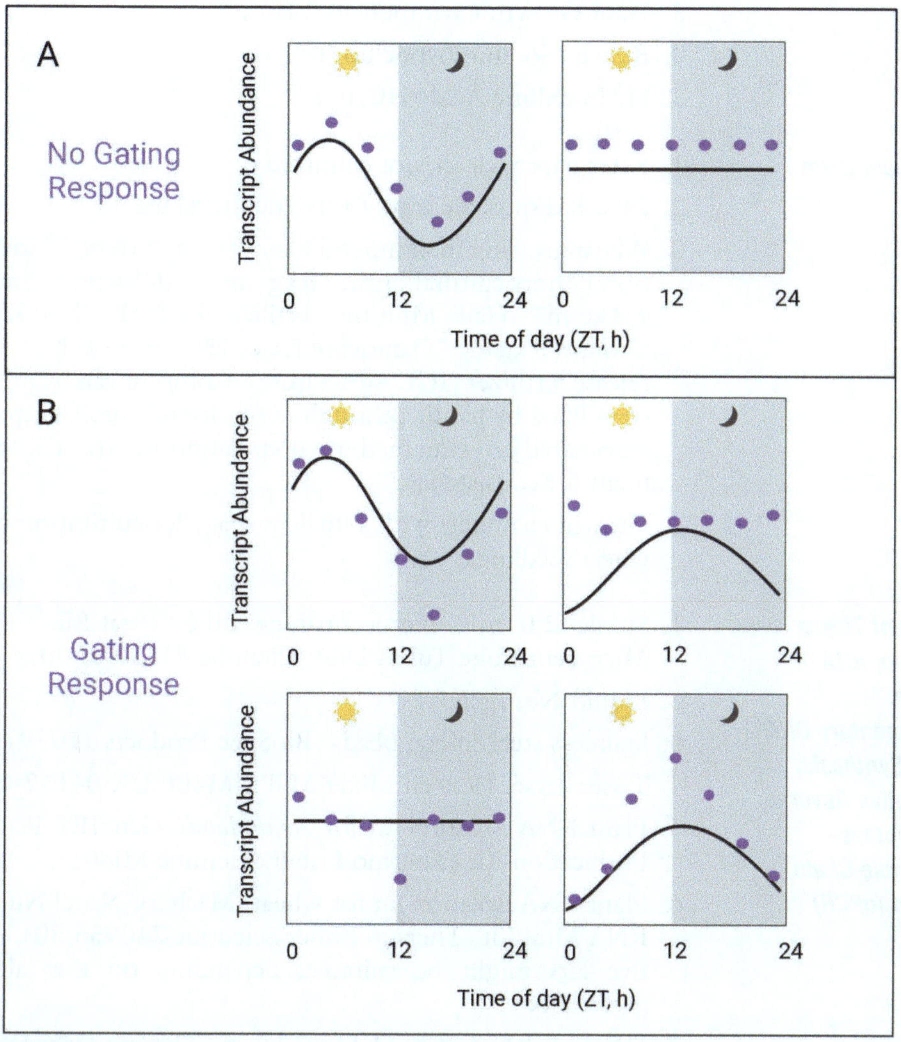

Fig. 1 Schematic representations of circadian gating responses. (**a**) Transcripts that either have a circadian rhythm (left panel) or do not have a circadian rhythm (right panel), and do not undergo circadian gating of their response to a stimulus. (**b**) Examples of circadian gating of a response to a stimulus. In some examples, there is a circadian rhythm in the magnitude of the response to a stimulus, where the response also oscillates with a 24 h rhythm. Other examples do not have a cycling pattern of response. Black lines represent the transcript abundance under control conditions. Purple circles indicate the relative transcript abundance after a temperature stimulus, given separately at each time point. The white area represents the subjective day (ZT0-12, yellow sun), and the gray areas (ZT12-24, moon symbol) represent the subjective night, under conditions of constant light and constant temperature. Time of day is represented as Zeitgeber or hours (ZT, h). Image created with Biorender

2. Grade 1 circular qualitative filter paper standard grade, sterilized by autoclaving (90 mm, VWR 28450-081). To autoclave, wrap in aluminum foil and place in an autoclavable vessel. Working in a laminar flow hood, use sterile forceps to place a single filter paper on the MS plates (*see* **Note 2**).

3. Plant Growth Chambers for plates.
4. Bleach (Sodium hypochlorite).
5. Hydrochloric Acid (HCl).

2.2 Wheat Plant Growth

1. Filter paper (clean, not sterilized).
2. 24-cell disposable trays for horticultural use
3. Wheat growing medium: 65% (v/v) peat, 25% (v/v) loam, 10% (v/v) horticultural grit, 3 kg m^{-3} dolomitic limestone, 1.3 kg m^{-3} Haifa Multimix fertilizer 14:16:18 (N:P:K) (Haifa Group), 3 kg m^{-3} Osmocote Exact 15:9:11 3/4 (N:P:K) slow-release fertilizer (ICL Agriculture). Compost can be mixed and trays filled by hand, although at the John Innes Centre we use automated growing medium preparation and tray filling equipment for consistency.
4. Growth chambers with sufficient space for cultivation of bread wheat seedlings.

2.3 Plant Tissue Collection, RNA Isolation, Complementary DNA (cDNA) Synthesis, Quantitative Reverse Transcriptase-Polymerase Chain Reaction (qPCR)

1. Sterile 2.0 mL microcentrifuge tubes (Seal-Rite® 2.0 mL Microcentrifuge Tubes USA Scientific #1620-2700).
2. Liquid Nitrogen.
3. Stainless steel/metal beads (BioSpec Products 11079132ss).
4. Tissue Lyser (Retsch Mixer Mill MM400 UX-04182-09).
5. Plant RNA isolation kit for *Arabidopsis*: GeneJET Plant RNA Purification Kit (Thermo Fisher Scientific K0801).
6. Plant RNA isolation kit for wheat: Machery-Nagel Nucleospin RNA Mini Kit (Thermo Fisher Scientific 740955.50). Alternative kits might be suitable, depending on the laboratory preference.
7. DNase I, RNase-free (1 U/μL) (e.g., Thermo Fisher Scientific EN0521, or as supplied with Machery-Nagel kit).
8. cDNA synthesis kit for *Arabidopsis*: iScript™ cDNA Synthesis Kit, which includes oligo(dT) and random primers as part of the 5X iScript Reaction Mix (Biorad 1708890),
9. cDNA synthesis kit for wheat: High-Capacity cDNA Reverse Transcription Kit, with random primers (Thermo Fisher Scientific, 4368814),
10. qPCR reaction mix for *Arabidopsis*: Maxima SYBR Green mix (Thermo Fisher Scientific K0252),
11. qPCR reaction mix for wheat: qPCRBIO SyGreen LO-ROX RT-qPCR reagent (PCR Biosystems, selecting reagent according to qPCR machine model),
12. RNaseZap™ RNase Decontamination Solution (Thermo Fisher Scientific AM9780).

13. Filter Pipette Tips (e.g., USA Scientific TipOne RPT ultra-low retention filter tips, 1000uL #1182–1830; 200ul #1180–8810; 10ul #1181–3810).
14. Real-Time PCR Detection System (e.g. CFX96 or CFX384 Real-Time PCR Machine Biorad).
15. Hard-Shell® 96-Well PCR Plates, low profile, thin wall, skirted, white/clear (Biorad HSP9601) or TempPlate polypropylene 384-well PCR plate (USA Scientific #1438–4790).
16. Microseal "B" PCR Plate Sealing Film, adhesive, optical (Bio-rad MSB1001).
17. Arabidopsis qPCR housekeeping gene primers: *IPP2* (IPP2-F: 5′- CTCCCTTGGGACGTATGCTG-3′; IPP2-R: 5′- TTGA ACCTTCACGTCTCGCA-3′) and *PP2A* (PP2A-F: 5′-TAAC GTGGCCAAAATGATGC -3′; PP2A-F: 5′- GTTCTCCACAACCGATTGGT-3′).
18. Wheat qPCR housekeeping gene primers: *TaHK4* (TaHK4-F: 5′- TCTAAATGTCCAGGAAGCTGTTA -3′; TaHK4-R: 5′- -CCTGTGGTGCCCAACTATT-3′).

2.4 Candidate Marker Genes with Circadian-Gated Responses

1. First, it is critical to confirm that the appropriate rhythmicity is observed in the control samples following collection and tissue processing, using a couple of known clock genes and qPCR analysis as described below (*see* **Note 3**). To validate the successful implementation of a gating experiment, several marker genes can be used to confirm the dynamics of time of day and temperature-dependent changes in transcript accumulation. As an example, *Arabidopsis GIGANTEA (GI)* transcript abundance has a circadian gated response to heat stress (1 h at 37 °C; Fig. 2a). GI is clock-regulated and functions in the photoperiod flowering pathway [22, 23]. Furthermore, in the context of temperature and gating, *GI* transcript accumulation increases at high ambient temperatures (28 °C) as compared to the cooler temperatures of 12 °C at dawn and is also cold-responsive [24, 25]. In recent years, several studies have reported a functional role for GI in the gating of light signals, hormone sensitivity, and shade avoidance responses [26–28]. Of note, in fundamental studies, GI has been reported to participate in temperature compensation of the clock, primarily at higher temperatures [23, 29]. Together, this implicates an important role for GI as a marker for abiotic stress responses and environmental changes (*see* **Note 4**). Likewise, there are excellent marker genes for the validation of cold-temperature responses in wheat. One example is the orthologs of the *CBF* gene family in wheat, which are strongly cold-induced and undergo circadian gating of their response to cold [11]. In wheat, we have used *TaCBFIIId_D19* (Fig. 2b).

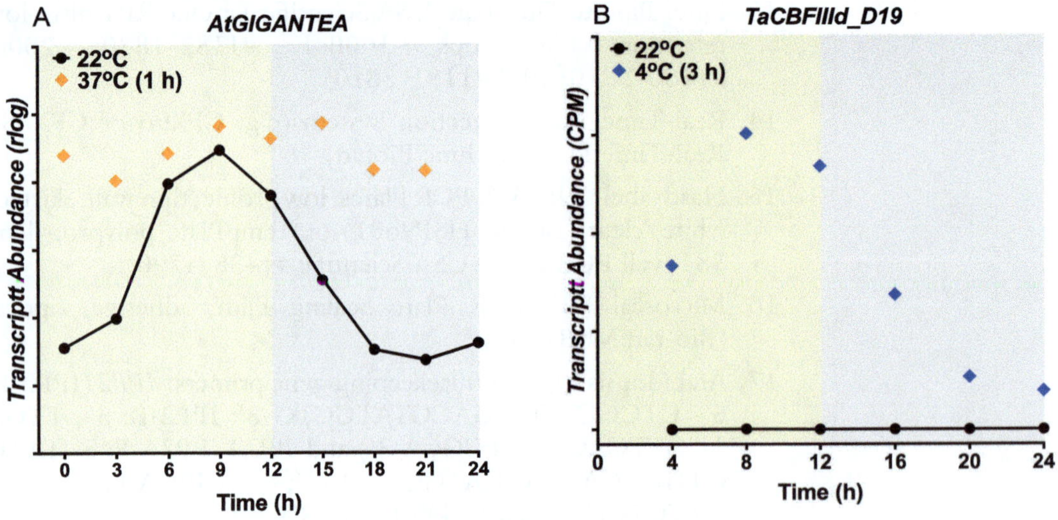

Fig. 2 Examples of genes that have circadian gating of their responses to acute temperature stimuli. (**a**) Arabidopsis *GIGANTEA* transcript abundance (*Rlog normalized counts*). Data are means for N = 3 biological replicates. 37 °C (orange diamond) and 22 °C (black line and circles). Each exposure to heat stress (37 °C) was for 1 h. Time (h) represents (ZT 48–77). *Arabidopsis* data reproduced from Bonnot et al., 2022 [18] with permission from Oxford Academic Press. (**b**) Wheat *CBF (TaCBFIIId_D19)* transcript abundance (*CPM, counts per million*). Data are means for N = 3 biological replicates. 4 °C (blue diamond) and 22 °C (black circles). Each exposure to cold stress (4 °C) was for 3 h and ended at the time shown on the x-axis. Time (h) represents (ZT 28-48). Data extracted from European Nucleotide Archive dataset PRJEB56923 (collected by Graham CA, Paajanen P, Edwards KJ, Dodd AN, within [19]). Yellow/gray shading = subjective day/night

3 Methods

Growth and experimental procedures are indicated where applicable. Otherwise, all procedures can be performed at room temperature.

3.1 Plant Growth and Tissue Collection

For *Arabidopsis*

1. Surface sterilize *Arabidopsis* seeds (100 mL of bleach +3–4 mL of Hydrochloric Acid (HCl)) in a glass desiccator jar for at least 3 h (*see* **Note 5**).
2. In a laminar flow hood, sprinkle the sterilized seeds onto the filter paper containing MS plates attempting to disperse the seeds evenly across the filter paper. Stratify the seeds for 2–4 days at 4 °C in the dark.
3. Transfer plates to a growth chamber (Percival, Perry, IA) with specified conditions depending on the plant. For *Arabidopsis*, grow seedlings in 12 h light (Fluorescent Tube, ~130 μmol m^{-2} s^{-1}): 12 h dark and constant 22 °C for 10 days. If using circadian conditions, on Day 11, transfer plates to conditions of

constant light (Fluorescent Tube, 80–100 μmol m^{-2} s^{-1}) and temperature for 2 days (*see* **Note 6**).

For Wheat

4. Place filter paper inside a petri dish and evenly spread wheat seeds over the filter paper. Moisten filter paper pieces with tap water so the entire paper is damp but seeds are not floating. The filter paper should be clean but does not need to be sterile.
5. Stratify the wheat seeds on filter paper at 4 °C for 3 days, then transfer to 22 °C in darkness for two days to germinate.
6. Pot germinated seedlings into a cereal cultivation mixture (see Materials and Methods for composition), in 24-cell horticultural trays.
7. Cultivate for 12 days under cycles of 12 h light/12 h darkness, at 22 °C, in MLR-352 growth chambers (Panasonic, Osaka, Japan; white light supplied by LED tubes). Rotate trays every three days to ensure even light distribution and consistent growth.

3.2 Temperature Stress and Sample Collection

For *Arabidopsis*

1. For heat shock gating of *Arabidopsis* seedlings, transfer plates to a growth chamber preset to 37 °C for 1 h every 3 or 4 h over a 24 h period. Control (non-stressed) plates should be kept at 22 °C (*see* **Note 7**).
2. Immediately collect whole seedlings in 2 mL centrifuge tubes from heat-stressed plates following the 1 h treatment and from the control plates. Snap freeze seedlings containing centrifuge tubes in liquid nitrogen (N_2). Tissue can either be immediately ground into powder or stored at −80 °C until processed. If using a grinding mill, include 3 metal beads in the tube before adding the sample (*see* **Note 8**).

For Wheat

3. For cold temperature treatment of wheat plants, transfer 14-day old seedlings (at Zadok Stage GS1.2) to conditions of continuous light and temperature (22 °C), 24 h before sampling begins. This ensures that the transitory effects of the final dawn are absent from the data.
4. After 25 h of constant light conditions, transfer three individual plants to 4 °C, with three control plants remaining at 22 °C. Both remain under white light (*see* **Note 9**).
5. After 3 h exposure to 4 °C, sample the tissue. Collect 3 cm of tissue from the second leaf into a pre-labeled Eppendorf tube

loaded with 2 or 3 metal (tungsten) beads. Freeze immediately in liquid N_2, and store at −80 °C for later processing. Repeat this process for the remaining two cold-treated individuals, and also for three control-temperature individuals.

6. Repeat this process after 29 h of constant light conditions and every further 4 h, for at least 24 h of sampling.

3.3 Total RNA Isolation

1. Clean the work area with RNaseZap prior to the RNA isolation procedure (*see* **Note 10**).
2. For *Arabidopsis*, grind seedlings using a tissue lyser (also known as a ball mill or bead beater) at a frequency rate of 28 Hz for 1 min or a mortar and pestle. Both the tissue lyser racks and mortar and pestle should be chilled with liquid nitrogen prior to grinding and kept cold throughout. For wheat, load Eppendorfs containing samples into a tissue lyser rack that has been stored at −80 °C. Grind wheat samples using a tissue lyser for 2 min at 28 Hz.
3. Isolate total RNA from *Arabidopsis* using the GeneJET Plant RNA Purification Mini Kit (*see* **Note 11**), and from wheat using the NucleoSpin RNA Mini Kit (Machery-Nagel, Düren, Germany), applying an on-column DNase treatment.
4. Check total RNA concentration using a nanodrop. A minimum of 150 ng/μL of total RNA is recommended for the next steps.
5. Check RNA integrity by running 1 μL of total RNA on a 1.5% agarose gel. Intact high-quality total RNA will have sharp and distinct 28S and 18S rRNA bands in a ~ 2:1 ratio (28S:18S) (*see* **Note 12**). Alternatively, RNA integrity can be assessed using a 2100 Bioanalyzer System (Agilent, Santa Clara, USA), confirming RIN scores >6.

3.4 Quantitative Real-Time PCR

1. For *Arabidopsis*, treat 1 μg of total RNA with DNase I to remove DNA contamination following the manufacturer's protocol. For Thermo Fisher DNase I, incubate the reaction at 37 °C for 30 min. Add 1 μL of 50 mM EDTA and incubate at 65 °C for 10 min (*see* **Note 13**). Alternatively, the wheat RNA isolation protocol uses an on-column DNA digest.
2. For quantitative real-time PCR, perform complementary DNA (cDNA) synthesis using the preferred kit. For iScript cDNA Synthesis Kit (Bio-Rad), add 4 μL of 5× iScript Reaction Mix, 1 μL of iScript RT, and 4.1 μL of kit provided H_2O to the 10.9 μl of DNase treated total RNA. Incubate the cDNA synthesis reaction in a PCR machine as follows: 5 min at 25 °C, 30 min at 42 °C, 5 min at 85 °C, and hold at 4 °C.
3. Dilute the cDNA by adding 180 μL of sterile H_2O.

4. Perform quantitative real-time PCR using an SYBR Green kit such as Maxima SYBR Green mix (Thermo Fisher Scientific). Include at least 2 known and well-established housekeeping genes such as *Protein Phosphate 2A (PP2A)* and *isopentyl-diphosphate delta-isomerase II (IPP2)* to be to normalize controls for relative transcript abundance (*see* **Note 14**). For wheat, *TraesCS5A02G015600 (Traes_5AS_019ECA143)* that encodes an ion exchange channel can be used as a stable housekeeping gene (as well as this gene, other candidates are also proposed in Borrill et al., 2016 [30]). Ensure appropriate tests for efficiency of amplification and primer specificity are conducted before embarking on the analysis of large numbers of samples.

3.5 RNA-Sequencing

1. For genome-wide studies and questions concerning circadian gating, mRNA sequencing can be performed on the collected samples using a variety of published approaches [16, 19, 31].

3.6 Data Interpretation

1. A variety of strategies can be considered for the interpretation of the data. First, it is important to confirm that transcripts known to be rhythmic under control temperature conditions have the expected circadian rhythm. This is necessary to ensure that the sampled plant material had been entrained to environmental conditions, before investing time and research funds in further sample analysis. To subsequently test the hypothesis that a transcript undergoes circadian gating of its response to a stimulus, it is necessary to test whether there is a 24 h fluctuation in its response to the stimulus (Fig. 1). At a straightforward level, it is possible to test whether the response to the stimulus has a circadian rhythm in its magnitude. This approach is informative when the number of time points tested is relatively small. For example, if the transcript abundance of a selected gene increases or decreases in magnitude at one or more but not all time points, it could be considered to be gated. This analysis can be elaborated, for example, to test whether the magnitude of response to the stimulus itself has a circadian rhythm. One such approach is to calculate the difference, for each time point, between the transcript abundance at the control temperature and after the temperature treatment. It is then possible to test whether this time series of differences has a circadian rhythm, using a tool such as MetaCycle or BioDare2 [32–34]. Downstream experiments can then be designed based on the gating of the temperature response of the gene(s) of interest, focusing on regulatory players or functional relevance. The approach also enables researchers to subsequently select appropriate times of day to perform a temperature treatment to obtain the most appropriate and reproducible response.

4 Notes

1. The inclusion of sucrose in cultivation media is optional and somewhat laboratory-dependent, but an important consideration and decision because exogenous sucrose can affect the functioning of the *Arabidopsis* circadian clock [35].
2. The use of filter paper allows easier collection of seedlings and prevents root damage during harvest. Users can choose to grow seedlings on a membrane rather than filter paper, and some users prefer to allow roots to grow into the agar and harvest the aerial parts of the seedlings.
3. Following the collection of the time course of samples and tissue processing, it is strongly recommended to examine by qPCR a morning-expressed and evening-expressed clock gene, to confirm that the samples were collected at the correct times of day, that known clock genes are peaking at the correct times in the control samples, and to confirm that abnormal samples are not present, before investing time in further analysis.
4. Although *GIGANTEA (GI)* is used as a marker gene for gating here, other genes may be more suitable depending on the plant species, developmental age, and nature of the temperature or stress treatment.
5. Other methods of seed sterilization can be used such as 70% (v/v) Ethanol/0.1% (v/v) Triton X-100 solutions. If using Bleach/HCl, it is important that tubes are labeled with a pencil rather than a pen, to prevent bleaching of the labels.
6. Different light and temperature regimes, growth duration, and control conditions can be used, depending on the plant species.
7. The temperature and duration used for heat stress may vary depending on the plant species. Alternative methods to apply heat stress can be used.
8. Other methods to grind tissue finely may be used. It is important that the tissue remains frozen until the start of the total RNA isolation procedure.
9. It is highly recommended that you include additional plants as replicates in the event that plants/samples are compromised during subsequent steps.
10. Other RNA decontamination methods can be used.
11. Other plant RNA isolation kits or Trizol methods can be used.
12. Checking RNA integrity is optional but recommended prior to cDNA synthesis.
13. cDNA synthesis kits from other manufacturers can be used.

14. SYBR green kits from other manufacturers may be used. Both *IPP2* (IPP2-F: 5′- CTCCCTTGGGACGTATGCTG-3′; IPP2-R: 5′- TTGAACCTTCACGTCTCGCA-3′) and *PP2A* (PP2A-F: 5′-TAACGTGGCCAAAATGATGC-3′; PP2A-F: 5′-GTTCTCCACAACCGATTGGT-3′) are commonly used in circadian studies in Arabidopsis. In wheat, *TaHK4* (TaHK4-F: 5′-TCTAAATGTCCAGGAAGCTGTTA-3′; TaHK4-R: 5′-CCTGTGGTGCCCAACTATT-3′) have been used. It is important to validate that your housekeeping gene does not display robust circadian cycling or has altered expression based on the treatment.

Acknowledgments

This work was supported by NSF Early Career Award IOS 1942949 to D.H.N. and BBSRC Institute Strategic Programmes GEN (BB/P013511/1) and BRiC (BB/X01102X/1) to A.N.D., and BBSRC SWBio DTP grants BB/J014400/1 and BB/M009122/1 to A.N.D. and C.A.G. The authors declare no competing interests.

References

1. Paajanen P, de Barros L, Dantas L, Dodd AN (2021) Layers of crosstalk between circadian regulation and environmental signalling in plants. Curr Biol 31:R399–R413. https://doi.org/10.1016/j.cub.2021.03.046
2. Harmer SL, Hogenesch JB, Straume M et al (2000) Orchestrated transcription of key pathways in *Arabidopsis* by the circadian clock. Science 290:2110–2113. https://doi.org/10.1126/science.290.5499.2110
3. Pittendrigh CS, Skopik SD (1970) Circadian systems, V. The driving oscillation and the temporal sequence of development. Proc Natl Acad Sci USA 65:500–507. https://doi.org/10.1073/pnas.65.3.500
4. Sweeney BM, Hastings JW (1958) Rhythmic cell division in populations of *Gonyaulax polyedra*. J Protozool 5:217–224. https://doi.org/10.1111/j.1550-7408.1958.tb02555.x
5. McWatters HG, Bastow RM, Hall A et al (2000) The ELF3 zeitnehmer regulates light signalling to the circadian clock. Nature 408: 716–720. https://doi.org/10.1038/35047079
6. Covington MF (2001) ELF3 modulates resetting of the circadian clock in *Arabidopsis*. Plant Cell 13:1305–1316. https://doi.org/10.1105/tpc.13.6.1305
7. Thines B, Harmon FG (2010) Ambient temperature response establishes ELF3 as a required component of the core *Arabidopsis* circadian clock. Proc Natl Acad Sci USA 107: 3257–3262. https://doi.org/10.1073/pnas.0911006107
8. Masuda K, Tokuda IT, Nakamichi N et al (2021) The singularity response reveals entrainment properties of the plant circadian clock. Nat Commun 12:864. https://doi.org/10.1038/s41467-021-21167-7
9. Millar AJ, Kay SA (1996) Integration of circadian and phototransduction pathways in the network controlling *CAB* gene transcription in *Arabidopsis*. Proc Natl Acad Sci USA 93: 15491–15496. https://doi.org/10.1073/pnas.93.26.15491
10. Salter MG, Franklin KA, Whitelam GC (2003) Gating of the rapid shade-avoidance response by the circadian clock in plants. Nature 426: 680–683. https://doi.org/10.1038/nature02174
11. Fowler SG, Cook D, Thomashow MF (2005) Low temperature induction of *Arabidopsis* CBF1, 2, and 3 is gated by the circadian clock. Plant Physiol 137:961–968. https://doi.org/10.1104/pp.104.058354

12. Legnaioli T, Cuevas J, Mas P (2009) TOC1 functions as a molecular switch connecting the circadian clock with plant responses to drought. EMBO J 28:3745–3757. https://doi.org/10.1038/emboj.2009.297
13. Wilkins O, Waldron L, Nahal H et al (2009) Genotype and time of day shape the *Populus* drought response. Plant J 60:703–715. https://doi.org/10.1111/j.1365-313X.2009.03993.x
14. Wilkins O, Bräutigam K, Campbell MM (2010) Time of day shapes *Arabidopsis* drought transcriptomes. Plant J 63:715–727. https://doi.org/10.1111/j.1365-313X.2010.04274.x
15. Greenham K, Guadagno CR, Gehan MA et al (2017) Temporal network analysis identifies early physiological and transcriptomic indicators of mild drought in *Brassica rapa*. eLife 6: e29655. https://doi.org/10.7554/eLife.2965
16. Blair EJ, Bonnot T, Hummel M et al (2019) Contribution of time of day and the circadian clock to the heat stress responsive transcriptome in *Arabidopsis*. Sci Rep 9:4814. https://doi.org/10.1038/s41598-019-41234-w
17. Grinevich DO, Desai JS, Stroup KP et al (2019) Novel transcriptional responses to heat revealed by turning up the heat at night. Plant Mol Biol 101:1–19. https://doi.org/10.1007/s11103-019-00873-3
18. Bonnot T, Nagel DH (2021) Time of the day prioritizes the pool of translating mRNAs in response to heat stress. Plant Cell 33:2164–2182. https://doi.org/10.1093/plcell/koab113
19. Graham CA, Paajanen P, Edwards KJ et al (2022) Genome-wide circadian gating of a cold temperature response in bread wheat. PLoS Genet 19:e1010947. https://doi.org/10.1371/journal.pgen.1010947
20. Bonnot T, Somayanda I, Jagadish SVK et al (2023) Time of day and genotype sensitivity adjust molecular responses to temperature stress in *sorghum*. Plant J 116:1081–1096. https://doi.org/10.1111/tpj.16467
21. Medina-Chávez L, Camacho C, Martínez-Rodríguez JA et al (2023) Submergence stress alters the expression of clock genes and configures new zeniths and expression of outputs in *Brachypodium distachyon*. Int J Mol Sci 24: 8555. https://doi.org/10.3390/ijms24108555
22. Mizoguchi T, Wright L, Fujiwara S et al (2005) Distinct roles of *GIGANTEA* in promoting flowering and regulating circadian rhythms in *Arabidopsis*. Plant Cell 17:2255–2270. https://doi.org/10.1105/tpc.105.033464
23. Gould PD, Locke JCW, Larue C et al (2006) The molecular basis of temperature compensation in the *Arabidopsis* circadian clock. Plant Cell 18:1177–1187. https://doi.org/10.1105/tpc.105.039990
24. Fowler S, Thomashow MF (2002) *Arabidopsis* transcriptome profiling indicates that multiple regulatory pathways are activated during cold acclimation in addition to the CBF cold response pathway. Plant Cell 14:1675–1690. https://doi.org/10.1105/tpc.003483
25. Paltiel J, Amin R, Gover A et al (2006) Novel roles for GIGANTEA revealed under environmental conditions that modify its expression in *Arabidopsis* and *Medicago truncatula*. Planta 224:1255–1268. https://doi.org/10.1007/s00425-006-0305-1
26. Nohales MA, Kay SA (2019) GIGANTEA gates gibberellin signaling through stabilization of the DELLA proteins in *Arabidopsis*. Proc Natl Acad Sci USA 116:21893–21899. https://doi.org/10.1073/pnas.1913532116
27. Nohales MA, Liu W, Duffy T et al (2019) Multi-level modulation of light signaling by GIGANTEA regulates both the output and pace of the circadian clock. Dev Cell 49:840–851.e8. https://doi.org/10.1016/j.devcel.2019.04.030
28. Martínez-Vasallo C, Cole B, Gallego-Bartolomé J et al (2023) Epidermal Gigantea adjusts the response to shade at dusk by directly impinging on Phytochrome Interacting Factor 7 function. bioRxiv. 2023.03.21.533699. https://doi.org/10.1101/2023.03.21.533699
29. Edwards KD, Lynn JR, Gyula P et al (2005) Natural allelic variation in the temperature-compensation mechanisms of the *Arabidopsis thaliana* circadian clock. Genetics 170:387–400. https://doi.org/10.1534/genetics.104.035238
30. Borrill P, Ramirez-Gonzalez R, Uauy C (2016) expVIP: a customizable RNA-seq data analysis and visualization platform. Plant Physiol 170: 2172–2186. https://doi.org/10.1104/pp.15.01667
31. Rees H, Rusholme-Pilcher R, Bailey P et al (2022) Circadian regulation of the transcriptome in a complex polyploid crop. PLoS Biol 20:e3001802. https://doi.org/10.1371/journal.pbio.3001802
32. Moore A, Zielinski T, Millar AJ (2014) Online period estimation and determination of rhythmicity in circadian data, using the BioDare data infrastructure. Methods Mol Biol 1158:13–44.

https://doi.org/10.1007/978-1-4939-0700-7

33. Zielinski T, Moore AM, Troup E et al (2014) Strengths and limitations of period estimation methods for circadian data. PLoS One 9: e96462. https://doi.org/10.1371/journal.pone.0096462
34. Wu G, Anafi´RC, Hughes ME et al (2016) MetaCycle: an integrated R package to evaluate periodicity in large scale data. Bioinformatics 32:3351–3353. https://doi.org/10.1093/bioinformatics/btw405
35. Haydon MJ, Mielczarek O, Robertson FC et al (2013) Photosynthetic entrainment of the *Arabidopsis thaliana* circadian clock. Nature 502:689–692. https://doi.org/10.1038/nature12603

Chapter 21

RuBisCO Depletion and Subcellular Fractionation for Enhanced Florigen Detection in *Arabidopsis*

Katarzyna Gawarecka and Ji Hoon Ahn

Abstract

In plants, complex signaling networks monitor and respond to environmental cues to determine the optimal time for the transition from the vegetative to reproductive phase. Understanding these networks requires robust tools to examine the levels and subcellular localization of key factors. The florigen FLOWERING LOCUS T (FT) is a crucial regulator of flowering time and occurs in soluble and membrane-bound forms. At low ambient temperatures, the ratio of these forms of FT undergoes a significant shift, which leads to a delay in the onset of flowering. To investigate these changes in FT localization, epitope-tagged FT protein can be isolated from plants by subcellular fractionation and its localization examined by immunoblot analysis of the resulting fractions. However, the highly abundant protein ribulose-1,5-bisphosphate carboxylase/oxygenase (RuBisCO) can interfere with methods to detect and characterize low-abundance proteins such as FT. In this chapter, we present a method for analyzing the ratio of HA-tagged FT (HA:FT) in different subcellular fractions while mitigating the interference from RuBisCO by using protamine sulfate (PS) to deplete RuBisCO during protein purification, thereby enhancing HA:FT detection in fractionated samples.

Key words Flowering time, FT, RuBisCO depletion, Fractionation, *Arabidopsis*

1 Introduction

Extensive research conducted over the past century has shed light on the diverse environmental factors [1, 2] and genetic factors [3] that influence the timing of the transition from vegetative to reproductive growth in plants. In response to environmental conditions, a multitude of activators and repressors intricately regulate plant growth and development [4].

One key player in regulating flowering time is FLOWERING LOCUS T (FT), a highly conserved member of the phosphatidylethanolamine-binding protein (PEBP) family. FT plays a pivotal role in regulating flowering time across diverse plant species [5]. FT is considered to function as a florigen, a crucial mobile signaling molecule that facilitates long-distance

Meng Chen (ed.), *Thermomorphogenesis: Methods and Protocols*, Methods in Molecular Biology, vol. 2795, https://doi.org/10.1007/978-1-0716-3814-9_21,

communication between leaves and the shoot apical meristem, where floral initiation occurs [6]. FT integrates signals from environmental cues and coordinates the transition from vegetative to reproductive growth, ensuring successful reproduction and facilitating plant adaptation to varying environmental conditions. For example, changes in ambient temperature have significant effects on the flowering time signaling network; by interacting with phospholipids (particularly phosphatidylglycerol) embedded in the cell membrane, FT finely orchestrates the timing of phase transitions in response to changes in ambient temperature [7, 8].

Temperature fluctuations strongly influence the different forms of FT. As the temperature decreases, the level of soluble FT decreases, while the level of membrane-bound FT increases, thus decreasing the availability of FT as a mobile signal. This dynamic shift in FT forms represents a crucial regulatory mechanism that fine-tunes flowering time in response to temperature changes. Understanding the balance between the soluble and membrane-bound forms of FT is therefore of paramount importance in elucidating the molecular mechanisms underlying plant development. Consequently, comprehensive analysis of total FT protein levels and FT distribution within different subcellular fractions is essential for investigating the intricate interplay between FT forms during the regulation of flowering time.

Previous studies have described various techniques for subcellular fractionation [9], which have been adapted for the analysis of FT protein in *Arabidopsis thaliana*, as described in this chapter. To minimize interference from larger protein tags such as green fluorescent protein (GFP), researchers commonly use a small HA tag to detect FT in plant cells. However, detecting the 20 kDa FT protein by immunoblotting with anti-HA antibody often yields nonspecific signals, particularly in membrane fractions (Fig. 1). Using protamine sulfate (PS) precipitation to remove the highly abundant protein RuBisCO, which accumulates in leaves [10, 11], can significantly enhance proteomic analyses [12] and improve the selectivity of the HA antibody toward FT (Fig. 2).

This chapter provides a detailed protocol for subcellular fractionation combined with RuBisCO depletion, along with instructions for detecting HA-tagged FT (HA:FT) on a polyvinylidene fluoride (PVDF) membrane. This protocol has been developed to analyze the level of HA:FT in 8–10-day-old *Arabidopsis* plants overexpressing HA:FT and grown under long-day conditions at 23 °C, or in 14–17-day-old *Arabidopsis* plants overexpressing HA:FT and grown under long-day conditions at 16 °C (*see* **Note 1**). By following this protocol, researchers will be able to assess the levels of HA:FT in different subcellular fractions and gain insight into the dynamic changes in FT forms during the regulation of flowering time.

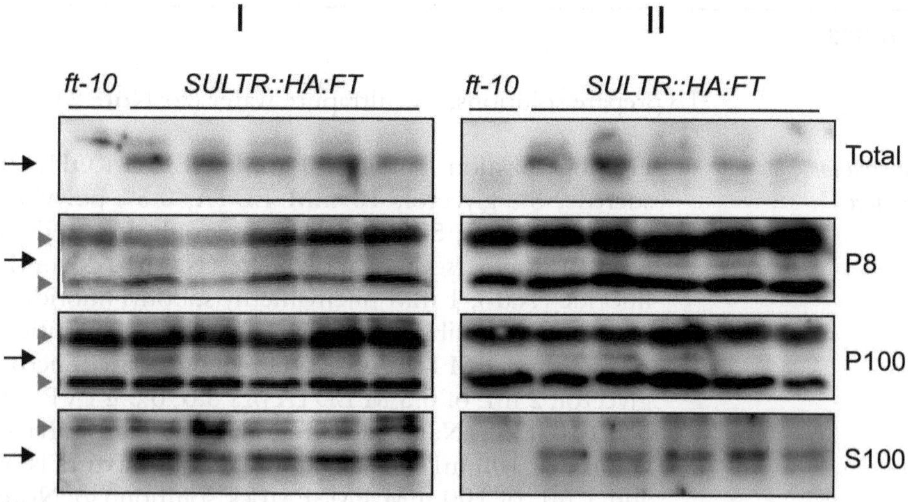

Fig. 1 Representative immunoblots showing two biological replicates (I and II) collected from plants overexpressing HA:FT under the control of the *SULTR* promoter without RuBisCO precipitation. The *ft-10* mutant was used as a control. Arrows indicate the signal from HA:FT, while arrowheads indicate nonspecific signals observed after incubation with anti-HA antibody. The subcellular fractions are labeled as Total, P8, P100, and S100

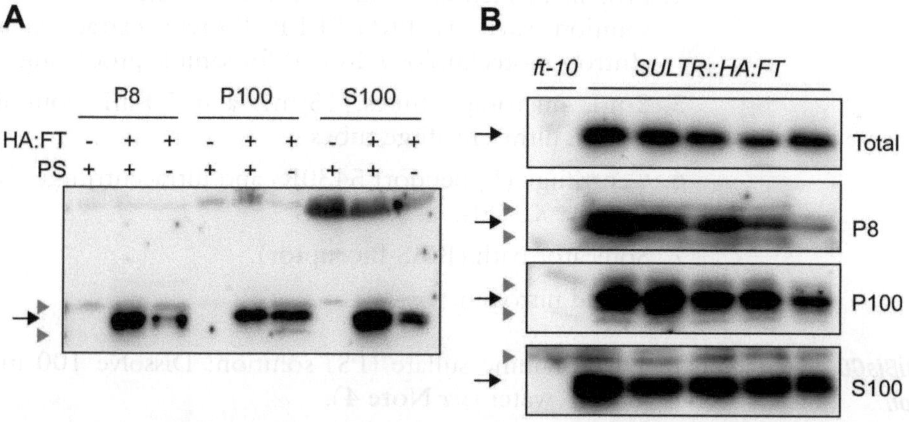

Fig. 2 Immunoblots obtained after combining subcellular fractionation with RuBisCO depletion. A: Comparison of immunoblots produced after subcellular fractionation with (+) and without (−) RuBisCO precipitation using protamine sulfate (PS). B: Immunoblot analysis of samples obtained from the same plants as in Fig. 1 after RuBisCO depletion using PS. The *ft-10* mutant was used as a control. Arrows indicate the signal from HA:FT, while arrowheads indicate nonspecific signals observed after incubation with anti-HA antibody. The subcellular fractions are labeled as Total, P8, P100, and S100

2 Materials

To prepare solutions, use ultrapure water (*see* **Note 2**).

2.1 Subcellular Fractionation

1. Homogenization buffer: 50 mM HEPES-KOH, 250 mM sucrose, 5% glycerol, 10 mM EDTA, 0.5% polyvinylpyrrolidone (PVP-10), 50 mM sodium pyrophosphate decahydrate (NaPP), 1 mM sodium molybdate (NaMo), 25 mM sodium fluoride (NaF), 1 mM phenylmethylsulfonyl fluoride (PMSF), 1× protease inhibitor mix. To make homogenization buffer, add 5 mL of 1 M HEPES-KOH pH 7.7, 8.56 g sucrose, 5 mL glycerol, 2 mL of 0.5 M EDTA pH 8.0, 0.5 g PVP-10, 1.32 g NaPP, 0.026 g NaMo, 0.105 g NaF, and add water to a final volume of 100 mL. Once a homogenous mixture is obtained, add 1 mL of 100 mM PMSF stock solution (*see* **Note 3**) and 2 tablets of protease inhibitor cocktail (cOmplete, Roche).
2. Solution for concentrating the S100 fraction: 1/4/1/3 (v/v/v/v) sample/methanol/chloroform/water; for each sample, prepare 24 mL methanol, 6 mL chloroform, and 18 mL water.
3. Miracloth (Merck): cut the Miracloth sheet into 12 × 12-cm squares for use in subsequent steps.
4. Protein extraction solution: use a suitable protein extraction solution such as PRO-PREP Protein extraction solution (Intron Biotechnology, Korea) for sample processing.
5. 2-mL microfuge tubes, 15-mL and 50-mL conical tubes, 13-mL ultracentrifuge tubes
6. Centrifuge (Eppendorf 5430R) and ultracentrifuge (Beckman Optima XE-90).
7. Sonicator bath (BMS Bioruptor).
8. Liquid nitrogen.

2.2 RuBisCO Depletion

1. 1% protamine sulfate (PS) solution: Dissolve 100 mg PS in 9.9 mL water (*see* **Note 4**).

2.3 Gel Electrophoresis and Transfer

1. 1.5 M Tris–HCl (pH 8.8) buffer: Dissolve 90.85 g Tris base in 400 mL of double-distilled water. Adjust the pH of the solution to 8.8 using 12 M HCl (*see* **Note 5**) and add water to bring the final volume to 500 mL.
2. 1.0 M Tris–HCl (pH 6.8) buffer: Dissolve 60.6 g Tris base in 400 mL double-distilled water. Adjust the pH of the solution to 6.8 using 12 M HCl (*see* **Note 5**) and add water to bring the final volume to 500 mL.
3. 10% sodium dodecyl sulfate (SDS) solution: Dissolve 10 g of SDS in 90 mL of water. Store at room temperature.

4. 10% ammonium persulfate (APS) solution: Dissolve 1 g of APS in 9 mL of water. Store at 4 °C
5. N,N,N,N-tetramethyl-ethylenediamine (TEMED).
6. Acrylamide running gel: 12% acrylamide, 0.39 M Tris–HCl, 0.1% SDS, 0.1% APS, 0.04% TEMED; For a 1.5 mm–thick mini gel, thoroughly mix 3.3 mL water, 4 mL of 30% 29:1 acrylamide mix, 2.5 mL of 1,5 M Tris–HCl (pH 8.8) buffer, 0.1 ml of 10% SDS solution, 0.1 mL of 10% APS solution, and 0.004 mL TEMED.
7. Acrylamide stacking gel: 5% acrylamide, 0.13 M Tris-HCl, 0.1% SDS, 0.1% APS, 0.1% TEMED; For a 1.5 mm–thick mini gel, thoroughly mix 2.08 mL water, 0.5 mL of 30% 29:1 acrylamide mix, 0.38 mL of 1.0 M Tris–HCl (pH 6.8) buffer, 0.03 mL of 10% SDS, 0.03 mL of 10% APS, and 0.003 mL TEMED.
8. 4× SDS-PAGE loading buffer: 0.25 M Tris–HCl, 10% SDS, 50% glycerol, 0.001% bromophenol blue, 20% β-mercaptoethanol; Combine 2 mL of 1 M Tris–HCl (pH 6.8), 0.8 g SDS, 4 mL glycerol, a pinch of bromophenol blue, and water to a final volume of 8 mL. Store the loading buffer at room temperature. Before use, add 0.2 mL β-mercaptoethanol to 1 mL (final volume) of 4× SDS-PAGE loading buffer (*see* **Note 6**).
9. 10× Running buffer: 0.25 M Tris, 1.9 M glycine, 1% SDS; Dissolve 30.3 g Tris base, 144.1 g glycine, and 10 g SDS in 800 mL of water. Adjust the final volume to 1 L and store at room temperature.
10. 1× Running buffer: 0.025 M Tris, 0.19 M glycine, 0.1% SDS; Mix 100 mL of 10× running buffer with 900 mL water before use.
11. 20× Transfer buffer: 0.2 M Tris, 2 M glycine; Dissolve 24.2 g Tris base and 150.1 g glycine in 800 mL of water. Adjust the final volume to 1 L and store at 4 °C.
12. 1× Transfer buffer: 0.01 M Tris, 0.1 M glycine, 10% methanol; mix 50 mL of 20× running buffer with 100 mL methanol and 850 mL water before use (*see* **Note 7**).
13. PVDF membrane (*see* **Note 8**).

2.4 Immunoblotting

1. 10× TBS-Tween-20 solution: 0.2 M Tris–HCl, 1.5 M sodium chloride (NaCl), 1% Tween-20; Dissolve 24 g of Tris base and 88 g of NaCl in 900 mL of water. Adjust the pH to 7.6 with 12 M HCl (*see* **Note 5**) and bring the final volume to 1 L. Add 10 mL of Tween-20 and store at room temperature.

2. 1× TBS-Tween-20 solution: 0.02 M Tris–HCl, 0.15 M NaCl, 0.1% Tween-20; Dilute 100 ml of the 10× TBS-Tween-20 solution with 900 mL of water. Store at room temperature.
3. 5% skim milk blocking solution (for one membrane): 5% skim milk, 0.02 M Tris–HCl, 0.15 M NaCl, 0.1% Tween-20; Dissolve 500 mg of skim milk powder in 8.5 mL of water and add 1 mL of 10× TBS-Tween buffer.
4. 1.5% skim milk solution (for one membrane): 1.5% skim milk, 0.02 M Tris–HCl, 0.15 M NaCl, 0.1% Tween 20; Dissolve 150 mg of skim milk powder in 8.85 mL of water and add 1 mL of 10× TBS-Tween buffer.
5. To detect HA:FT protein: Dissolve 5 μL of anti-HA stock solution (Roche) in 10 mL of 1.5% skim milk solution (1: 2000). Dissolve 1.4 μL of anti-Rat stock solution (ThermoFisher) in 10 mL of 1.5% skim milk solution (1:7000) (*see* **Note 9**).
6. Total protein fraction loading control: Dissolve 1 μL of anti-Actin stock solution (Agrisera) in 10 mL of 1.5% skim milk solution (1:10,000). Dissolve 0.3 μL of anti-Rabbit stock solution (ThermoFisher) in 10 mL of 1.5% skim milk solution (1: 30,000) (*see* **Note 9**).
7. P8 fraction loading control: Dissolve 3.3 μL of anti-TIC40 stock solution (Agrisera) in 10 mL of 1.5% skim milk solution (1:3000). Dissolve 1.25 μL of anti-Rabbit stock solution (ThermoFisher) in 10 mL of 1.5% skim milk solution (1: 8000) (*see* **Note 9**).
8. P100 fraction loading control: Dissolve 1 μL of anti-H^+ATPase stock solution (Agrisera) in 20 mL of 1.5% skim milk solution (1:10,000). Dissolve 0.6 μL of anti-Rabbit stock solution (ThermoFisher) in 10 mL of 1.5% skim milk solution (1: 15,000) (*see* **Note 9**).
9. S100 fraction loading control: Dissolve 0.4 μL of anti-UGPase stock solution (Agrisera) in 20 mL of 1.5% skim milk solution (1:25,000). Dissolve 0.2 μL of anti-Rabbit stock solution (ThermoFisher) in 10 mL of 1.5% skim milk solution (1: 60,000) (*see* **Note 9**).
10. PicoEPD Western Reagent (ELPIS BIOTECH, Korea).
11. Immunoblot stripping buffer (LPS Solution, Korea).
12. Membrane Imaging system or X-ray film.

3 Methods

3.1 Protein Extraction and Subcellular Fractionation with RuBisCO Depletion (Fig. 3)

1. Collect 3 g of tissue from fresh, healthy seedlings grown under the selected growth conditions in a growth chamber. Grind the tissue in a prechilled mortar and pestle on ice in 3 mL of homogenization buffer for 3 min (*see* **Note 10**). Freeze 0.05 g of the sample in liquid nitrogen and extract proteins with 0.07 mL of protein extraction solution. The extracted proteins will serve as the total protein fraction.

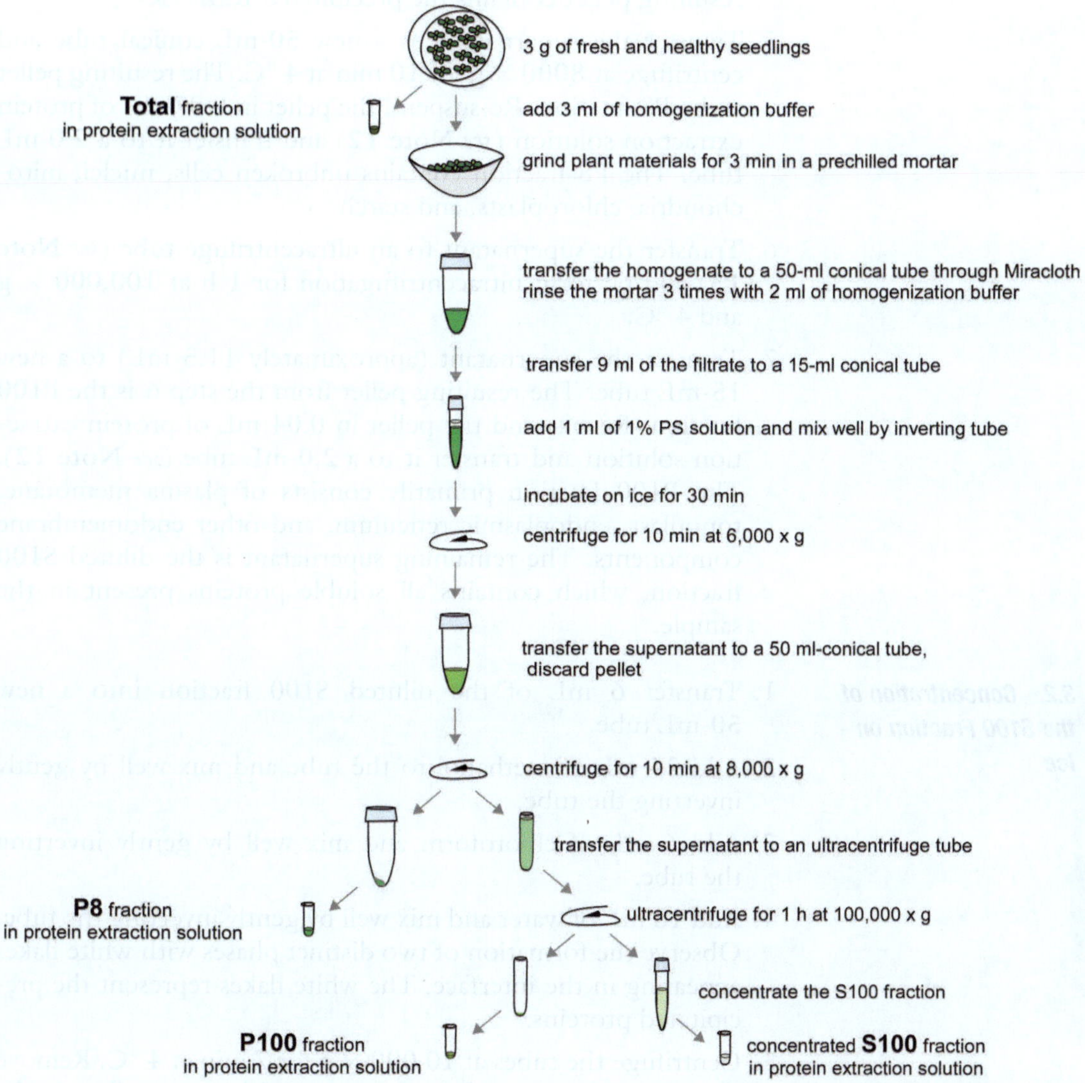

Fig. 3 Workflow illustrating subcellular fractionation with RuBisCO depletion using protamine sulfate

2. Transfer the homogenate to a 50-mL conical tube lined with two layers of Miracloth squares and squeeze the homogenate through the cloth. Rinse the mortar and pestle 3 times with 2 mL of homogenization buffer and add it to the tubes through the Miracloth (*see* **Note 11**).
3. Transfer the filtrate to a new 15-mL conical tube and adjust the volume to 9 mL with homogenization buffer if necessary. Add 1 mL of 1% PS solution to precipitate RuBisCO. Mix well and incubate the samples on ice for 30 min.
4. Centrifuge the tube at 6000 × g for 10 min at 4 °C. The resulting pellet contains the precipitated RuBisCO.
5. Transfer the supernatant to a new 50-mL conical tube and centrifuge at 8000 × g for 10 min at 4 °C. The resulting pellet is the P8 fraction. Re-suspend the pellet in 0.07 mL of protein extraction solution (*see* **Note 12**) and transfer it to a 2.0-mL tube. The P8 fraction contains unbroken cells, nuclei, mitochondria, chloroplasts, and starch.
6. Transfer the supernatant to an ultracentrifuge tube (*see* **Note 13**) and perform ultracentrifugation for 1 h at 100,000 × g and 4 °C.
7. Transfer the supernatant (approximately 11.5 mL) to a new 15-mL tube. The resulting pellet from the step 6 is the P100 fraction. Re-suspend the pellet in 0.04 mL of protein extraction solution and transfer it to a 2.0-mL tube (*see* **Note 12**). The P100 fraction primarily consists of plasma membrane, tonoplast, endoplasmic reticulum, and other endomembrane components. The remaining supernatant is the diluted S100 fraction, which contains all soluble proteins present in the sample.

3.2 Concentration of the S100 Fraction on Ice

1. Transfer 6 mL of the diluted S100 fraction into a new 50-mL tube.
2. Add 24 mL of methanol to the tube and mix well by gently inverting the tube.
3. Add 6 mL of chloroform and mix well by gently inverting the tube.
4. Add 18 mL of water and mix well by gently inverting the tube. Observe the formation of two distinct phases with white flakes appearing in the interface. The white flakes represent the precipitated proteins.
5. Centrifuge the tubes at 10,000 × g for 7 min at 4 °C. Remove the upper aqueous phase, being careful not to disturb the precipitated pellet at the interface (*see* **Note 14**).

6. Add 22 mL of methanol to the lower organic phase and mix it well by gently inverting the tube. The white protein flakes will quickly settle to the bottom of the tube.
7. Use a cut 1-mL pipette tip to carefully transfer all the protein sediment to a new 2-mL tube.
8. Centrifuge the tubes at maximum speed at 4 °C and carefully remove the supernatant. The resulting pellet represents the concentrated S100 fraction.
9. Allow the pellet to air dry at room temperature for 15 min in a fume hood. Re-suspend the pellet in 0.2 mL of protein extraction solution (*see* **Note 12**).
10. Sonicate at 60 kHz the P8, P100, and S100 fractions using 11 cycles, with 30-second intervals per cycle, to ensure the complete dissolution of the proteins in protein extraction buffer.
11. Store all fractions at −80 °C until further use for SDS-PAGE and immunoblot analysis.

3.3 Protein Analysis Via SDS-PAGE and Immunoblotting

1. Measure the protein concentration in each collected fraction using the appropriate method (*see* **Note 15**).
2. Prepare 1.5 mm–thick mini gels with acrylamide running gel and acrylamide stacking gel following any standard protocol.
3. Prepare the desired amount of each sample (*see* **Note 12**), add 4× SDS-PAGE loading buffer with β-mercaptoethanol in 4/1 (v/v) sample/buffer ratio and perform electrophoresis at 100 V using 1× running buffer until the bromophenol blue dye has migrated off the gel (*see* **Note 16**).
4. Transfer the proteins from the acrylamide gel onto a PVDF membrane (*see* **Note 8**) using 1× transfer buffer for 105 min at 100 V.
5. After transfer, carefully dry the membrane with a paper towel before proceeding to the blocking step (*see* **Note 17**).
6. Block the membranes with 5% skim milk blocking solution at room temperature for 45–60 min.
7. Remove the blocking solution and incubate the membranes in anti-HA antibody solution overnight at 4 °C.
8. Remove the anti-HA antibody solution and wash the membranes four times for 5 min each time with 10 mL of 1× TBS-Tween-20 solution.
9. Add the anti-Rat antibody solution and incubate the membranes for 2 days at 4 °C.
10. Remove the anti-Rat antibody solution and wash the membranes four times for 5 min each time with 10 mL of 1× TBS-Tween-20 solution.

11. Analyze the protein signal using a Membrane Imaging system or X-ray film, utilizing PicoEPD Western Reagent (*see* **Note 18**).
12. After protein analysis, strip the membranes using stripping buffer and repeat **steps 7–11** for each loading control, using the appropriate antibody solutions mentioned in the Materials and Methods. For the loading control, the incubation time with anti-Rabbit solution can be reduced to 2 h at room temperature instead of 2 days at 4 °C (*see* **Note 19**).

4 Notes

1. Older plant tissues contain a significant amount of starch, which can affect fraction separation and the results of protein analysis.
2. Ultrapure water prepared by purifying deionized water to attain a resistivity of 18 MΩ-cm at 25 °C. Prepare fresh solutions before use at room temperature or use stock solutions when indicated. Use analytical grade reagents for all experiments. Ensure compliance with waste disposal regulations when disposing of waste materials.
3. Dissolve 350 mg of PMSF in 27.7 mL of 100% propan-2-ol to create a 100 mM stock solution. Divide the solution into 1-mL aliquots and store them at −20 °C until use. Note that PMSF may precipitate under this condition, so ensure that it dissolves completely at room temperature before use.
4. It may take some time to dissolve PS in water. Consider placing the tube on a shaking device overnight at room temperature before the experiment.
5. When working with concentrated HCl, always wear gloves and safety glasses and work in a fume hood.
6. 4× SDS-PAGE loading buffer with β-mercaptoethanol can be stored for a short time at 4 °C.
7. The 1× transfer buffer can be reused for a second run if desired.
8. Activate the PVDF membrane by immersing it in 100% methanol for 5 min prior to transfer.
9. Antibody solutions can be reused multiple times. Store the working antibody solution at −20 °C.
10. Maintain consistent grinding times for each sample to ensure reproducibility.
11. Perform all extraction steps on ice to preserve protein integrity.
12. Adjust the protein extraction solution volume according to the pellet size and protein analysis requirements. Aim for at least

0.08 mg protein for the Total and P8 fractions, or 0.05 mg protein for the P100 and S100 fractions. Do not exceed 0.110 mg of protein.

13. Fill the tubes to up to 90% of their volume, and ensure that the weights of the tubes filled with samples are consistent before ultracentrifugation.
14. Exercise caution to avoid removing proteins when discarding the upper phase, as it contains chloroform. Handle it according to the waste disposal regulations in your facility.
15. A bicinchoninic acid (BCA) test can be used to measure protein concentration.
16. Electrophoresis at 100 V typically takes approximately 2 h 15 min to complete.
17. During membrane drying, the transferred proteins may produce a visible black signal on the membrane surface.
18. The signals from HA:FT in the P8 and P100 fractions may be low. To improve protein signal visualization on the membranes, consider using a more sensitive chemiluminescence mixture.
19. As the loading control proteins are larger than 40 kDa, the membranes can be cut in half. Each membrane half can be incubated with the appropriate antibody solution simultaneously to reduce the analysis time.

Acknowledgments

We thank Dr. Hendry Susila for his initial contribution to this work. This work was supported by a National Research Foundation of Korea grant (NRF-2022R1A3B1078180 and RS-2023-00221182) to J.H.A.

References

1. Gawarecka K, Ahn JH (2021) Isoprenoid-derived metabolites and sugars in the regulation of flowering time: does day length matter? Front Plant Sci 12:765995
2. Renau-Morata B, Carrillo L, Dominguez-Figueroa J et al (2020) CDF transcription factors: plant regulators to deal with extreme environmental conditions. J Exp Bot 71(13): 3803–3815
3. Kinoshita A, Richter R (2020) Genetic and molecular basis of floral induction in Arabidopsis thaliana. J Exp Bot 71(9):2490–2504
4. Jin S, Ahn JH (2021) Regulation of flowering time by ambient temperature: repressing the repressors and activating the activators. New Phytol 230(3):938–942
5. Jin S, Nasim Z, Susila H, Ahn JH (2021) Evolution and functional diversification of FLOWERING LOCUS T/TERMINAL FLOWER 1 family genes in plants. Semin Cell Dev Biol 109:20–30
6. Corbesier L, Vincent C, Jang S et al (2007) FT protein movement contributes to long-distance signaling in floral induction of Arabidopsis. Science 316(5827):1030–1033
7. Susila H, Jurić S, Liu L et al (2021) Florigen sequestration in cellular membranes modulates

temperature-responsive flowering. Science 373(6559):1137–1142

8. Susila H, Nasim Z, Gawarecka K et al (2021) PHOSPHORYLETHANOLAMINE CYTIDYLYLTRANSFERASE 1 modulates flowering in a florigen-independent manner by regulating SVP. Development 148(1): dev193870
9. LaMontagne ED, Collins CA, Peck SC, Heese A (2016) Isolation of microsomal membrane proteins from Arabidopsis thaliana. Curr Protoc Plant Biol 1(1):217–234
10. Sharkey TD (2023) The discovery of rubisco. J Exp Bot 74(2):510–519
11. Prywes N, Phillips NR, Tuck OT et al (2023) Rubisco function, evolution, and engineering. Annu Rev Biochem
12. Kim YJ, Lee HM, Wang Y et al (2013) Depletion of abundant plant RuBisCO protein using the protamine sulfate precipitation method. Proteomics 13(14):2176–2179

Chapter 22

Identifying Causal Genes for Thermo-Responsive Rosette Growth in *Arabidopsis* by Genome-Wide Association Study

Zhixue Wang and Jian Hua

Abstract

Temperature responsive plant growth, such as thermo-morphogenesis, varied greatly among *Arabidopsis* natural accessions. Here we describe a procedure to identify causal genes for thermo-morphogenesis variation by a genome-wide association study (GWAS). It includes methods of measuring thermo-responsive rosette growth architecture, using GWA-Portal webtool for GWAS, analyzing haplotypes, and validating functional natural variations. In addition, we discuss key factors that affect the success of identifying causal genes from GWAS.

Key words Thermo-morphogenesis, GWAS, Rosette growth architecture, *Arabidopsis*

1 Introduction

High ambient temperature induces plant thermo-morphogenesis by promoting hypocotyl and petiole elongation and decreasing leaf thickness. *Arabidopsis* natural accessions exhibit great diversity in growth thermo-morphogenesis. Genome-wide association study (GWAS) has been used to investigate the molecular genetic basis of natural variations in growth thermo-morphogenesis by identifying the regulators such as *SMALL AUXIN UP RNA 26* (*SAUR26*) subfamily genes [1]. Multiple growth parameters, such as hypocotyl elongation and rosette compactness [1–3], could be measured for thermo-morphogenesis. Here, we describe a method for GWAS on thermo-morphogenesis using *Arabidopsis* thermo-responsiveness of rosette compactness as an example. Rosette compactness index was defined as the ratio of rosette weight to rosette outgrowth area. Thermo-responsiveness is calculated as the differences between indices of 22 °C and 28 °C relative to 22 °C. GWAS is performed with thermo-responsiveness of rosette compactness index using GWA-Portal [4]. Causal genes are then identified based

Meng Chen (ed.), *Thermomorphogenesis: Methods and Protocols*, Methods in Molecular Biology, vol. 2795, https://doi.org/10.1007/978-1-0716-3814-9_22,

on haplotype/allele analysis and mutant characterization. Lastly, the effect of natural variations is assessed by quantifying growth thermo-morphogenesis of transgenic plants carrying different haplotypes of the causal genes.

2 Materials

2.1 Plant Materials

1. *Arabidopsis* natural accessions are used. Until now, more than one thousand *Arabidopsis* natural accessions have been fully sequenced (https://1001genomes.org/). Choosing a suitable set of accessions for thermo-morphogenesis is the first and important step for GWAS [5]. In a previous work, we used 69 accessions from the first publicly available panel of 195 accessions collected worldwide to identify *SAUR26* subfamily genes for thermo-responsive rosette growth [1]. Martínez-Berdeja and collogues used a collection of 559 accessions from a wide climate range to identify *DOG1* as a regulator for seed-chilling responses and seasonal life history [6]. Regional collections were also used for GWAS to increase the chance of identifying rare alleles. For example, a panel of 174 accessions collected from the Iberian Peninsula was used for GWAS of environmental factors and flowering time [7].

   Below are three key factors to consider when selecting accessions for GWAS:

2. The first factor is the number of accessions. In general, a larger number of accessions (in hundreds) will increase the resolution and the probability of the genetic architecture captured by GWAS. However, a large number of accessions require more labor and resources, and it may also introduce variations when accessions are assayed in multiple batches. In practice, panels of dozens to a hundred accessions could lead to the successful identification of causal genes.
3. The second factor is the phenotypic variation of accessions. It is important that accessions in the panel have large diversity in the trait/phenotype (such as thermo-morphogenesis) analyzed. Using a small but representative collection of accessions with diverse geographic distribution could be effective and time saving for GWAS.
4. The third factor is the population structure of the accessions. Strong population structure generates spurious associations between unlinked SNPs and phenotypes. Many GWAS algorithms, such as mixed models, accounts for population structures, but the false positives arising from population structure may not be fully controlled [8]. Thus, a panel of accessions with a lower population structure is preferred.

2.2 Growth Conditions

1. Set the growth chamber conditions: 24 h constant light, a light intensity of approximately 100 μmol m^{-2} s^{-1}, 50–70% humidity, and a temperature of either 22 °C or 28 °C.

2.3 Equipment and Materials

1. ImageJ software [9].
2. Precision balance.
3. GWA-Portal webtools (https://gwas.gmi.oeaw.ac.at/).
4. R program (https://www.r-project.org/).
5. pCR™8/GW/TOPO™ TA Cloning Kit (ThermoFisher, Catalog: K250020).

3 Methods

3.1 Analyses of Plant Thermo-Responsive Rosette Growth Architecture

1. Propagate seeds of *Arabidopsis* natural accessions from the stock, and sow fresh seeds directly on the soil with six seeds in a pot. Cover the pots with a transparent plastic dome and keep them at 4 °C for 4–7 days according to the instructions for the stock (*see* **Note 1**).
2. After stratification, transfer the pots to a 22 °C growth chamber for germination. This will take 2–3 days.
3. After germination, maintain the plants in the 22 °C growth chamber or transfer them to a 28 °C growth chamber (*see* **Note 2**).
4. Analyze the rosette of each plant for both area and fresh weight at the time of bolting (Fig. 1).
5. Quantify the thermo-responsiveness of rosette architecture as following (*see* **Note 3**):

$$\text{Compactness index (CI)} = \frac{\text{Weight}}{\text{Area}}$$

$$\text{Thermo} - \text{responsiveness} = \frac{\text{CI}_{22^\circ\text{C}} - \text{CI}_{28^\circ\text{C}}}{\text{CI}_{22^\circ\text{C}}}.$$

3.2 Genome-Wide Association Study

1. Perform GWAS using the GWA-Portal webtool (*see* **Note 4**) with a browser. A detailed protocol for running GWAS with GWA-Portal has been reported [4]. Critical steps are briefly described here.
2. To initiate the GWAS analysis, create a new study and upload the phenotype measurement as a comma-separated values file (.csv). In the phenotype file, the first column should be ecotype (or accession) id, which is a four-digit name (*see* **Note 5**), while the second and subsequent columns are the phenotypes.

Fig. 1 Quantification of rosette area using ImageJ. The circle defines the area

	A	B	C	D	E	F	G
1	ecotype_id	phenotype1	phenotype2	phenotype3	phenotype4	phenotype5	phenotype6
2	108						
3	139						
4	159						
5	265						
6	350						
7	351						
8	403						
9	410						
10	1002						
11	1066						
12	1158						

Fig. 2 A sample phenotype file for GWAS

Ensure the column names are included in the first row of the file and avoid using spaces within the column names (Fig. 2).

3. Next, choose a genotype dataset for GWAS analysis based on the accessions used. There are four genotype datasets in GWA-Portal: the 250 k SNP dataset, 1001 full sequence dataset, Swedish genomes dataset, and imputed full sequence dataset. Choose a genotype dataset with the highest overlap with the accessions used.

4. Choose a transformation method to make the phenotype data as close to normal distribution as possible. GWA-Portal provides three transformation methods: logarithmic (LOG), square root (SQRT), and Box-Cox; it calculates the Shapiro-Wilk scores for each transformation. The desired transformation is selected based on data distribution.

5. In the last step, choose a GWAS algorithm and run the analysis. There are three algorithms in GWA-Portal, but only accelerated mixed model (AMM) accounts for the population structure. Thus, AMM is generally chosen (*see* **Note 6**).

3.3 Data Processing

1. Download the result in csv format and filter out rare SNPs by setting minor allele frequency (MAF) at 5%, equal to minor allele count (MAC) at 10 for 200 accessions (*see* **Note 7**), using "subset" function in R program with the following script.

```
>mydata <- read.csv("filename.csv")
>mydata <- subset(mydata, mac>=10)
```

2. Format the data and generate the Manhattan plot using "qqman" package in R program (https://cran.r-project.org/web/packages/qqman/).

```
>library("qqman")
##In the original file, p values for each SNP are stored as
-log10(P) scores. Convert scores to p values.
>mydata <- transform(mydata, P = 10^-(score))
##Convert data in integer format to numeric format.
>mydata$chr <- as.numeric(mydata[,1])
>mydata$pos <- as.numeric(mydata[,2])
##Rename column names.
>colnames(mydata) <- c("CHR","BP","score","maf","mac","G-
VE","P")
##Add a new row to store the SNP number.
>mydata2$SNP <- seq.int(nrow(mydata2))
##Generate Manhattan plot
>manhattan(mydata2, col=c("gray10","gray60"), suggestiveli-
ne=F, genomewideline = F)
```

3. In a GWAS analysis, millions of statistical tests are performed, which introduce false positives. To minimize the effect from multiple comparisons, consider employing multiple comparison correction methods to adjust the *p* value threshold. Several correction methods can be used, including Bonferroni correction, Benjamino-Hochberg correction and šidák correction (*see* **Note 8**). According to Bonferroni correction, *p* value threshold is calculated as follows:

$$P_{\text{threshold}} = \frac{0.05}{N}, \text{where } N \text{ is the total number of SNPs tested}$$

4. Due to the linkage disequilibrium, the truly causal genetic loci are generally identified as peaks comprising multiple highly significant SNPs. A peak with only a single SNP is often indicative of a false positive.

3.4 Causal Genes Identification

1. Genes linked with the significant SNPs are considered as candidate genes. To identify the causal gene among the candidates, the loss of function mutant and sometimes overexpressing of the genes can be characterized for thermo-morphogenesis.
2. T-DNA insertion mutants are available for most of the genes in Arabidopsis and can be requested from Arabidopsis Biological Resource Center (ABRC, https://abrc.osu.edu/). If T-DNA insertion mutants are not available, the loss of function mutants can o be generated by CRISPR/Cas9 genome editing tools [10].
3. To generate overexpression lines, PCR amplify the coding regions of these genes from the genomic DNAs or cDNA library and clone them into the pCR8/GW/TOPO vector using a TA cloning kit. Clone the DNA fragments into the binary vector, which contains a 35S promoter [such as pMDC32 [11]]. Transform the overexpression construct into Agrobacterium tumefaciens strain GV3101 (pMP90) and then into plants using floral dipping method [12].
4. Grow the mutant and overexpression plants, as well as the wild-type control, at both 22 °C and 28 °C and calculate the thermo-responsiveness of rosette architecture. Compare the thermo-morphogenesis of the mutant and overexpression plants with the wild type to determine the function of the candidate gene in thermos-morphogenesis.

3.5 Functional Assessment of Natural Variations of the Causal Gene

1. Calculate the linkage disequilibrium of the associated SNPs from the same peak and identify different haplotypes for the causal gene. Associate the alleles with phenotypes to identify the high and low thermo-responsive haplotypes.
2. PCR amplify the genomic DNA fragments of the high and low thermo-responsive haplotypes of the causal gene from the representative accessions. Clone the genomic DNA fragments into the pCR8/GW/TOPO vector using a TA cloning kit and then clone them into the binary vector pMDC99 [11]. Transform the desired construct into Agrobacterium tumefaciens strain GV3101 (pMP90) and generate transgenic plants using floral dipping method in the loss of function mutant background.
3. Grow the transgenic plants at both 22 °C and 28 °C and calculate the thermo-responsiveness of rosette architecture. Compare the thermo-morphogenesis of the transgenic plants carrying different haplotypes to validate the effects of these natural variations.

4 Notes

1. Cold treatment helps the seed to synchronize germination. Different *Arabidopsis* natural accessions may require different length of treatment, and this information can be found at ABRC website.
2. To minimize the positional effect on plant growth, rotate the pots in random position once a day in a flat.
3. The average weight and the average area of six plants from the same accession are used to calculate the compactness index.
4. Many algorithms have been developed for GWAS. The GWA-Portal webtool is designed for *Arabidopsis* but not other organisms.
5. The accession ID can be found by searching the accession name in germplasm catalog in The Arabidopsis Information Resource (TAIR, https://www.arabidopsis.org/) or GWA-Portal.
6. The accelerated mixed model corrects the false positives caused by population structure.
7. SNPs that are rare will more likely have false positive association. To reduce the false positive chance from the rare SNPs, the MAF is typically defined at 1 or 5%. However, this elimination could also remove true positive rare SNPs.
8. Because SNPs are not always independent due to linkage disequilibrium, Bonferroni correction is conservative and sometimes causes false negatives. Other correction methods may also be considered [13]. Alternatively, the most significant SNPs could be tested without correction.

References

1. Wang Z, Yang L, Liu Z et al (2019) Natural variations of growth thermo-responsiveness determined by SAUR26/27/28 proteins in Arabidopsis thaliana. New Phytol 224:291–305
2. Sanchez-Bermejo E, Zhu W, Tasset C et al (2015) Genetic architecture of natural variation in thermal responses of Arabidopsis. Plant Physiol 169:647–659
3. Zhang LL, Shao YJ, Ding L et al (2021) XBAT31 regulates thermoresponsive hypocotyl growth through mediating degradation of the thermosensor ELF3 in Arabidopsis. Sci Adv 7:eabf4427
4. Seren Ü (2018) GWA-portal: genome-wide association studies made easy. In: Ristova D, Barbez E (eds) Root development. Methods in molecular biology, vol 1761. Humana Press, New York, pp 303–319
5. Belzile F, Torkamaneh D (2022) Designing a genome-wide association study: Main steps and critical decisions. In: Torkamaneh D, Belzile F (eds) Genome-wide association studies. Methods in molecular biology, vol 2481, Humana, New York, pp 3–12
6. Martínez-Berdeja A, Stitzer MC, Taylor MA et al (2020) Functional variants of DOG1 control seed chilling responses and variation in seasonal life-history strategies in *Arabidopsis thaliana*. Proc Natl Acad Sci USA 117:2526–2534
7. Tabas-Madrid D, Méndez-Vigo B, Arteaga N et al (2018) Genome-wide signatures of flowering adaptation to climate temperature:

regional analyses in a highly diverse native range of Arabidopsis thaliana. Plant Cell Environ 41:1806–1820

8. Yano K, Yamamoto E, Aya K et al (2016) Genome-wide association study using whole-genome sequencing rapidly identifies new genes influencing agronomic traits in rice. Nat Genet 48:927–934
9. Schneider CA, Rasband WS, Eliceiri KW (2012) NIH image to ImageJ: 25 years of image analysis. Nat Methods 9:671–675
10. Wang ZP, Xing HL, Dong L et al (2015) Egg cell-specific promoter-controlled CRISPR/Cas9 efficiently generates homozygous mutants for multiple target genes in Arabidopsis in a single generation. Genome Biol 16:144
11. Curtis MD, Grossniklaus U (2003) A gateway cloning vector set for high-throughput functional analysis of genes in planta. Plant Physiol 133:462–469
12. Clough SJ, Bent AF (1998) Floral dip: a simplified method for agrobacterium-mediated transformation of Arabidopsis thaliana. Plant J 16:735–743
13. Lee T, Lee I (2021) Genome-wide association studies in Arabidopsis thaliana: statistical analysis and network-based augmentation of signals. In: Sanchez-Serrano JJ, Salinas J (eds) Arabidopsis protocols. Methods in molecular biology, vol 2200, Humana, New York, pp 187–210

Chapter 23

Mathematical Modeling of Photo- and Thermomorphogenesis in Plants

Gabriel Rodriguez-Maroto, Pablo Catalán, Cristina Nieto, Salomé Prat, and Saúl Ares

Abstract

Increased day lengths and warm conditions inversely affect plant growth by directly modulating nuclear phyB, ELF3, and COP1 levels. Quantitative measures of the hypocotyl length have been key to gaining a deeper understanding of this complex regulatory network, while similar quantitative data are the foundation for many studies in plant biology. Here, we explore the application of mathematical modeling, specifically ordinary differential equations (ODEs), to understand plant responses to these environmental cues. We provide a comprehensive guide to constructing, simulating, and fitting these models to data, using the law of mass action to study the evolution of molecular species. The fundamental principles of these models are introduced, highlighting their utility in deciphering complex plant physiological interactions and testing hypotheses. This brief introduction will not allow experimentalists without a mathematical background to run their own simulations overnight, but it will help them grasp modeling principles and communicate with more theory-inclined colleagues.

Key words Mathematical modeling, Ordinary differential equations (ODEs), Law of mass action, Systems biology, Dynamical systems, Numerical simulation, Bifurcation analysis, Model building, Simulated annealing

1 Introduction

After the discovery that warm temperatures modify light-grown seedling architecture and the coining of the "thermomorphogenesis" biological term, much recent work has focused on understanding the effects of both light and temperature on plant development [1]. To disentangle the complex network of molecular actors involved in sensing and responding to light and temperature, development of mathematical models is a very helpful asset. In this chapter, we outline the basic procedure to develop a

Authors Gabriel Rodriguez-Maroto and Pablo Catalán are the co-first authors for this chapter.

Meng Chen (ed.), *Thermomorphogenesis: Methods and Protocols*, Methods in Molecular Biology, vol. 2795, https://doi.org/10.1007/978-1-0716-3814-9_23,

mathematical model of plant systems biology. We will focus on systems of ordinary differential equations (ODEs), and in what follows, we will explain the basic concepts for their derivation.

Mathematical models help us understand our system of interest by allowing us to outline its temporal behavior given certain initial conditions. They also help to test different hypotheses and check if they are consistent with our data or our previous knowledge. Finally, once they are correctly built, they allow us to make predictions about the behavior of our system outside the range of experimental conditions previously measured.

Modeling is an iterative process. First, several hypotheses about the system's behavior are made. Then, a combination of those hypotheses, the available experimental data, and theoretical knowledge (physical and chemical laws) are combined to build the model. The next step is to verify whether the model is able to capture and mimic the experimental results. If the model's results match the observed data, the model has been successfully built. On the other hand, if there are inconsistencies between the model's results and the experimental ones, we will have to review the initial hypotheses and adjust the model. In fact, this process can be done even when the results match. For instance, it might be possible to reduce the complexity of the model without omitting any biologically relevant parts by simply ignoring certain variables, grouping parameters, or simplifying some intermediate steps.

The primary objective is not to develop a mathematical model that incorporates the maximum number of variables. Even if such a comprehensive model were feasible, deriving biological insights from it would prove exceedingly challenging due to its inherent complexity [2]. Instead, what mathematical modeling pursues in biology is to develop a model able to capture the essence of the significant interactions, thus providing an outcome that can be fully understood from a biological perspective.

A mathematical model can be defined as a set of equations describing the behavior of a system. When modeling plants' responses to light and temperature, the state variables will very likely represent the abundance of specific molecular species as a function of time, as, for instance, would be the concentration of active phytochrome B (phyB) in the nucleus [3]. If the biological system under study is a biochemical network, the components of that network will be the various molecules involved, and the interactions will correspond to the chemical reactions occurring between them. Assuming we know which chemical reactions occur, they can be mathematically expressed as [4]:

$$R_j : \sum_{i=1}^{n} \alpha_{ij} X_i \xrightarrow{k_j} \sum_{i=1}^{n} \beta_{ij} X_i,$$

where R_j represents each of the reactions included in the network; α_{ij} and β_{ij} are the stoichiometric coefficients determining the number of molecules of reactants and products (respectively) involved in the reaction; the k_j are the reaction constants, and X_i identifies each of the species participating in the reaction. The left- and right-hand side terms represent the reactants and products, respectively, and the arrow indicates the direction of the reaction.

Once we have identified the chemical reactions, the next step is to derive the mathematical equations describing the temporal behavior of the system. From a mathematical point of view, this is equivalent to finding the temporal derivative of whatever magnitude is used to describe the state of the system's molecules. As we will later see, these rates depend on the species participating in the reaction, but they can also be functions of environmental conditions, such as light or temperature: many reactions happen only during the daytime [5], or reaction rates can be accelerated when temperature increases [6].

The magnitude we will use to describe the state of the system is the concentration of molecules of interest [2]. This way, the mathematical model will consist of a set of ODEs describing the temporal evolution of the concentration of molecules of each species involved in the network (*see* **Note 1** for an alternative modeling framework). Furthermore, when modeling chemical reaction networks, two assumptions are frequently considered. The first is the lack of spatial dependence in the reaction rates, which can be justified on the grounds of a homogeneous distribution of the reactants. The second assumption refers to the use of molecular concentrations instead of discrete numbers of molecules. This is possible as long as the number of molecules is high enough, and it is designated as the *continuum hypothesis* [2].

If we can properly justify the two previous assumptions, the equations modeling the overall network will arise from the application of the Law of Mass Action. This law states that the rate of any chemical reaction is proportional to the product of the concentrations of the reactants [2, 4]. Consider, for instance, the simple reaction:

$$A + B \xrightarrow{k} C.$$

The rate of production of the species C can be easily obtained by applying the Law of Mass Action, and it is equal to $k[A][B]$, where $[A]$ and $[B]$ denote the concentrations of each reactant, and k is a constant of proportionality (rate constant). Importantly, the units of this rate constant will depend on the number of reactants in each reaction, while the number of products does not play any role in the reaction rate. A further important factor to take into consideration is the kinetic order of each reactant [2], which sets the exponent to which each concentration has to be powered. In the previous reaction, the kinetic order of A and B is one. If we had the reaction:

$$2A + B \xrightarrow{k} C,$$

the kinetic order of A would be two, and hence the reaction rate of C would be expressed as $k[A]^2[B]$.

Our aim is to obtain an ODE describing the temporal variation of the concentration of the different species participating in the reaction. Thus, time will be the independent variable, and the species' concentrations will become the dependent variables. In the case of a species participating in a single reaction, what we chemically called the "reaction rate" is mathematically nothing but a temporal derivative. Thus, the reaction rate of the production of C is just the temporal variation of C's concentration:

$$\frac{d[C]}{dt} = k[A][B].$$

Note also that if a species participates in several reactions, the temporal derivative will be obtained from the addition of the different reaction rates.

In studying plants' responses to light and temperature, we will need to model the transcription and translation of genes of interest, as well as the protein-protein and protein-gene interaction events comprising the gene regulatory network (GRNs). The general scheme follows the same guidelines explained above, but for the sake of clarity, we will provide some examples of modeling GRNs.

Basically, what makes these GRNs different from the previously explained chemical networks is their complexity. Instead of a set of single reactions, we have complex chains of reaction processes. For instance, both transcription and translation involve a very large number of individual biochemical reactions. However, it is not just a matter of the high number of reactions involved—these also occur in metabolic or pure chemical networks—but the fact that many of those reactions are not fully characterized [7]. This clearly prevents us from knowing the exact chemical reaction that takes place, a main feature that differentiates biological systems with respect to other chemical networks.

In order to circumvent this problem, we often make the simplifying assumption that both transcription and translation happen at phenomenological—"effective," in modeling jargon—rates, therefore modeling more abstract reactions. As such, the gene expression description will not be as precise as that of the previous biochemical networks—it will be more coarse-grained—but it will still be useful and manageable. However, detailed models of these processes also exist, see for example [8].

A second problem associated with modeling gene regulatory networks has to do with the *continuum hypothesis*, since often the number of molecules involved in the regulation of gene expression is not that high. For instance, in the case of proteins regulating gene expression, the number of copies per species is in the range of several hundred or even less [9]. Regarding encoding genes, the numbers are even lower—rarely more than dozens and typically just one or two. Surprisingly, the mass-action formalism is still applicable in systems with low numbers of molecules; however, this entails a small change in the perspective of the model [2]. Now, the differential equations should be understood as descriptors of the average behavior of the system over a large population of cells. By contrast, to capture the behavior of individual cells, we would better employ a different modeling scheme (*see* **Note 1**).

2 Materials

In order to perform numerical simulations of the model, you will need a computer. Personal computers (including laptops) are enough for most tasks, while access to high-performance computing clusters may be necessary for specific aspects such as parameter fitting, especially if the model is very complex or involves many parameters.

Numerical simulations of ODEs can be written in most programming languages. We favor C++ and Python 3, both open source and available for free, and MATLAB (Mathworks, Inc.), which requires a license. Tutorials on how to use these languages and how to simulate dynamical systems are readily available online [10–12].

We utilize AUTO [13] for bifurcation analysis, while Mathematica (Wolfram Research) aids us in handling complex analytical calculations.

3 Methods

In this section, we report on how to build a mathematical model from scratch.

3.1 Read the Literature

To start building the model, you will need to investigate the known interactions between the molecular actors involved in your system of interest, by performing a thorough literature review. In modeling plants' responses to light and temperature, you will probably have to study the effects of photoreceptors (typically phytochromes, but also cryptochromes or others [14]) on your system. You will also probably have to include the effect of the circadian clock [15]. Identify the main genes and proteins involved in your process. Write down all relevant interactions, as well as those that you hypothesize are possible but have not been measured. And, of course, do not reinvent the wheel: there are several beautiful models out there [16–19]. Take inspiration from or recycle parts of published models as needed. Building a mathematical model from the gathered information is the key part of the procedure, and the most difficult one. You will have to turn the previously noted interactions into ODEs. This can be done either using a bottom-up or a top-down approach.

3.2 Build the Mathematical Model Using a Bottom-Up Approach

Start by writing down the chemical reactions in which all the molecular actors are involved: transcription, translation, dimerization, degradation, and so on. Use the Law of Mass Action to derive a system of ODEs for each molecular species (these will include intermediate complexes as well as the molecules of interest). By inspecting the scales of the different parameter values, you can introduce steady-state approximations that simplify the system. For example, the process of a transcription factor binding to DNA is generally assumed to occur faster than protein translation. Consequently, the equation representing a TF-DNA complex can be considered to be in equilibrium. A simple example is as follows: we know that *constitutive photomorphogenic* 1 (COP1) is involved in degradation of both *phytochrome* B (phyB) and *early flowering* 3 (ELF3) [20, 21]. When we started to model this interaction, we hypothesized that phyB and ELF3 might be competing for a finite pool of COP1, so that higher levels of ELF3 would result in less phyB degradation since ELF3 would be sequestering COP1. In terms of reactions, this translates into:

$$E + C \xrightarrow{k_1} D_1 \xrightarrow{k_2} C,$$

$$D_1 \xrightarrow{k_{-1}} E + C,$$

$$B + C \xrightarrow{q_1} D_2 \xrightarrow{q_2} C,$$

$$D_2 \xrightarrow{q_{-1}} B + C,$$

where E represents ELF3, C is COP1, B is phyB, D_1 and D_2 are the respective ELF3-COP1 and phyB-COP1 complexes, and k_1, k_{-1}, k_2, q_1, q_{-1}, q_2 are the corresponding rate constants of the reactions. For simplicity, we can assume that both ELF3 and phyB will be created at rates p_E and p_B, respectively.

Translating this into equations, we obtain:

$$\frac{d[E]}{dt} = p_E - k_1[E][C] + k_{-1}[D_1],$$

$$\frac{d[B]}{dt} = p_B - q_1[B][C] + q_{-1}[D_2],$$

$$\frac{d[C]}{dt} = -k_1[E][C] + (k_{-1} + k_2)[D_1] - q_1[B][C] + (q_{-1} + q_2)[D_2],$$

$$\frac{d[D_1]}{dt} = k_1[E][C] - (k_{-1} + k_2)[D_1],$$

$$\frac{d[D_2]}{dt} = q_1[B][C] - (q_{-1} + q_2)[D_2],$$

where the brackets, as before, indicate concentrations. Considering that the formation and break-down of these protein complexes is fast, we can assume that the derivatives of $[D_1]$ and $[D_2]$ are close to zero (the steady-state assumption), and therefore express their concentrations in terms of E, B and C. Also, the sum of the last three equations is zero. This implies that the sum of the concentrations of C, D_1 and D_2 is constant and equal to the initial concentration of COP1, $[C]_0$ (conservation of mass). After some algebra, we can express the change in $[E]$ and $[B]$ by only using their rate constants and $[C]_0$:

$$\frac{d[E]}{dt} = p_E - \frac{\frac{k_1}{k_{-1}+k_2} k_2[C]_0[E]}{\frac{k_1}{k_{-1}+k_2}[E] + \frac{q_1}{q_{-1}+q_2}[B] + 1},$$

$$\frac{d[B]}{dt} = p_P - \frac{\frac{q_1}{q_{-1}+q_2} q_2[C]_0[B]}{\frac{k_1}{k_{-1}+k_2}[E] + \frac{q_1}{q_{-1}+q_2}[B] + 1}.$$

Figure 1 shows the comparison between the true solution (solid lines) and the approximate solution (dashed lines). Note that, after an initial transient, both lines quickly converge once the intermediate complexes reach equilibrium.

3.3 Build the Mathematical Model Using a Top-Down Approach

When multiple molecular actors are involved and, more importantly, the details of their molecular interactions are not well known, an alternative approach to model building is to write interactions from scratch. For instance, if we know that the expression of the *phytochrome interacting factor* 4 (PIF4) target *arabidopsis thaliana homeobox2* (ATHB2) is inhibited by phyB [22], ELF3 [23] and *elongated hypocotyl* 5 (HY5) [24], we can write the ODE for ATHB2 as follows:

$$\frac{d[\mathrm{ATHB2}]}{dt} = \frac{p_P[P]}{1 + p_P[P] + p_E[E] + p_B[B] + p_H[H]},$$

where $[P]$ is the concentration of PIF4 and $[H]$ is the concentration of HY5, the p_X indicate the binding strength of species X, while the rest are as above. Once you get a sense of how these models work, you can come up with a model based on these types of equations without all the intermediate derivation.

In many cases, the top-down approach results in simpler equations than the ones obtained using the bottom-up approach. For instance, when we derived our hypocotyl growth model [25] in the

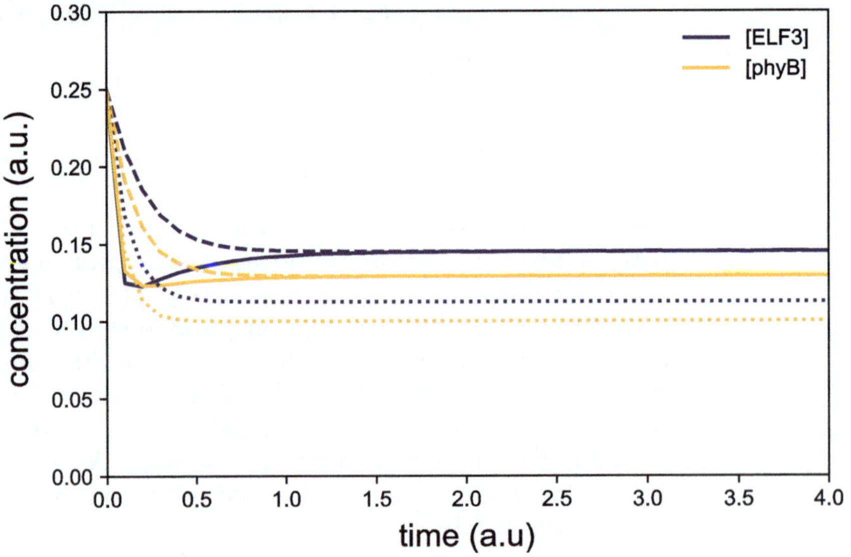

Fig. 1 Time evolution of ELF3 and phyB concentrations, using different equations. Solid lines represent the complete numerical solution, while dashed lines represent the approximate solution using the steady-state approximation for the intermediate complexes. Dotted lines show the solution when there is no competition for COP1 between ELF3 and phyB. Parameters: $p_E = 1$, $p_P = 1.2$, $k_1 = 10$, $k_{-1} = 5$, $k_2 = 4$, $q_1 = 9$, $q_{-1} = 3$, $q_2 = 6$, $[E](0) = 0.25$, $[B](0) = 0.25$, $[C](0) = 2$, $[D_1](0) = 0$, $[D_2](0) = 0$

example of the previous section, we soon realized that the ODEs had too many parameters. A model without considering the competition between ELF3 and phyB for COP1 interaction was actually equally capable of reproducing the experimental results. The equations for ELF3 and COP1 nuclear abundance without such a competition effect become:

$$\frac{d[E]}{dt} = p_E - \frac{k_1}{k_{-1} + k_2} k_2 [C]_0 [E],$$

$$\frac{d[B]}{dt} = p_P - \frac{q_1}{q_{-1} + q_2} q_2 [C]_0 [B].$$

These are plotted in Fig. 1 as dotted lines. Note that while this solution is not the same as the complete (bottom-up) model, for high enough values of $[C]_0$ it is close to the experimental data. More importantly, the qualitative behavior of the curves is very similar, which allows fitting the experimental data with these simplified equations (*see* Subheading 3.7 and **Note 2**).

Once you have decided on a final set of ODEs, you should study the properties of your dynamical system. In order to do this, you will have to use a combination of analytical and numerical methods. A good starting point is to just run some simulations of the equations, playing with the parameter values, to get a feel for what the model can do. By studying the properties of the model, you will be able to see if the hypotheses you used for developing it are consistent or not. If not, you should now modify the equations in an iterative process until you get the desired results.

3.4 Find the Fixed Points of Your System

In many cases, it is very useful to find the fixed points of the system of ODEs, i.e., the set of points that make all derivatives equal to zero. If they exist and are stable (more on stability in the next subsection), these are the stationary values where the system variables will settle once the initial transients have receded. If the system is simple enough, you will be able to obtain analytical expressions for these. More frequently, however, the nonlinearities of the system will prevent you from obtaining closed analytical expressions. In this case, you can try testing approximations for when some parameters are small or large using methods from perturbation theory [26], or find the fixed points by solving the nonlinear algebraic equations numerically. Often, the initial exploratory study of the system through simulations (more on how to do this below) will find sets of values where the variables stabilize.

Studying fixed points means focusing on the system's long-term behavior, excluding its transient states. In many cases, transients are not relevant for our purposes, and so the analysis of the model will be restricted to the final equilibrium. In other cases, we may also be interested in the transient behavior, and numerical simulations will then be necessary. Finding analytical solutions, that is, closed formulas, for the time dependence of the variables is almost for certain an intractable problem for most models in plant systems biology.

3.5 Study the Stability of the Fixed Points

The fixed points we do obtain might be stable, meaning that the state of the system robustly stays at the fixed point once it has been reached. It is important to study this stability, not only to find out whether the fixed point is a valid stationary state of the system but also to determine if more complex behaviors, like multistability or oscillations, may arise. This is usually done by calculating the eigenvalues of the Jacobian matrix associated with the system [27]. When all the eigenvalues of the Jacobian evaluated at a fixed point have negative real part, the fixed point is stable; otherwise, it is unstable. As before, you will sometimes be able to do this by hand or with the help of a symbolic programming language such as Mathematica. In many other cases, you will need to analyze the stability of your fixed point by using numerical methods. For this, a very good tool is AUTO [13] or its derived package XPPAUT [28]. These allow you to study the stability of your system as the values of the parameters change—for instance, by drawing bifurcation diagrams—(the use of XPPAUT is nevertheless not straightforward; *see* **Note 3**).

3.6 Simulate the Model Using Numerical ODE Solvers

This step helps you visualize the dynamics of the system, even when it is far from the fixed points. The information gathered throughout the previous steps is crucial here, since numerical routines can be unstable and yield erroneous solutions. Knowing where the fixed points should be, and their stability, allows you to check your routines.

Common routines to solve ODEs are ode45 in MATLAB [29], integrate.solve_ivp in SciPy (Python) [30] and odeint::integrate in C++ [31]. They all offer the user a wide variety of numerical solvers (both implicit and explicit, with adaptive and fixed step sizes) that suit different problems. Knowing when to use a particular solver is not easy, but having them already implemented in these languages assists in the trial-and-error process. A common dilemma faced with the modeling of large systems of equations is deciding which numerical integrator to use. Here, different aspects like the efficiency, accuracy, and stability of the methods need to be carefully analyzed. A common strategy is to assume that the ODE problem is in principle non-stiff, while it can be defined as stiff if very different timescales play a role in the system. Some clever approximations may eliminate fast timescales when formulating the problem, but this is not always possible. For stiff systems, some numerical methods will force us to use an extremely small step size to find the desired solution—explicit methods—with implicit methods performing in general much better in this case [32]. You can begin by using a non-stiff solver and then, depending on the behavior of the integrator used—accuracy of the found solution, computational time, and so on—decide whether to change it or not (*see* **Note 4**).

3.7 Parameter Fitting

If experimental data is available, you should fit the model to it. When studying the response of plants to temperature, data spanning a range of temperatures will be helpful in order to make predictions, as the reaction rates will change with different temperatures, and you will need to perform separate fits for each of them. Similarly, when studying light, either a range of daylengths or light intensities will help you discern the effect of light on the problem you are studying. Ideally, a careful experimental design should be considered when developing a mathematical model.

As the number of parameters in these models is quite high, a manual trial-and-error strategy in which we try to adjust the model predictions to the experimental results is not feasible. We need to turn to optimization methods, which try to minimize the value of a so-called energy (or cost) function. Often, this function will be the sum of the squares of the differences between the experimental values and the model predictions. Typical methods to obtain parameter fits are Monte Carlo algorithms [33], where you start with an initial arbitrary set of parameters and make small modifications to those while accepting the new parameters if the energy function satisfies particular conditions. Monte Carlo methods vary widely in their algorithms of acceptance, as well as their implementations (*see* **Note 5**). Our method of choice is simulated annealing [34].

At each step, the simulated annealing algorithm evaluates some neighboring state of the current parameter state. By modifying one of the parameters slightly and comparing the energy function associated with it (E_{new}) with the one associated with the unperturbed parameter set (E_{old}), it decides whether to accept the new parameter values or not, based on the acceptance probability:

$$P_{acc} = \begin{cases} 1 \text{ if } E_{\text{old}}/E_{\text{new}} < 1, \\ \frac{E_{\text{old}}}{E_{\text{new}}} T_{\text{A}} \text{ if } E_{\text{old}}/E_{\text{new}} < 1 \end{cases},$$

where T_A is the "temperature" of the annealing, which decreases with each step. The idea is to use probabilities ensuring that the system moves to lower energy states. The particularity of simulated annealing is that the probability of accepting a parameter change resulting in an increase of the energy function decreases with each annealing step, using a particular function T_A that requires to be fine-tuned through trial-and-error. We often use:

$$T_{\text{A}} = \frac{0.8}{\sqrt{1+i}}.$$

The algorithm will be repeated until either a preset maximum number of iterations has been achieved or it successfully meets an error tolerance criterion (*see* **Note 6**).

3.8 Making Predictions

Once the model is fitted to the data, you can make predictions of the behavior of the system by varying relevant parameters. If your model includes the effect of temperature, you can predict the behavior of the plant for temperatures out of the set that has been measured. Equally, with light, we were able to predict the growth of plants at 4 h of day lengths, an experimental condition that had not been measured for our system [25].

Finally, you can also assess how changing parameters will change the behavior of the system. Perhaps a fixed point becomes unstable, thus leading to completely different dynamics. Combining this last step with the previous ones will help you to understand the behavior of the system, and gain entirely novel biological insights concerning the studied functions.

4 Notes

1. ODEs are not the only modeling framework available. If you need to introduce spatial information into your model (for

instance, because you want to understand how different parts of a stem or a leaf react to different stimuli), you will need to use partial differential equations (PDEs), which we have not mentioned here. Some relevant plant modeling studies using this framework can be found in [35].

Also, it may be that stochasticity is relevant to the system. This happens when the number of molecules involved is small (and the continuum approximation does not work), and also when the response of the cells to environmental changes is not deterministic. When this is the case, one may use stochastic processes to model the system [36]. Alternatively, stochastic differential equations (SDEs), such as the Langevin eq. [37], are very common in modeling biological systems, see for instance [38, 39].

It is worth noting that the simulation and analysis techniques differ for these methods, but detailing them is outside the scope of this chapter.

2. In some cases, simplification of the model using steady-state assumptions is not straightforward. It is not until you perform numerical simulations that you realize, for instance, that some variables reach equilibrium much faster than others; you can then assume those to be constants, consequently simplifying the simulations. Also, depending on the range of some parameter values, corresponding terms in the ODEs can be neglected. This does not mean that these interactions do not exist, but rather that they are not relevant for the system under the experimental conditions being studied.

   Most models of biological systems are "sloppy" [40], meaning that some parameters are very important to the dynamics while others are not and can thus be removed. For large models, this procedure is not easy to do by hand, and automated methods making use of the geometry of the parameter-to-function manifolds [41] will help us understand the basic structure behind these models.

   Sloppiness is a consequence of not having enough data to determine the structure of the model. This means that, for a given problem, many putative models will be compatible with the data, and so the search for the "best" model is futile. We can then only aim to understand how the data restricts the kinds of models we can build.

3. It is convenient to emphasize that despite the usefulness of XPPAUT for carrying out bifurcation analysis, its handling is very demanding. Understanding how to use the program is a complex and trial-and-error process. Despite the existing documentation about this software, the practical examples are mostly reduced to prototypical and ideal cases. For real problems, it can be highly complex not only to solve a particular

error but just to understand and find out the type of error. Nevertheless, XPPAUT and its underlying software, AUTO, are considered among the best—if not the best— computational tools for this type of analysis.

4. Numerical solvers (especially in Python) often yield numerical errors, and great care has to be taken with the parameters of the algorithm: step size, solver, and stopping conditions. Given that this process can be time-consuming, having approximate analytical forms of the solutions or fixed points, regardless of their roughness, can prove helpful. You do not want to report numerical artifacts as the solution.
5. Many Monte Carlo methods using Bayesian frameworks exist for parameter fitting, and most of them are implemented in libraries in Python (for instance, the pyABC library) and Matlab (the Markov Chain Monte Carlo Toolbox), making this step easier. However, in our experience, these libraries do not work well with the kind of problems we have worked with, perhaps because of our own limited experience with them. Therefore, we have preferred to code our own optimization algorithms. Particularly, simulated annealing has worked very well for us in many cases.
6. Model building can be carried out in a more sophisticated manner. Moreover, additional steps can be considered, such as studying the identifiability of parameters. We refer the reader to reference (42) for more details on this approach.

Acknowledgments

Research has been supported by Spanish MCIN/AEI/10.13039/501100011033/ and FEDER, EU (grant nos. PGC2018-098186-B-100 and PID2022-142185NB-C22 to P.C., PID2022-142185NB-C21 to S.A., and BIO2017-90056-R/PID2020-119758RB-I00 to S.P.), besides grant BADS, no. PID2019-109320GB-100, to S.A. and P.C. The CNB and CRAG Institutes also received the "Severo Ochoa" Centers of Excellence SEV 2017-0712 (CNB) and CEX2019-000902-S (CRAG) awards from the Spanish Ministerio de Ciencia e Innovación. We thank Jorge J. Casal for enlightening discussions.

References

1. Legris M, Nieto C, Sellaro R, Prat S, Casal JJ (2017) Perception and signalling of light and temperature cues in plants. Plant J 90:683–697
2. Ingalls BP (2013) Mathematical modeling in systems biology: an introduction. MIT press
3. Rausenberger J, Hussong A, Kircher S, Kirchenbauer D, Timmer J, Nagy F, Schafer E, Fleck C (2010) An integrative model for phytochrome B mediated photomorphogenesis: from protein dynamics to physiology. PLoS One 5:e10721

4. Feinberg M (2019) Applied mathematical sciences: foundations of chemical reaction network theory. Springer, New York
5. Pay ML, Kim DW, Somers DE, Kim JK, Foo M (2022) Modeling of plant circadian clock for characterizing hypocotyl growth under different light quality conditions. silico Plants 4: diac001
6. Avello PA, Davis SJ, Ronald J, Pitchford JW (2019) Heat the clock: entrainment and compensation in Arabidopsis circadian rhythms. J Circadian Rhythms 17:5
7. Reed MC (2004) Why is mathematical biology so hard. Not Am Math Soc 51:338–342
8. Weiße AY, Oyarzún DA, Danos V, Swain PS (2015) Mechanistic links between cellular trade-offs, gene expression, and growth. Proc Natl Acad Sci USA 112:E1038–E1047
9. Milo R, Phillips R (2015) Cell biology by the numbers. Garland Science, New York
10. C++ language. https://cplusplus.com/doc/tutorial/. Accessed July 7 2023
11. Mathworks (2023) Learn with Matlab tutorials. https://mathworks.com/support/learn-with-matlab-tutorials.html. Accessed July 7 2023
12. The Python Software Foundation (2023) The Python Tutorial. https://docs.python.org/3/tutorial/. Accessed July 7 2023
13. Doedel EJ, Oldeman B (1998) Auto-07p: continuation and bifurcation software. Concordia University Canada, Montreal, QC
14. Paik I, Huq E (2019) Plant photoreceptors: multi-functional sensory proteins and their signaling networks. Semin Cell and Dev Biol 92: 114–121
15. Sanchez SE, Kay SA (2016) The plant circadian clock: from a simple timekeeper to a complex developmental manager. Cold Spring Harb Perspect Biol 8:a027748
16. Johansson H, Jones HJ, Foreman J, Hemsted JR, Stewart K, Grima R, Halliday KJ (2014) Arabidopsis cell expansion is controlled by a photothermal switch. Nat Commun 5:4848
17. Legris M, Klose C, Burgie ES, Rojas CCR, Neme M, Hiltbrunner A, Wigge PA, Schäfer E, Vierstra RD, Casal JJ (2016) Phytochrome B integrates light and temperature signals in Arabidopsis. Science 354:897–900
18. Jung JH, Domijan M, Klose C, Biswas S, Ezer D, Gao M, Khattak AK, Box MS, Charoensawan V, Cortijo S, Kumar M (2016) Phytochromes function as thermosensors in Arabidopsis. Science 354:886–889
19. De Caluwé J, Xiao Q, Hermans C, Verbruggen N, Leloup JC, Gonze D (2016) A compact model for the complex plant circadian clock. Front Plant Sci 7:74
20. Yu JW, Rubio V, Lee NY, Bai S, Lee SY, Kim SS et al (2008) COP1 and ELF3 control circadian function and photoperiodic flowering by regulating GI stability. Mol Cell 32:617–630
21. Jang IC, Henriques R, Seo HS, Nagatani A, Chua NH (2010) Arabidopsis phytochrome interacting factor proteins promote phytochrome B polyubiquitination by COP1 E3 ligase in the nucleus. Plant Cell 22:2370–2383
22. Park E, Kim Y, Choi G (2018) Phytochrome B requires PIF degradation and sequestration to induce light responses across a wide range of light conditions. Plant Cell 30:1277–1292
23. Nieto C, López-Salmerón V, Davière JM, Prat S (2015) ELF3-PIF4 interaction regulates plant growth independently of the evening complex. Curr Biol 25:187–193
24. Osterlund MT, Wei N, Deng XW (2000) The roles of photoreceptor systems and the COP1-targeted destabilization of HY5 in light control of Arabidopsis seedling development. Plant Physiol 124:1520–1524
25. Nieto C, Catalán P, Luengo LM, Legris M, López-Salmerón V, Davière JM, Casal JJ, Ares S, Prat S (2022) COP1 dynamics integrate conflicting seasonal light and thermal cues in the control of Arabidopsis elongation. Sci Adv 8:eabp8412
26. Bender CM, Orszag SA (1999) Advanced mathematical methods for scientists and engineers I: asymptotic methods and perturbation theory. Springer, New York
27. Strogatz SH (2018) Nonlinear dynamics and chaos with student solutions manual: with applications to physics, biology, chemistry, and engineering. CRC Press, Boca Raton
28. Ermentrout B, Mahajan A (2003) Simulating, analyzing, and animating dynamical systems: a guide to XPPAUT for researchers and students. Appl Mech Rev 56:B53
29. Mathworks (2023) ode45: Solve nonstiff differential equations — medium order method. https://mathworks.com/help/matlab/ref/ode45.html. Accessed July 7 2023
30. SciPy documentation (2023) scipy.integrate.solve_ivp. https://docs.scipy.org/doc/scipy/reference/generated/scipy.integrate.solve_ivp.html. Accessed July 7 2023
31. Boost C++ Libraries (2023) Odeint. https://www.boost.org/doc/libs/1_74_0/libs/numeric/odeint/doc/html/index.html. Accessed July 7 2023
32. Spijker MN (1996) Stiffness in numerical initial-value problems. J Comput Appl Math 72:393–406

33. Liu JS, Liu JS (2001) Monte Carlo strategies in scientific computing, vol 75. Springer, New York
34. Kirkpatrick S, Gelatt CD Jr, Vecchi MP (1983) Optimization by simulated annealing. Science 220:671–680
35. Prusinkiewicz P, Rolland-Lagan AG (2006) Modeling plant morphogenesis. Curr Opin Plant Biol 9:83–88
36. Karlin S, Taylor HM (1975) A first course in stochastic processes. Academic Press
37. Gillespie DT (2000) The chemical Langevin equation. J Chem Phys 113:297–306
38. Muñoz-García J, Ares S (2016) Formation and maintenance of nitrogen-fixing cell patterns in filamentous cyanobacteria. Proc Natl Acad Sci USA 113:6218–6223
39. Casanova-Ferrer P, Ares S, Muñoz-García J (2022) Terminal heterocyst differentiation in the Anabaena patA mutant as a result of post-transcriptional modifications and molecular leakage. PLOS Comp Biol 18:e1010359
40. Transtrum MK, Machta BB, Brown KS, Daniels BC, Myers CR, Sethna JP (2015) Perspective: sloppiness and emergent theories in physics, biology, and beyond. J Chem Phys 143:010901
41. Transtrum MK, Qiu P (2014) Model reduction by manifold boundaries. Phys Rev Lett 113:098701
42. Villaverde AF, Pathirana D, Fröhlich F, Hasenauer J, Banga JR (2022) A protocol for dynamic model calibration. Brief Bioinform 23:bbab387

Correction to: Chromatin Immunoprecipitation to Investigate H2A.Z Dynamics in Response to Environmental Changes

J. Paola Saldierna Guzmán, Mark Zander, and Björn C. Willige

Correction to:
Chapter 17 in: Meng Chen (ed.), *Thermomorphogenesis: Methods and Protocols*, Methods in Molecular Biology, vol. 2795,
https://doi.org/10.1007/978-1-0716-3814-9_17

This book was inadvertently published with one of the author names in Chapter 17 mentioned incorrectly, which has now been corrected in the chapter and contributors list as shown below.

First: J. Paola
Last: Saldierna Guzmán

The updated version of this chapter can be found at
https://doi.org/10.1007/978-1-0716-3814-9_17

Meng Chen (ed.), *Thermomorphogenesis: Methods and Protocols*, Methods in Molecular Biology, vol. 2795,
https://doi.org/10.1007/978-1-0716-3814-9_24,

Index

A

Arabidopsis 3, 18, 26, 38, 45, 56, 65, 76, 86, 96, 105, 113, 151, 162, 169, 184, 198, 214, 228, 239
Arabidopsis thaliana 4, 18, 19, 26, 27, 30, 32, 44, 56, 86–88, 124, 151, 169, 214, 228, 253

B

Bifurcation analysis 251, 258
Bioluminescence 46, 48, 49, 51, 52
Biomolecular condensates 95, 106–109

C

CBF1 .... 184, 185, 188, 190, 191
Cell length 78, 80
Chlorophyll 26, 28–32, 35, 44, 51, 62, 71, 86, 92
Chromatin 135, 136, 169–181
Chromatin immunoprecipitation (ChIP) 135, 170, 172, 177, 178, 181
Circadian oscillator 44
Circular dichroism (CD) 151–156

D

Dynamical systems 251, 254

E

EARLY FLOWERING 3 (ELF3) 18, 38, 123–132, 137, 140, 196, 252–254
EARLY FLOWERING 4 (ELF4) 124, 137, 138, 140–142, 146
Electrophoretic mobility shift assay (EMSA) 143–145
Entrainment 46, 49, 213
Evening complex (EC) 124, 137, 139, 142, 145, 146

F

Flowering time (FT) 4, 63, 170, 227–229, 232, 237, 240
Fluorescence-activated particle sorting (FAPS) 114–116, 118, 121
Fluorescence recovery after photobleaching (FRAP) 107, 108, 110, 124, 127, 129–132
Förster resonance energy transfer (FRET) 151–157
Fractionation 114, 116, 120, 227–237

G

Gating 119, 121, 213–223
Genome-wide association study (GWAS) 239–245
Growth rate analyses 4
GST pull down 196, 197, 200, 203, 205, 207, 208

H

Hairpins 150–152, 154–156
H2A.Z 169–172, 181
High ambient temperature 26, 38, 66, 162, 217, 239
High temperature 17, 56, 65–72
Histones 169–171
Hypocotyl length 5, 8, 20, 21, 75, 78
Hypocotyl measurement 19–21

I

Imaging of stomata 67, 70
Immunoblots 119, 191, 196, 197, 206–208, 210, 229, 232, 235
In vitro kinase assay 162, 163
In vivo spectroscopy 86–88, 92

L

Law of mass action 249, 252
Leaf senescence 25, 26, 35
Light signal 217
LUX ARRYTHMO (LUX) 124, 137–142, 144–146

M

Mathematical modeling 248
Model building 253, 259

Meng Chen (ed.), *Thermomorphogenesis: Methods and Protocols*, Methods in Molecular Biology, vol. 2795, https://doi.org/10.1007/978-1-0716-3814-9,

N

Numerical simulation 251, 254, 258

O

Ordinary differential equations (ODEs) 248, 251–254, 256–258

P

Pathogen growth 56, 60, 61, 63
Period analysis 49, 52
Phase 43, 105–110, 179, 181, 189, 210, 213, 228, 234, 237
Phase diagrams 124, 125, 127, 129
Phase separation 105, 106, 113, 124–132
Phenotypic analysis 65, 66, 162
Phenotyping 5, 40
Photobody 95–103, 105, 113–121
Photoperiod 4, 59, 68, 71, 217
Phytochrome 29, 85, 113, 137, 151, 162, 170, 184, 251
Phytochrome B (phyB) 17, 18, 26, 38, 85–92, 95, 96, 105–110, 113–121, 162, 184, 185, 188, 248, 252–254
Phytochrome interacting factor 4 (PIF4) 11, 18, 26, 38, 124, 137, 162, 163, 165, 170, 184, 185, 188, 191, 195–211, 253
Plant immunity 55, 56, 63
Plant tissue clearing and fixation 66, 72
Prion-like domain (PrD) 123, 124
Protein-DNA binding 138, 140
Protein-protein interaction 162, 196, 208
Protein purification 138, 140, 141
Pst DC3000 56–63

R

Ratiospectrophotometer 86–88, 90
RNA structures 150, 151, 156
RNA thermometer 150
Root 3–5, 7–9, 13, 15, 37–41, 222
RootDetection 5, 9–12, 15
rootdetectR 4, 5, 9–15
Root phenotyping 40
Rosette growth architecture 241
RT-qPCR 27, 215
RuBisCO depletion 227–237

S

Senescence marker gene 26, 27, 29, 33, 35
Shade avoidance response 217
Simulated annealing 257, 259
Stomatal development 37, 65, 66, 162, 195
SUPPRESSOR OF PHYA-105 1 (SPA1) 162, 163, 165, 166
Systems biology 248, 254

T

Temperature 3, 17, 26, 43, 55, 66, 76, 86, 95, 105, 114, 124, 135, 150, 162, 169, 184, 195, 213, 228, 241, 247
Temperature buffering 143, 231
Temperature dependence 86
Temperature stress 219, 220
Thermal reversion 17, 85–92, 95
Thermomorphogenesis 3–15, 17–20, 37, 38, 40, 162, 170, 195, 196, 239, 240, 244, 247–259
Transcript abundance 214, 215, 217, 218, 221
Turbidity assays 124, 129

Y

Yeast three-hybrid (Y3H) 185, 191

MIX
Papier aus verantwortungsvollen Quellen
Paper from responsible sources
FSC® C105338

If you have any concerns about our products,
you can contact us on
ProductSafety@springernature.com

In case Publisher is established outside the EU,
the EU authorized representative is:
Springer Nature Customer Service Center GmbH
Europaplatz 3, 69115 Heidelberg, Germany

Printed by Libri Plureos GmbH
in Hamburg, Germany